MySQL 8.0

运维与优化 （微课视频版）

姚远 ◎ 编著

清华大学出版社
北京

内容简介

本书全面介绍 MySQL 数据库的管理、监控、备份恢复和高可用等方面的知识，并在此基础上讨论如何优化 MySQL 的实例和 SQL 语句，书中还包括大量的实战案例。

全书分五部分：第一部分（第 1～4 章）为管理部分，包括安装和运行、账号和权限、日志和安全等内容；第二部分（第 5～7 章）为监控部分，介绍通过 MySQL 自带的 3 个系统数据进行监控的方法；第三部分（第 8～11 章）为备份恢复部分，介绍常用的逻辑备份和物理备份工具，还介绍在没有备份的情况下如何进行数据救援；第四部分（第 12～14 章）为高可用部分，包括 MySQL Shell、复制和 InnoDB 集群；第五部分（第 15～18 章）为优化部分，介绍基准测试工具和优化 MySQL 实例和 SQL 方法。

本书适合具有一定 IT 基础知识的 MySQL 数据库爱好者阅读，也可以作为准备 MySQL OCP 考试的备考书。

图书在版编目（CIP）数据

MySQL 8.0 运维与优化：微课视频版/姚远编著. —北京：清华大学出版社，2022.5（2022.11重印）
ISBN 978-7-302-60268-2

Ⅰ. ①M… Ⅱ. ①姚… Ⅲ. ①SQL 语言－数据库管理系统 Ⅳ. ①TP311.138

中国版本图书馆 CIP 数据核字（2022）第 036839 号

责任编辑：安　妮　李　燕
封面设计：刘　键
责任校对：胡伟民
责任印制：沈　露

出版发行：清华大学出版社
网　　址：http://www.tup.com.cn，http://www.wqbook.com
地　　址：北京清华大学学研大厦 A 座　　**邮　　编**：100084
社 总 机：010-83470000　　**邮　　购**：010-62786544
投稿与读者服务：010-62776969，c-service@tup.tsinghua.edu.cn
质量反馈：010-62772015，zhiliang@tup.tsinghua.edu.cn
课件下载：http://www.tup.com.cn，010-83470236
印 装 者：北京同文印刷有限责任公司
经　　销：全国新华书店
开　　本：185mm×260mm　　**印　张**：27.5　　**字　　数**：688 千字
版　　次：2022 年 5 月第 1 版　　**印　　次**：2022 年 11 月第 2 次印刷
印　　数：2001～3500
定　　价：99.80 元

产品编号：092222-01

推荐序

姚远老师是我在鼎甲科技的同事，我和他是 2019 年经朋友介绍认识的，当时我们在网罗数据库方面的专家。第一次见面我觉得姚老师对加入鼎甲科技是有疑虑的，“这个池子似乎小了一点，我来鼎甲工作有什么发展机会呢？”，毕竟鼎甲是做数据备份的公司，而他是国内知名的数据库专家，还是中国为数不多的 Oracle ACE，金融和互联网企业也急需像姚老师这样的人才。然而当我告诉姚老师，鼎甲是当时全球唯一一家实现了 MySQL InnoDB 集群在线备份的厂商，且率先在 Oracle 数据库上实现基于连续日志保护专利技术的数据零丢失方案，他开始对鼎甲产生了兴趣。我继续告诉他，鼎甲还在包括 MySQL、PostgreSQL、MongoDB 以及衍生的一系列国内外数据库产品进行持续和深入的研究时，姚老师的态度又发生了比较大的变化。他想不到一个做数据备份的公司竟会投入这么多精力研究各种开源数据库，于是他委婉地提出了想和我们研发同事聊聊的请求。聊完之后，我感觉他已经不再怀疑我们对各种数据库技术都做过深度研究，因为我们的研发同事们已经展现了他们在数据库方面的能力。姚老师接下来的疑问则变成，“既然鼎甲已经有一群数据库高手，还需要我做什么？”

实际上，当初我们找姚老师来也是情势所逼。鼎甲在 2015 年将自己潜心研发 6 年的 DBackup 备份软件及基于 DBackup 的鼎甲备份一体机 Infokist 推向市场，仅仅用了 3 年时间，鼎甲就成为中国备份一体机市场国产品牌的第 1 名。随之而来的是大量的用户需求，特别是对一些国产数据库的备份需求，需要我们研发团队迅速地给出解决方案。然而大多数国产数据库都是基于各种开源数据库二次开发出来的，由于这些数据库大都只是在界面和表达层进行修改，很少对数据库文件系统、日志系统以及数据库引擎进行优化和改进，因此要想备份这些数据库，与其去找这些厂商要接口资料，还不如直接研究原生态的开源数据库，从源头解决问题。但是一方面，国内在开源数据库方面的中文资料通常是翻译国外开源英文资料或外版书籍，这样就会出现中文资料隔代延迟的情况，即等中文版书籍出来的时候，开源数据库已经升级换代了。另一方面，为了提高中文资料的时效性，很多翻译人员直接翻译开源社区的英文资料，由于他们没有研究过相关数据库，翻译出来的资料不仅承袭了开源资料原有的问题，还可能由于理解不够带来更多谬误。因此，鼎甲的研发人员一直都持之以恒地直接研究国外开源数据库的原生资料。由于业务迅速增长和国产数据库层出不穷，我们的研发人员需要花大量时间编写代码，这导致他们研究和跟踪国外最新技术的时间越来越少；并且随着政府对国外技术社区无差别地管控，和国外同行进行技术咨询和交流也变得越来越困难。所以我们需要像姚老师这样的专家，将最新的数据库技术系统性地传授给我们的研发、测试、售前、售后团队。出于对鼎甲自主创新的认可，姚老师最终拒绝了金融机构和互联网企业抛出的橄榄枝，毅然加入了鼎甲。

加入鼎甲以来，姚老师负责向同事和客户提供数据库方面的培训和技术支持，他先后主讲了 Linux、Oracle、MySQL、SQL Server 和 DB2 的培训课程。在 MySQL 方面，他主讲了面向初学者的基础培训、面向开发人员的 SQL 优化培训、面向运维人员的实例优化和健康检查培训。他还应邀在墨天轮、ITPUB、爱可生开源社区、华为云社区进行了技术分享，参加了 2021 Oracle Groundbreakers 亚太巡演。在鼎甲服务的两万多个客户里，除了有号称宇宙第一大银行的金融机构以及全球最大的移动通信公司，也有许多中小型企业和政府单位，虽然鼎甲是一个备份厂商，但很多客户遇到了数据库方面的难题还是会第一时间咨询我们，因为有姚老师的鼎甲几乎没有解决不了的数据库问题。姚老师帮助过许多同事和客户，慢慢地，大家给他封了一个“鼎甲首席布道师”的头衔。现在，姚老师将他这几年在鼎甲研究、教学和咨询 MySQL 数据库的内容整理出版，和更多的同行分享，这得到了公司和同事的支持，也是我们鼎甲的荣幸。本书将是国内在 MySQL 方面最新、最全、最可靠的参考文档。我希望姚老师的新作能够给蓬勃发展的国产数据库事业带来贡献。我相信这只是开始，假以时日，姚老师会有更多、更好的数据库相关著作问世。

王子骏

鼎甲科技总裁、创始人、首席科学家

2022 年 2 月

前言

写这本书的原因

MySQL是世界上排名第一的开源数据库，实际上，MySQL加上其分支已经超过了Oracle数据库的排名，成了事实上的业界王者。自从2009年Oracle公司通过收购SUN公司将MySQL置于旗下后，MySQL数据库的发展更加迅猛，特别是MySQL 8.0版本的推出，使MySQL数据库和其他开源数据库的差距进一步拉大。在开源数据库越来越流行的今天，学习MySQL数据库成了很多数据库爱好者的首选。

我从21世纪初开始使用MySQL数据库，刚开始做开发，后来做运维。和很多MySQL的爱好者一样，我在遇到问题时喜欢从网上搜索答案，但这样得到的知识是碎片化的，要想真正精通MySQL必须进行系统、全面的学习。我学习MySQL的资料来源主要有两个：第一个来源是MySQL的官方文档，但MySQL数据库的官方文档和Oracle数据库的官方文档比起来相当简陋，很难让人相信是出自同一家公司之手；第二个资料来源是MySQL方面的书籍，我基本上把国内所有的Oracle和MySQL的书籍都看过了。感觉MySQL方面的中文书还是太少，特别是涉及MySQL 8.0新特性和优化方面的书籍更少。于是我在学习过程中就萌生了一个想法，要自己写本MySQL方面的书，为开源数据库在中国的推广做点贡献。

这些年来，我不断地把自己实际工作中积累的经验整理成对外的博客和对内的工作笔记。随着自己博客和工作笔记的积累，再加上在广州鼎甲公司和互联网上的授课，输出一本MySQL的书就成为水到渠成的事情了。

这本书中的很多内容是我工作经验的总结，例如，我在工作中处理最多的故障是MySQL无法启动，针对MySQL系统参数设置错误而造成的无法启动，我总结了一种排除法找出阻止MySQL启动的错误系统参数，又总结了SELinux和AppArmor阻止MySQL启动的解决办法，还总结了MySQL无法启动时进行数据救援的方法。如果只是采用平铺直叙的方式把技术要点描述清楚，这样的书市面上很多，我的目标是要写出自己的东西、自己的特色、自己对技术的理解和应用，这样便增加了写书的难度和出错的可能，但这样做才更有意义，至于做得如何就交给读者来评价吧。

学习数据库的方法

因为工作的关系，我对目前所有流行的数据库都有所研究，特别是Oracle和MySQL数据库，很多朋友问我是如何学好数据库的，这里和大家做一个交流。

要学好任何一门技术都要有一种静下心来做学问的态度，数据库的技术相对比较复杂，短期内无法速成，只有经过长时间的钻研才能精通这门技术。我把学习数据库技术的过程大致分为如下三步。

第一步是理论学习，主要是阅读官方文档。在理论学习过程中，首先要掌握概念和原理，因此我学习 Oracle 的时候，看得最多的文档是 *Database Concepts*，后来我把这个文档的电子版发给淘宝的卖家，他们帮忙打印和装订成两册，大约花了 70 元钱，我看了很多遍，所有的概念和原理搞得烂熟。不光是数据库技术，所有的 IT 技术都要先把理论搞清楚再动手操作，所有的理工科的学习都是这样，理论清楚了再谈操作。绕过理论学习操作并不是走捷径，因为不久就发现实际上是在走弯路。广州鼎甲公司有两万多个客户，我在日常的工作中处理了很多数据库的疑难杂症，这些故障表面上看起来各不相同，但实际上只要理论清楚，所有的故障都能解决。

顺便说到，如果要掌握高端的 IT 技术，必须具备相当的英文阅读能力，虽然我经过专门的英语学习，托业考试成绩是 890 分，但并不建议 IT 从业人员专门学习英语，因为阅读自己专业的英文文档既能学习英语，又能学习技术。一些初学者畏惧英文文档，想绕开走，实际上只要硬着头皮去啃，一两年的时间就能翻过这个坎。

第二步是动手实验。做实验是提高自己动手能力的最有效的方法。我曾经花了两年的时间准备 Oracle 10g 的 OCM 考试，那个时候没有培训班，我就是依据考试大纲阅读 Oracle 的官方文档，然后猜测考试时会遇到什么样的场景，自己设计了数百个实验题，反复练习，把理论知识练成了直觉。业界的一个常见现象是某位技术大咖被提拔成了领导，减少了亲自动手操作的机会，越不动手，越不自信，就越不敢动手，形成了一个恶性循环，最后退化成了光说不练的领导，很可惜！

第三步是在生产系统的运维中磨炼提高。实验做得再熟，和在生产系统上操作还是有差距的，因为数据量和业务量的关系，一些在开发、测试环境执行得很顺畅的命令到生产系统上却会遇到各种问题。我在生产上进行实施或故障排除中有两个原则：一是对生产系统要有敬畏之心，在我的职业生涯中只闯了一次大祸，却永远铭记，每次进行生产变更都小心翼翼。业界有一种人是“无知无畏”，实际上越是水平高的人，在生产系统上进行变更时越谨慎，毕竟 80%的生产事故是由变更引起的；二是思路要清楚，其实不光是解决技术问题，人生中的所有事情都要基于这个原则，你要知道你的目标是什么，怎么达到你的目标，你采取每个步骤的后果是什么。

这三步是一个多次循环的过程，理论知识的欠缺会导致操作的错误，操作中的出错也可以反映出对理论知识的掌握不足，只有从理论到实践的多次反复才能磨砺出一个真正的数据库专家。

本书的主要内容

本书的内容基于 MySQL 8.0 的版本，全书分为 18 章。

第 1 章介绍 MySQL 的安装、升级和启动过程中的排错。

第 2 章介绍账号、权限、密码、角色等内容，还介绍如何解决忘记 root 密码的问题。

第 3 章介绍 MySQL 的 4 种日志：错误日志、通用查询日志、慢查询日志和二进制日志。

第 4 章介绍安全方面的内容，包括账号和密码的安全、数据加密和审计等。

第 5～7 章介绍 3 个 MySQL 的系统数据库：information_schema、performance_schema 和 sys。

第 8 章介绍逻辑备份，包括 4 种常见的逻辑备份工具及其对比，还讨论了备份集的一致性和加快数据导入速度的问题。

第 9、10 章分别介绍常用的物理备份工具：XtraBackup 和 MySQL Enterprise Backup。

第 11 章介绍在没有备份而且 MySQL 无法启动时如何进行数据恢复。

第 12 章介绍 MySQL 的新客户端工具 MySQL Shell。

第 13、14 章分别介绍 MySQL 的两个高可用解决方案：复制和 InnoDB 集群。

第 15 章介绍 3 种基准测试工具。

第 16 章介绍数据库优化的重要性和实例优化的方法。

第 17 章介绍 SQL 优化的基础知识。

第 18 章从实战角度介绍如何进行 SQL 优化。

本书讲述的 MySQL，除非有特别说明，默认都是 Linux 环境的 MySQL 8.0，存储引擎默认是 InnoDB。本书适合具有一定 IT 基础知识的 MySQL 数据库爱好者学习，也可以作为准备 MySQL OCP 考试的备考书。

本书的配套资源

我针对书中的重点和难点设计了 30 多个实验，并录制了实验的操作视频，读者可以通过扫描书中的二维码在线观看。读者还可以通过扫描下方二维码下载本书的全部脚本。

脚本

致谢

感谢我的父母、妻子、妹妹和两个小宝贝雯雯和咚咚对我永远的支持，你们的支持是我写书的动力，这本书同时也是献给你们的礼物。

感谢广州鼎甲公司的创始人和总裁王子骏教授，这里我想引用阿达在 2020 年年会上的话对王总表示感谢：“感谢王总创立了鼎甲，使我们有了一个施展自己才能的平台。”

感谢广州鼎甲的领导和同事：马总、樊总、吕总、汤总、大侠、阿达、冷工、武总、常馨、水仙、婷娟、肖廷楷、范华，以及全体鼎甲的同仁，感谢你们在工作上对我的帮助，每天和你们这些业界精英在一起工作对自己技术水平的提高大有裨益，这本书也是和你们思想碰撞的产物。广州鼎甲成为国内容灾行业龙头的历史同时也是你们这些技术精英的成长史。

感谢各位数据库领域的朋友，每天浏览你们的博客和朋友圈学到了很多东西，不仅限于技术方面。

由于自己水平有限，书中不当之处在所难免，欢迎广大同行和读者批评指正。

姚 远

2021 年 8 月

目录

第一部分 管理

第二部分　监控

第三部分　备份恢复

第四部分 高可用

第五部分 优化

第一部分

管　理

第1章

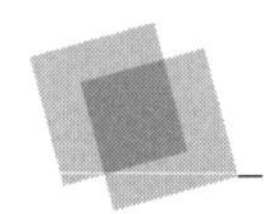

安装和运行

本章的内容是 MySQL 用户需要最先掌握的知识。主要介绍 MySQL 的安装，包括 MySQL 8.0 社区版的安装和 MySQL 的一个重要分支——Percona 的安装；还介绍了 Sakila 示例数据库的安装(后面章节的例子几乎都基于该数据库)；并介绍了检查 MySQL 服务和升级到 MySQL 8.0 的方法。

几乎所有的人安装 MySQL 后都会遇到无法启动的问题，本章还介绍了如何排除启动过程中的故障，包括 SELinux 和 AppArmor 阻止 MySQL 启动的问题，参数设置错误造成 MySQL 无法启动的问题，以及客户端连接不上的问题等。

1.1 MySQL 8.0 社区版的安装

Oracle 公司的 MySQL 版本分社区版和企业版两种，企业版比社区版功能强，但要付费，通常使用社区版即可满足一般要求。

Linux 平台上的 MySQL 版本主要分两类：一类是 Oracle Linux、Red Hat、CentOS 和 Fedora 平台上的 MySQL 版本；另一类是 Ubuntu 和 Debian 平台上的 MySQL 版本。安装方式有多种，常用的有软件仓库(YUM 源或 APT 源)安装、RPM 或 DEB 包安装、二进制文件安装、源码编译安装等，其对应关系如表 1.1 所示。

表 1.1

<table>
<tr><th rowspan="2">安装方式</th><th colspan="2">Linux 平台</th></tr>
<tr><th>Oracle Linux、Red Hat、CentOS 和 Fedora</th><th>Ubuntu 和 Debian</th></tr>
<tr><td>软件仓库</td><td>YUM 源</td><td>APT 源</td></tr>
<tr><td>安装包</td><td>RPM</td><td>DEB</td></tr>
<tr><td>二进制文件</td><td colspan="2">通用</td></tr>
<tr><td>源码编译</td><td colspan="2">通用</td></tr>
</table>

MySQL 版本按照 CPU 的架构分为 ARM 和 x86，CPU 的位数有 64 和 32 位两种。

这些 MySQL 版本的安装方式大同小异，本章介绍其中最常用的两种安装方式：RPM 包安装和通用二进制文件安装。

1.1.1 使用 RPM 包安装

1. 下载安装包

登录 MySQL 的官方网站，下载完整的 MySQL 8.0 社区版 RPM 安装包，注意选择对应的平台和操作系统版本，这里选择的是 Red Hat Enterprise（红帽）Linux，CPU 架构是 64 位的 x86，如图 1.1 所示。

图 1.1

如上下载的安装包里包括的 RPM 包如下：

```
$ tar tvf mysql-8.0.25-1.el7.x86_64.rpm-bundle.tar|awk '{print $6}'
mysql-community-client-8.0.25-1.el7.x86_64.rpm
mysql-community-client-plugins-8.0.25-1.el7.x86_64.rpm
mysql-community-common-8.0.25-1.el7.x86_64.rpm
mysql-community-devel-8.0.25-1.el7.x86_64.rpm
mysql-community-embedded-compat-8.0.25-1.el7.x86_64.rpm
mysql-community-libs-8.0.25-1.el7.x86_64.rpm
mysql-community-libs-compat-8.0.25-1.el7.x86_64.rpm
```

```
mysql-community-server-8.0.25-1.el7.x86_64.rpm
mysql-community-test-8.0.25-1.el7.x86_64.rpm
```

如上包的命名规则是 packagename-version-distribution-arch.rpm，其中：

(1) packagename 代表包名。

(2) version 代表版本号。

(3) distribution 代表 Linux 平台。

(4) arch 代表 CPU 类型。

例如，mysql-community-server-8.0.25-1.el7.x86_64.rpm 这个 RPM 包，其中 mysql-community-server 是包名；8.0.25-1 是版本号；el7 中的 el 代表 Enterprise Linux，包括 Oracle Linux、Red Hat Enterprise Linux 和 CentOS，7 是 Linux 的大版本号；x86_64 是 CPU 的类型。

2. 检查是否已经安装 MySQL 或 MariaDB

首先，检查默认的配置文件是否已经存在，命令如下：

```
$ ll /etc/my.cnf
-rw-r--r--. 1 root root 570 Mar 6 2014 /etc/my.cnf
```

然后，检查是否有 MySQL 或 MariaDB 的安装包，命令如下：

```
$ rpm -qa|grep -i mysql
$ rpm -qa|grep -i mariadb
mariadb-libs-5.5.35-3.el7.x86_64
```

Red Hat 和 CentOS 上通常已经安装了 mariadb -libs 的包，该包需要卸载，特别是/etc/my.cnf 文件，一定要先删除，如果不删除这个文件，新安装的 MySQL 会使用这个配置文件，届时会出现莫名其妙的错误。运行如下命令可卸载 mariadb -libs 包：

```
$ rpm -e mariadb-libs
$ ll /etc/my.cnf
ls: cannot access /etc/my.cnf: No such file or directory
```

卸载了 mariadb -libs 的包后，/etc/my.cnf 文件就一并被删除了。

3. 安装 RPM 包

服务端大部分时候只需要安装如下 5 个包：

(1) mysql-community-server。

(2) mysql-community-client。

(3) mysql-community-libs。

(4) mysql-community-common。

(5) mysql-community-libs-compat。

安装命令如下：

```
$ sudo yum install mysql-community-{server,client,common,libs}-*
```

客户端的安装命令如下：

```
$ sudo yum install mysql-community-{client,common,libs}-*
```

也可以把 yum install 替换成 rpm -Uvh，但因为 rpm -Uvh 不能自动解决包的依赖关系的问题而更容易出错。

4. 初始化数据库

安装完成后，先不要启动 MySQL 数据库，因为 MySQL 数据库在第一次启动时会自动进行初始化，在初始化之前要根据自己的需求对相关系统参数进行修改（默认的参数文件是/etc/my.cnf）。

(1) 最有可能修改的系统参数是数据文件目录 datadir，该系统参数的默认值是/var/lib/mysql，它通常在本地硬盘，在生产环境中通常是不合适的，可以将其改成其他实际生产中用到的目录，如/disk1/data。

(2) 系统参数 innodb_log_file_size 指定 InnoDB 重做日志文件的大小，默认值是 48MB，对于生产环境通常也小了，第 16 章会说明如何计算这个系统参数的大小，这里可以将其设置成 100MB。

(3) 系统参数 innodb_data_home_dir 指定 InnoDB 系统表空间的目录，默认值是 datadir，如果有必要，也可以调整该系统参数。

(4) 系统参数 innodb_log_group_home_dir 指定 InnoDB 重做日志文件的目录，默认值也是 datadir，如果有必要，也可以调整该系统参数。

参数文件准备好后，启动 mysqld 服务，命令如下：

```
$ sudo systemctl start mysqld
```

启动过程中会自动进行数据库的初始化工作，在初始化过程中会自动为超级用户 root@localhost 生成一个密码，可以在 MySQL 的错误日志里查询自动生成的密码，命令如下：

```
$ sudo grep 'temporary password' /var/log/mysqld.log
2020-12-22T08:08:27.273629Z 6 [Note] [MY-010454] [Server] A temporary password is generated for root@localhost: otvx.-q92,/N[
```

使用初始密码登录 MySQL 数据库并修改密码，这里需要注意，如果密码里面有特殊字符，密码应用单引号或双引号括起来，命令如下：

```
$ sudo mysql -uroot -p'otvx.-q92,/N'
mysql: [Warning] Using a password on the command line interface can be insecure.
Welcome to the MySQL monitor. Commands end with ; or \g.
Your MySQL connection id is 8
Server version: 8.0.25......
```

登录后，可以使用 set password 的命令修改密码，命令如下：

```
mysql> set password = 'Dingjia123!';
```

修改完成后使用新的密码重新登录，命令如下：

```
$ sudo mysql -uroot -pDingjia123!
```

1.1.2　使用二进制通用包安装

1. 下载

登录 MySQL 官网，下载二进制通用安装包，下载页面如图 1.2 所示。

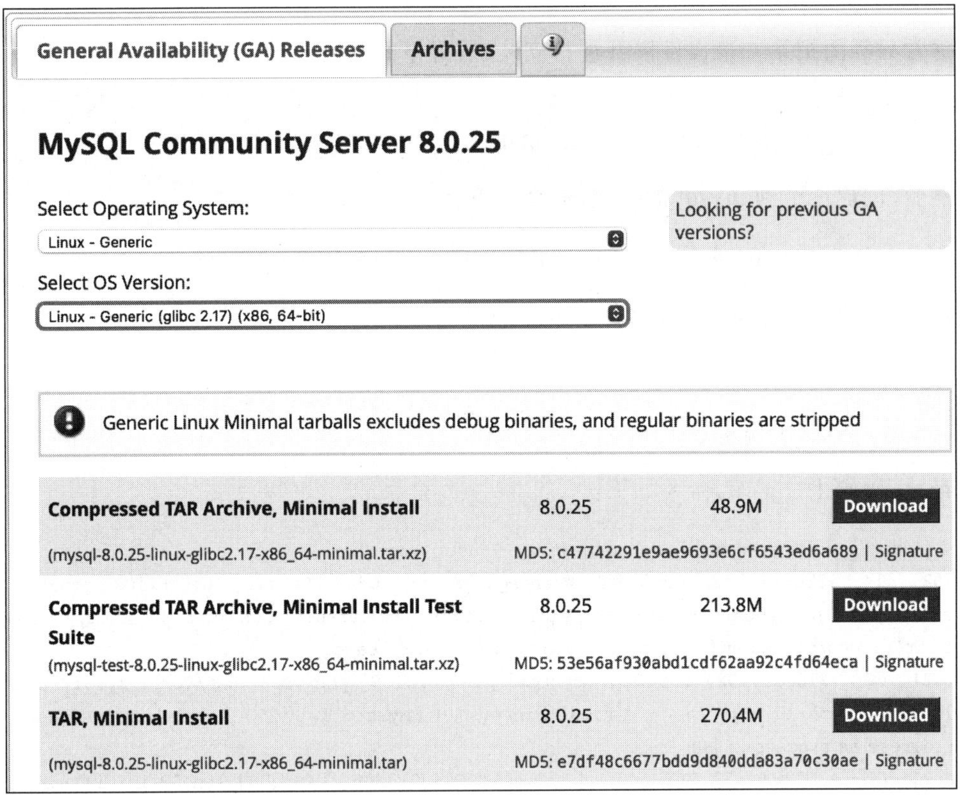

General Availability (GA) Releases　Archives

MySQL Community Server 8.0.25

Select Operating System:
Linux - Generic

Looking for previous GA versions?

Select OS Version:
Linux - Generic (glibc 2.17) (x86, 64-bit)

Generic Linux Minimal tarballs excludes debug binaries, and regular binaries are stripped

Compressed TAR Archive, Minimal Install (mysql-8.0.25-linux-glibc2.17-x86_64-minimal.tar.xz)	8.0.25	48.9M	Download MD5: c47742291e9ae9693e6cf6543ed6a689 | Signature
Compressed TAR Archive, Minimal Install Test Suite (mysql-test-8.0.25-linux-glibc2.17-x86_64-minimal.tar.xz)	8.0.25	213.8M	Download MD5: 53e56af930abd1cdf62aa92c4fd64eca | Signature
TAR, Minimal Install (mysql-8.0.25-linux-glibc2.17-x86_64-minimal.tar)	8.0.25	270.4M	Download MD5: e7df48c6677bdd9d840dda83a70c30ae | Signature

图　1.2

64 位 Linux 平台上不包括 debug 二进制安装包的包名是 mysql-8.0.25-linux-glibc2.17-x86_64-minimal.tar，该安装包的大小约为 270MB。

2. 安装 libaio 包

安装 libaio 包的命令如下：

```
$ yum search all libaio
$ sudo yum install libaio
```

3. 创建 MySQL 组和用户

创建一个 MySQL 组和用户的命令如下：

```
$ sudo groupadd mysql
$ sudo useradd -r -g mysql -s /bin/false mysql
```

该用户只作为 MySQL 文件的属主，不能登录。

4. 安装

下载的安装包中包括3个以tar. xz结尾的压缩文件，如下所示：

```
$ sudo tar xvf /root/mysql-8.0.25-linux-glibc2.17-x86_64-minimal.tar
mysql-test-8.0.25-linux-glibc2.17-x86_64-minimal.tar.xz
mysql-8.0.25-linux-glibc2.17-x86_64-minimal.tar.xz
mysql-router-8.0.25-linux-glibc2.17-x86_64-minimal.tar.xz
```

通常只需要把第二个文件展开进行安装，展开下载安装包的命令如下：

```
$ cd /usr/local
$ sudo tar xvf mysql-8.0.25-linux-glibc2.17-x86_64-minimal.tar.xz
```

然后，创建一个到mysql的链接，命令如下：

```
$ sudo ln -s mysql-8.0.25-linux-glibc2.17-x86_64-minimal mysql
```

接着，创建一个参数文件/etc/my. cnf，例如：

```
$ cat /etc/my.cnf
[mysqld]
datadir=/var/lib/mysql
```

最后，初始化数据库，命令如下：

```
$ sudo./mysql/bin/mysqld --initialize --user=mysql
2021-07-12T02:17:52.759935Z 0 [Warning] [MY-011070] [Server] 'Disabling symbolic links using --skip-symbolic-links (or equivalent) is the default. Consider not using this option as it' is deprecated and will be removed in a future release.
2021-07-12T02:17:52.760040Z 0 [System] [MY-013169] [Server] /usr/local/mysql-8.0.25-linux-glibc2.17-x86_64-minimal/bin/mysqld (mysqld 8.0.25) initializing of server in progress as process 25044
2021-07-12T02:17:52.768344Z 1 [System] [MY-013576] [InnoDB] InnoDB initialization has started.
2021-07-12T02:17:53.267734Z 1 [System] [MY-013577] [InnoDB] InnoDB initialization has ended.
2021-07-12T02:17:54.547118Z 6 [Note] [MY-010454] [Server] A temporary password is generated for root@localhost: =8qIdVoMbRqv
```

初始化数据库时使用参数--initialize会为root@localhost生成一个初始化密码，如果使用参数--initialize-insecure代替--initialize则不会生成密码，后续可以不用密码登录MySQL，然后手动设置密码。

初始化完成后，数据库就创建好了，但在启动MySQL前还要使用如下命令生成相关证书和密钥：

```
$ cd mysql
$ sudo bin/mysql_ssl_rsa_setup
```

一切准备就绪，现在可以使用mysqld_safe启动MySQL的服务进程了，命令如下：

```
$ sudo bin/mysqld_safe --user=mysql &
```

MySQL 服务启动后，客户端就可以连接进来了，使用前面的初始化密码进行登录，命令如下：

```
$ ./bin/mysql -uroot -p" = 8qIdVoMbRqv"
```

1.2 Percona Server for MySQL 的安装

MySQL 有很多分支，其中 Percona Server for MySQL 是一个非常优秀的 MySQL 分支，它完全兼容官方的 MySQL。

1.2.1 Percona Server for MySQL 和 MySQL 的对比

1. Percona Server for MySQL 和 MySQL 的关系

(1) Percona Server for MySQL 是基于 MySQL 社区版的源码进行二次开发而产生的。

(2) Percona Server for MySQL 修复了 MySQL 中一些已知的 Bug。

(3) Percona Server for MySQL 进行了功能和性能的增强。

(4) 除非启动了 Percona Server for MySQL 的某些特殊的功能，否则 Percona Server for MySQL 和 MySQL 可以无缝地进行切换。

2. Percona Server for MySQL 的优势

Percona Server for MySQL 通常发布的比对应的 MySQL 版本晚，它在 MySQL 的基础上进行了如下四方面的增强。

(1) 新增了针对写进行优化的存储引擎。MySQL 的 InnoDB 引擎在性能上偏向于读，Percona Server for MySQL 新增的引擎提高了写和压缩的性能，读的性能也不差。

(2) 增强了数据加密特性。

(3) 增强了灵活性和可管理性。

(4) 增加了一些企业版的特性，如线程池、审计插件、PAM 验证插件等，这些本来是 MySQL 企业版的功能，现已被迁移到 Percona Server for MySQL，以前需要付费才能使用的功能，现在可以免费使用了。

1.2.2 安装 Percona Server for MySQL 8.0

检查是否安装了 MySQL 和 MariaDB，如果已安装就将其卸载，这一步和其他安装方法相同。

登录 Percona 官网，下载 Percona Server for MySQL 的安装包，注意选择自己需要的版本和平台，下载页面如图 1.3 所示。

下载整个 Percona Server 的安装包需要占用 1.05GB 存储空间，如果只安装代替 MySQL 的 Percona Server for MySQL 8.0，只需要如下 4 个 RPM 包，占用空间大小约为 70MB。

(1) percona-server-client-8.0.22-13.1.el7.x86_64.rpm。

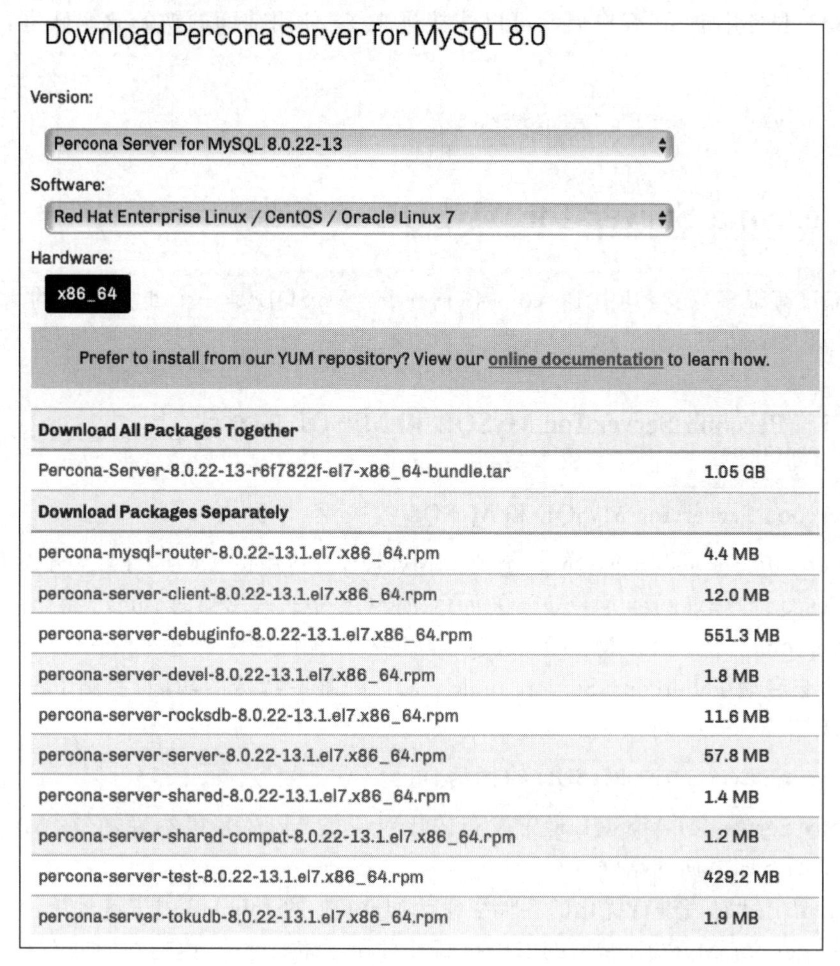
Download Percona Server for MySQL 8.0

Version:

Percona Server for MySQL 8.0.22-13

Software:

Red Hat Enterprise Linux / CentOS / Oracle Linux 7

Hardware:

x86_64

Prefer to install from our YUM repository? View our online documentation to learn how.

Download All Packages Together	
Percona-Server-8.0.22-13-r6f7822f-el7-x86_64-bundle.tar	1.05 GB
Download Packages Separately	
percona-mysql-router-8.0.22-13.1.el7.x86_64.rpm	4.4 MB
percona-server-client-8.0.22-13.1.el7.x86_64.rpm	12.0 MB
percona-server-debuginfo-8.0.22-13.1.el7.x86_64.rpm	551.3 MB
percona-server-devel-8.0.22-13.1.el7.x86_64.rpm	1.8 MB
percona-server-rocksdb-8.0.22-13.1.el7.x86_64.rpm	11.6 MB
percona-server-server-8.0.22-13.1.el7.x86_64.rpm	57.8 MB
percona-server-shared-8.0.22-13.1.el7.x86_64.rpm	1.4 MB
percona-server-shared-compat-8.0.22-13.1.el7.x86_64.rpm	1.2 MB
percona-server-test-8.0.22-13.1.el7.x86_64.rpm	429.2 MB
percona-server-tokudb-8.0.22-13.1.el7.x86_64.rpm	1.9 MB

图 1.3

（2）percona-server-shared-8.0.22-13.1.el7.x86_64.rpm。

（3）percona-server-server-8.0.22-13.1.el7.x86_64.rpm。

（4）percona-server-shared-compat-8.0.22-13.1.el7.x86_64.rpm。

安装代码如下：

```
$ sudo rpm -ivh percona-server-client-8.0.22-13.1.el7.x86_64.rpm
percona-server-server-8.0.22-13.1.el7.x86_64.rpm
percona-server-shared-8.0.22-13.1.el7.x86_64.rpm
percona-server-shared-compat-8.0.22-13.1.el7.x86_64.rpm
```

后续安装步骤和其他类型的安装基本相同。

1.3 安装 Sakila 示例数据库

Sakila 示例数据库是 MySQL 官方提供的一个学习 MySQL 的很好的素材，Sakila 是 MySQL 数据库海豚标志的名字。这个数据库模拟租借 DVD 的业务，它有 23 个表，用于替

代之前的 world 数据库，world 数据库比较简单，只包含三个表，分别为国家、城市、语言。该数据库的下载地址是 https://dev.mysql.com/doc/index-other.html。本书的后续章节中有很多例子都会用到这个数据库。

下载后是一个 sakila-db.zip 文件，展开后有如下三个文件：

（1）sakila-schema.sql 文件。它包括了所有的创建语句，用于创建 Sakila 数据库的结构，如表、视图、存储过程和触发器。

（2）sakila-data.sql 文件。它包括所有用于插入数据库中的实际数据和触发器的定义。

（3）sakila.mwb 文件。它是一个用于 MySQL Workbench 的数据模型。

安装的方法很简单，登录 mysql 后依次执行两个 sql 文件即可，代码如下：

```
$ sudo mysql -uroot -p
mysql> source sakila-schema.sql
mysql> source sakila-data.sql
```

安装完成后检查 Sakila 数据库，运行如下代码后，发现共生成了 23 个表。

```
$ mysqlshow -vv sakila;
Database: sakila
+----------------------------+----------+------------+
|           Tables           | Columns  | Total Rows |
+----------------------------+----------+------------+
| actor                      |        4 |        200 |
| actor_info                 |        4 |        200 |
| address                    |        9 |        603 |
| category                   |        3 |         16 |
| city                       |        4 |        600 |
| country                    |        3 |        109 |
| customer                   |        9 |        599 |
| customer_list              |        9 |        599 |
| film                       |       13 |       1000 |
| film_actor                 |        3 |       5462 |
| film_category              |        3 |       1000 |
| film_list                  |        8 |        997 |
| film_text                  |        3 |       1000 |
| inventory                  |        4 |       4581 |
| language                   |        3 |          6 |
| nicer_but_slower_film_list |        8 |        997 |
| payment                    |        7 |      16049 |
| rental                     |        7 |      16044 |
| sales_by_film_category     |        2 |         16 |
| sales_by_store             |        3 |          2 |
| staff                      |       11 |          2 |
| staff_list                 |        8 |          2 |
| store                      |        4 |          2 |
+----------------------------+----------+------------+
23 rows in set.
```

Sakila 数据库的 EER 图如图 1.4 所示。

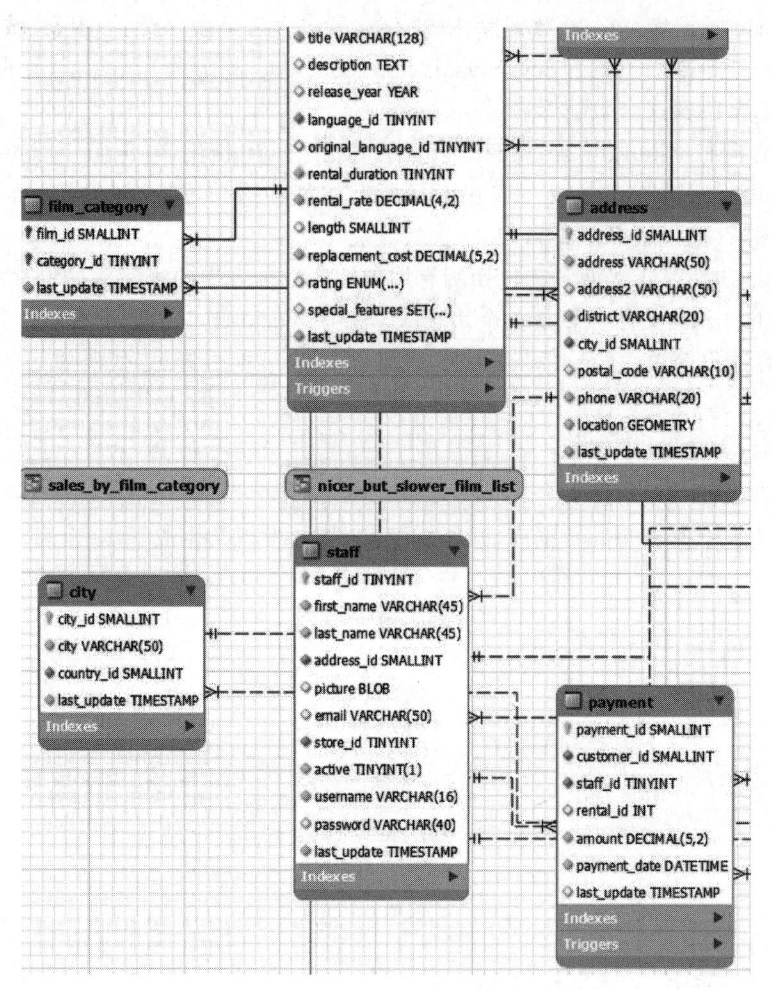

图 1.4

1.4 检查 MySQL 服务

MySQL 安装完成后要对 MySQL 服务进行检查，以验证安装的正确性。

1.4.1 启动和停止 MySQL 服务

MySQL 安装完成后可以使用 service 或 systemctl 命令来管理 MySQL 服务，MySQL 服务在 Red Hat 平台上的服务名是 mysqld，在 Debian 平台上的服务名是 mysql，这里以 Red Hat 平台为例说明常见的管理 MySQL 服务的命令，在 Debian 平台上需要把 mysqld 换成 mysql。

（1）启动 MySQL 服务：$ sudo service mysqld start 或 $ sudo systemctl start mysqld。

（2）检查 MySQL 服务的运行状态：$ sudo service mysqld status 或 $ sudo

systemctl status mysqld。

(3) 停止 MySQL 服务：$ sudo service mysqld stop 或 $ sudo systemctl stop mysqld。

(4) 重新启动 MySQL 服务：$ sudo service mysqld restart 或 $ sudo systemctl restart mysqld。

(5) 也可以使用 Linux 操作系统的 kill 命令“杀死”mysqld 进程来实现关闭 MySQL 的功能，默认“杀死”mysqld 的信号是 SIGTERM，mysqld 会正常关闭。如果用-9 参数“杀死”mysqld 进程，mysqld 进程会被强制终止，不会进行正常地关闭，下次启动时会自动进行崩溃恢复。如果 mysqld 是使用 mysqld_safe 启动的，或被 Linux 的 systemd 启动为服务，mysqld_safe 或 systemd 会自动重新启动没有正常关闭的 mysqld 进程。

1.4.2 使用 mysqladmin 工具检查 MySQL 服务

mysqladmin 是一个执行管理操作的 MySQL 客户端程序，它可以用来执行常用的管理 MySQL 的操作，也可以检查服务器的配置和当前状态。可以使用 mysqladmin 工具的 version 命令检查 MySQL 服务的状态，例如：

```
$ mysqladmin version
mysqladmin  Ver 8.0.25 for Linux on x86_64 (MySQL Community Server - GPL)
Copyright (c) 2000, 2020, Oracle and/or its affiliates. All rights reserved.

Oracle is a registered trademark of Oracle Corporation and/or its
affiliates. Other names may be trademarks of their respective
owners.

Server version      8.0.25
Protocol version    10
Connection          Localhost via UNIX socket
UNIX socket         /u01/mysql/mysql.sock
Uptime:             4 days 23 hours 40 min 46 sec

Threads: 2  Questions: 41  Slow queries: 0  Opens: 134  Flush tables: 3  Open tables: 55
Queries per second avg: 0.000
```

mysqladmin 工具的 version 命令显示的是 MySQL 数据库的一些重要信息，如版本号、连接信息、启动时间和 MySQL 服务的负载等。

1.4.3 使用 mysqlshow 检查已经存在的数据库

MySQL 安装完成后已经创建了四个自带的系统数据库，可以使用 mysqlshow 检查已经存在的数据库，相关命令及其运行结果如下：

```
$ mysqlshow -vv
```

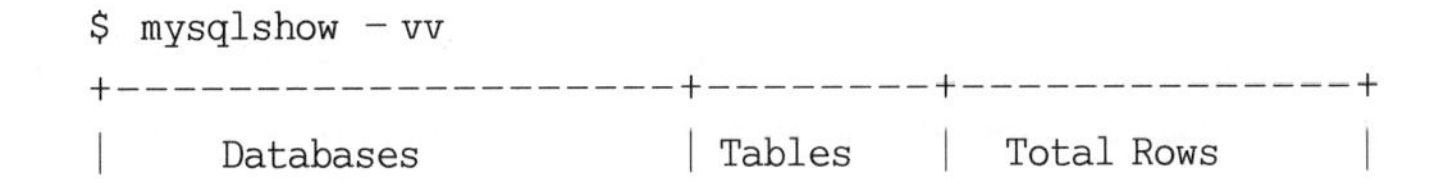

```
+--------------------+--------+--------------+
|     Databases      | Tables | Total Rows   |
```

```
+--------------------+--------+--------------+
| information_schema |     79 |        26031 |
| mysql              |     35 |         3822 |
| performance_schema |    110 |       212148 |
| sys                |    101 |         4594 |
+--------------------+--------+--------------+
4 rows in set.
```

1.5 升级到 MySQL 8.0

1.5.1 升级前检查

MySQL 可以从 5.7 升级到 8.0，但不能从 5.6 直接升级到 8.0，要先升级到 5.7 后再升级到 8.0。

为了解决不同版本数据的兼容问题，在升级前应使用 MySQL Shell 的升级检查工具对数据进行升级前的检查，例如：

```
$ mysqlsh -- util check-for-server-upgrade { --user=root --host=192.168.87.155 --port=3306 } --password='dingjia' --config-path=/etc/mysql/my.cnf
The MySQL server at 192.168.87.155:3306, version 5.7.33-0ubuntu0.16.04.1-log -
(Ubuntu), will now be checked for compatibility issues for upgrade to MySQL
8.0.23...

1) Usage of old temporal type
  No issues found

2) Usage of db objects with names conflicting with new reserved keywords
  No issues found

3) Usage of utf8mb3 charset
  No issues found
...
Errors:   0
Warnings: 1
Notices:  14

No fatal errors were found that would prevent an upgrade, but some potential issues were detected. Please ensure that the reported issues are not significant before upgrading.
```

这个工具将进行 21 项检查，逐一报告检查结果并给出进一步的检查或整改建议。

1.5.2 原地升级

MySQL 的升级方法分为原地升级和逻辑升级两种。原地升级是指直接在原来的数据库上进行升级，方法如下。

1. 关闭原 MySQL 实例

控制 MySQL 关闭时执行步骤的参数是 innodb_fast_shutdown，这个参数有 0、1 和 2 三个选项，分别被称为慢关闭、快关闭和冷关闭，应将该参数设置为 0 后再使用如下的代码关闭 MySQL 实例：

```
$ mysqladmin -u root -p shutdown
```

2. 升级 MySQL 的执行文件或包

以基于 YUM 源的安装包为例，先终止 MySQL 5.7 的 YUM 源，激活 MySQL 8.0 的源，代码如下：

```
$ sudo yum-config-manager --disable mysql57-community
$ sudo yum-config-manager --enable mysql80-community
```

然后升级 MySQL 的服务包，代码如下：

```
$ sudo yum update mysql-server
```

3. 启动安装好的 MySQL 8.0

启动时系统参数--datadir 指向原来的数据目录，在启动的过程中会自动进行数据升级，数据升级分为如下两步。

第一步：升级数据字典。包括：

- mysql 系统数据库中的数据字典表。
- 系统数据库 performance_schema 和 information_schema。

第二步：升级 MySQL 服务。包括：

- mysql 系统数据库中的系统表。
- 系统数据库 sys。
- 用户数据库。

这两步的升级操作由 mysqld 的选项--upgrade 决定，--upgrade 有 4 个值，它们和这两步的对应关系如表 1.2 所示。

表 1.2

--upgrade 的值	与第一步的对应关系	与第二步的对应关系
AUTO	根据需要执行	根据需要执行
NONE	不执行	不执行
MINIMAL	根据需要执行	不执行
FORCE	根据需要执行	执行

--upgrade 的默认值是 AUTO。由于 MySQL 8.0 中的数据字典已经放入 InnoDB 表中，MySQL 5.7 中的 frm 文件将在升级过程中被自动删除。

1.5.3 逻辑升级

逻辑升级是指把源库的数据逻辑导出到升级后的空库的升级方法，这种方法比原地升级安全性高，但因为需要进行数据导出和导入，用时较长。这种方法又可以分为全库升级和元数据升级两种。

1. 全库升级

全库升级是把整个库的数据导出，再用 MySQL 8.0 创建一个新的空库，最后把导出的数据导入新库中，完成升级的过程，具体步骤如下。

(1) 从源库中将数据导出。使用 mysqldump 命令的代码如下：

```
$ mysqldump -u root -p  --add-drop-table --routines --events
  --all-databases --force > data-for-upgrade.sql
```

(2) 关闭源数据库。

(3) 安装 MySQL 8.0 软件包。

(4) 数据库初始化，使用 MySQL 8.0 生成一个新的空库，代码如下：

```
$ mysqld --initialize --datadir=/path/to/8.0-datadir
```

在数据库初始化过程中，会在错误日志中保存生成的 root 的临时密码。

(5) 启动 MySQL 8.0。启动时系统参数--datadir 指向新的数据目录。

(6) 重置 root 的密码，代码如下：

```
$ mysql -u root -p
Enter password: ****    # 输入初始化时生成的 root 的临时密码
mysql> alter user user() identified by 'new password';
```

(7) 将前面导出的数据导入新库中，代码如下：

```
$ mysql -u root -p --force < data-for-upgrade.sql
```

(8) 数据升级。关闭 MySQL 实例，然后使用--upgrade=FORCE 启动 MySQL 实例。在启动过程中会自动完成数据升级。

2. 元数据升级

使用逻辑升级时，有的生产库数据量很大，进行数据导入和导出的时间很长，数据升级时的转换时间也比较长，可以先进行元数据升级，升级完成后再将实际数据导入，这种升级方法的数据导入和导出的主要步骤如下。

(1) 第一次导出时只导出生产实例的元数据，相应的 mysqldump 命令如下：

```
$ mysqldump --all-databases --no-data --routines --events > dump-defin.sql
```

其中的--no-data 表示只导出对象的定义语句，并不导出实际数据，然后再用如下代码将导出的数据导入升级实例中：

```
$ mysql -u root -p --force < dump-defin.sql
```

这个升级实例中并不包括实际数据，因此可以很快完成数据的导入和导出。接着可以使用--upgrade=FORCE 启动 MySQL 实例。在启动过程中会自动完成数据升级。待解决了所有的兼容问题后，再将实际数据导入升级实例中。

(2) 第二次导出不包括对象定义语句的实际数据，相应的 mysqldump 命令如下：

```
$ mysqldump --all-databases --no-create-info > dump-data.sql
```

其中的--no-create-info 表示只导出实际数据,并不导出元数据,然后再将数据导入升级实例中,代码如下:

```
shell> mysql -u root -p --force < dump-data.sql
```

接着检查表中的数据和进行业务测试,完成升级。

1.6 Linux 对 MySQL 的强制访问控制

为了提高 Linux 系统的安全性,在 Linux 系统上通常会使用 SELinux(Red Hat 类 Linux 上使用)或 AppArmor(Debian 类 Linux 上使用)实现强制访问控制(mandatory access control,MAC)。这两种安全机制都是一个内核级的安全机制,集成在 Linux 内核中,它们允许管理员细粒度地定义访问控制,而未经定义的访问一律禁止。对于 MySQL 数据库的强制访问控制策略通常是激活的,如果用户采用默认的配置,并不会感到强制访问控制策略对 MySQL 数据库的影响,但一旦用户修改了 MySQL 数据库的默认配置,如默认的数据目录或监听端口,MySQL 数据库的活动就会被 SELinux 或 AppArmor 阻止,数据库将无法启动。用户在新安装的 MySQL 数据库上会经常遇到这类故障。

1.6.1 SELinux 与 MySQL

1. SELinux 的简介

SELinux 有如下三种工作模式。

(1) enforcing:强制模式。任何违反策略的行为都会被禁止,并且产生警告信息。

(2) permissive:允许模式。违反策略的行为不会被禁止,只产生警告信息。

(3) disabled:关闭 SELinux。

使用 getenforce 命令来显示 SELinux 的当前模式。更改模式使用 setenforce 0(设置为允许模式)或 setenforce 1(强制模式)。这些设置重启后就会失效,可以编辑/etc/selinux/config 配置文件并设置 SELinux 变量为 enforcing、permissive 或 disabled,保存设置让其重启后也有效。

使用如下命令查看 SELinux 的状态:

```
$ sudo sestatus
SELinux status:                 enabled
SELinuxfs mount:                /sys/fs/selinux
SELinux root directory:         /etc/selinux
Loaded policy name:             targeted
Current mode:                   enforcing
Mode from config file:          enforcing
Policy MLS status:              enabled
Policy deny_unknown status:     allowed
Max kernel policy version:      28
```

2. 查看 MySQL 的 SELinux 的上下文

可以使用 ps -eZ 查看 mysqld 进程的 SELinux 的上下文，代码如下：

```
$ sudo ps -eZ | grep mysqld
system_u: system_r: mysqld_t: s0     2381 ?          00: 01: 00 mysqld
```

也可以使用 ls -dZ 查看 MySQL 数据目录的 SELinux 的上下文，代码如下：

```
$ sudo ls -dZ /var/lib/mysql
drwxr-x--x. mysql mysql system_u: object_r: mysqld_db_t: s0 /var/lib/mysql
```

参数说明如下：

（1）system_u 是系统进程和对象的 SELinux 用户标识。

（2）system_r 是用于系统进程的 SELinux 角色。

（3）objects_r 是用于系统对象的 SELinux 角色。

（4）mysqld_t 是与 mysqld 进程相关的 SELinux 类型。

（5）mysqld_db_t 是与 MySQL 数据目录相关的 SELinux 类型。

3. 修改对 MySQL 数据目录的访问控制

如果把 MySQL 数据目录从默认的/var/lib/mysql 改为其他目录，如/disk1/data，SELinux 将会阻止 mysqld 进程访问新的数据目录，从而造成 MySQL 无法启动，相关拒绝访问的信息记录在审计日志(/var/log/audit/audit.log)中，代码如下：

```
$ sudo tail -100 /var/log/audit/audit.log|grep mysqld
...
type=AVC msg=audit(1629100778.112: 5132): avc: denied  { write } for
pid=17855 comm="mysqld" name="data" dev="dm-0" ino=137756704
scontext=system_u: system_r: mysqld_t: s0
tcontext=unconfined_u: object_r: default_t: s0
tclass=dir
type=SYSCALL msg=audit(1629100778.112: 5132): arch=c000003e syscall=2
success=no exit=-13 a0=3fd69e0 a1=42 a2=1a0 a3=fffff000 items=0 ppid=1
pid=17855 auid=4294967295 uid=27 gid=27 euid=27
suid=27 fsuid=27 egid=27 sgid=27 fsgid=27 tty=(none) ses=4294967295 comm="mysqld"
exe="/usr/sbin/mysqld" subj=system_u: system_r: mysqld_t: s0 key=(null)
```

在审计日志中可以看到/usr/sbin/mysqld 命令被阻止了，这个进程号是 17855，在 MySQL 的错误日志里可以查到该进程号，如下所示：

```
$ sudo tail -100 /var/log/mysqld.log|grep 17855
2021-08-16T07: 59: 38.101354Z 0 [System] [MY-010116] [Server] /usr/sbin/mysqld (mysqld
8.0.23) starting as process 17855
```

在审计日志中还可以看到被阻止用户的用户号和组号都是 27，检查 mysql 用户的信息如下：

```
$ sudo id mysql
uid=27(mysql) gid=27(mysql) groups=27(mysql)
```

根据这些信息可以判断 mysql 用户启动 mysqld 进程时被 SELinux 阻止了，一个简单的解决方法是把 SELinux 关闭或改成允许模式后再启动 MySQL 数据库,但这样会把所有的 SELinux 的安全策略都终止了,留下了安全隐患。更专业的做法是把新的 MySQL 数据目录增加到 mysqld_db_t 这个 SELinux 类型中,如使用 semanage fcontext 命令的-a 选项增加一个目录为/disk1/data 的 MySQL 数据目录,代码如下:

```
$ sudo semanage fcontext -a -t mysqld_db_t "/disk1/data(/.*)?"
```

然后使用 restorecon 命令恢复这个数据目录对应的 SELinux 上下文,代码如下:

```
$ sudo restorecon -Rv /disk1/data
```

最后可以用 semanage fcontext 命令的-l 选项进行检查,代码如下:

```
$ sudo semanage fcontext -l|grep mysqld_db_t
/var/lib/mysql(/.*)?      all files      system_u:object_r:mysqld_db_t:s0
/disk1/data(/.*)?         all files      system_u:object_r:mysqld_db_t:s0
```

此时可发现 mysqld_db_t 类型现在有两条记录,分别是系统默认的和刚才增加的。再启动 mysqld 服务即可成功。

4. 修改对 MySQL 其他对象的访问控制

除了可以修改对 MySQL 数据目录的访问控制外,还可以采用类似的方法修改对其他 MySQL 对象的访问控制,例如,控制 MySQL 的错误日志的类型是 mysqld_log_t,采用如下命令增加 MySQL 的错误日志的记录:

```
$ sudo semanage fcontext -a -t mysqld_log_t "/path/to/my/custom/error.log"
$ sudo restorecon -Rv /path/to/my/custom/error.log
```

控制 MySQL 的 PID 文件的类型是 mysqld_var_run_t ,采用如下命令增加 MySQL 的 PID 文件的记录:

```
$ sudo semanage fcontext -a -t mysqld_var_run_t
"/path/to/my/custom/pidfile/directory/.*?"
$ sudo restorecon -Rv /path/to/my/custom/pidfile/directory
```

控制 MySQL 监听端口的类型是 mysqld_port_t ,采用如下命令增加一个 3307 的监听端口:

```
$ sudo semanage port -a -t mysqld_port_t -p tcp 3307
```

1.6.2　AppArmor 与 MySQL

1. AppArmor 简介

AppArmor 与 SELinux 的最大区别是 AppArmor 使用路径名作为安全标签,而 SELinux 是使用文件的 inode。AppArmor 通过一个配置文件(profile)来指定一个应用程序的相关权限。

AppArmor 配置文件有如下两种工作模式。

（1）Enforced/Confined：按照配置文件对进程的行为进行限制，并对于违反限制的进程进行日志记录。

（2）Complaining/Learning：不对进程的行为进行限制，只对进程的行为进行记录。这种模式可用于测试或开发新的 AppArmor 的配置文件。

apparmor-utils 包里的工具可以用于改变 AppArmor 配置文件的工作模式，查询配置文件的工作模式，创建新的配置文件等，例如，可以使用 apparmor_status 命令查询当前配置文件的状态，代码如下：

```
$ sudo apparmor_status
apparmor module is loaded.
7 profiles are loaded.
7 profiles are in enforce mode.
   /sbin/dhclient
   /usr/lib/NetworkManager/nm-dhcp-client.action
   /usr/lib/NetworkManager/nm-dhcp-helper
   /usr/lib/connman/scripts/dhclient-script
   /usr/sbin/mysqld
   /usr/sbin/ntpd
   /usr/sbin/tcpdump
0 profiles are in complain mode.
2 processes have profiles defined.
2 processes are in enforce mode.
   /usr/sbin/mysqld (6220)
   /usr/sbin/ntpd (12360)
0 processes are in complain mode.
0 processes are unconfined but have a profile defined.
```

可以看到，有两个进程的策略处于激活状态，其中包括 mysqld。

所有的 AppArmor 的配置文件都在/etc/apparmor.d 目录下，对该目录进行操作可以改变所有配置文件的模式，例如，可以使用如下的 aa-complain 命令将所有配置文件设置成 complaining 模式：

```
$ sudo aa-complain /etc/apparmor.d/*
```

也可以使用如下的 aa-cenforce 命令将所有配置文件设置成 enforced 模式：

```
$ sudo aa-enforce /etc/apparmor.d/*
```

apparmor_parser 命令可用于将配置文件装载到 Linux 的系统内核中，代码如下：

```
$ cat /etc/apparmor.d/profile.name | sudo apparmor_parser -a
```

如果修改了配置文件，可以使用如下代码重新装载：

```
$ cat /etc/apparmor.d/profile.name | sudo apparmor_parser -r
```

也可以使用 systemctl 命令重新装载所有的配置文件，代码如下：

```
$ sudo systemctl reload apparmor.service
```

也可以先将配置文件链接到/etc/apparmor.d/disable 目录，再用 apparmor_parser 的-

R 选项终止配置文件的 enforced 模式，代码如下：

```
$ sudo ln -s /etc/apparmor.d/profile.name /etc/apparmor.d/disable/
$ sudo apparmor_parser -R /etc/apparmor.d/profile.name
```

对于一个已经被终止了 enforced 模式的配置文件，可以通过删除/etc/apparmor.d/disable 目录下的链接，再用 apparmor_parser 的-a 选项将配置文件重新置于 enforced 模式，代码如下：

```
$ sudo rm /etc/apparmor.d/disable/profile.name
$ cat /etc/apparmor.d/profile.name | sudo apparmor_parser -a
```

2. MySQL 的 AppArmor 配置文件

MySQL 的 AppArmor 配置文件是/etc/apparmor.d/usr.sbin.mysqld，其中 usr.sbin.mysqld 代表执行程序/usr/sbin/mysqld，也就是 MySQL 服务的主程序，查看该配置文件的代码如下：

```
$ cat /etc/apparmor.d/usr.sbin.mysqld
#include <tunables/global>

/usr/sbin/mysqld {
  #include <abstractions/base>
  #include <abstractions/nameservice>
  #include <abstractions/user-tmp>
  #include <abstractions/mysql>
  #include <abstractions/winbind>

# Allow system resource access
  /sys/devices/system/cpu/ r,
  /sys/devices/system/node/ r,
  /sys/devices/system/node/** r,
  /proc/*/status r,
  capability sys_resource,
  capability dac_override,
  capability setuid,
  capability setgid,
  capability sys_nice,

# Allow network access
  network tcp,

  /etc/hosts.allow r,
  /etc/hosts.deny r,

# Allow config access
  /etc/mysql/** r,

# Allow pid, socket, socket lock file access
  /var/run/mysqld/mysqld.pid rw,
  /var/run/mysqld/mysqld.sock rw,
```

```
  /var/run/mysqld/mysqld.sock.lock rw,
  /var/run/mysqld/mysqlx.sock rw,
  /var/run/mysqld/mysqlx.sock.lock rw,
  /run/mysqld/mysqld.pid rw,
  /run/mysqld/mysqld.sock rw,
  /run/mysqld/mysqld.sock.lock rw,
  /run/mysqld/mysqlx.sock rw,
  /run/mysqld/mysqlx.sock.lock rw,

# Allow systemd notify messages
  /{,var/}run/systemd/notify w,

# Allow execution of server binary
  /usr/sbin/mysqld mr,
  /usr/sbin/mysqld-debug mr,

# Allow plugin access
  /usr/lib/mysql/plugin/ r,
  /usr/lib/mysql/plugin/*.so* mr,

# Allow error msg and charset access
  /usr/share/mysql/ r,
  /usr/share/mysql/** r,
  /usr/share/mysql-8.0/ r,
  /usr/share/mysql-8.0/** r,

# Allow data dir access
  /var/lib/mysql/ r,
  /var/lib/mysql/** rwk,

# Allow data files dir access
  /var/lib/mysql-files/ r,
  /var/lib/mysql-files/** rwk,

# Allow keyring dir access
  /var/lib/mysql-keyring/ r,
  /var/lib/mysql-keyring/** rwk,

# Allow log file access
  /var/log/mysql/ r,
  /var/log/mysql/** rw,

# Allow access to openssl config
  /etc/ssl/openssl.cnf r,

   # Site-specific additions and overrides. See local/README for details.
   #include <local/usr.sbin.mysqld>
}
```

从配置文件可以看到AppArmor对mysqld进程的访问控制粒度非常细，包括所有类型的文件、网络、系统资源等，这大大提高了安全性。但如果需要访问非默认的资源，就会被

AppArmor阻止。例如,MySQL默认的数据目录是/var/lib/mysql,如果把这个目录改成其他目录,如/disk1/data,因为AppArmor会阻止mysqld进程对新数据目录的访问,MySQL将无法启动,在Ubuntu的系统日志 /var/log/syslog 中会出现如下所示的记录:

```
Jan  1 23:13:44 scutech kernel: [3445518.170016] audit: type=1400
audit(1609555424.049:144): apparmor="DENIED" operation="mknod"
profile="/usr/sbin/mysqld"
name="/disk1/data/mysqld_tmp_file_case_insensitive_test.lower-test"
pid=10740 comm="mysqld" requested_mask="c" denied_mask="c" fsuid=0 ouid=0
Jan  1 23:13:44 scutech kernel: [3445518.178482] audit: type=1400
audit(1609555424.057:145): apparmor="DENIED" operation="mknod"
profile="/usr/sbin/mysqld"
name="/disk1/data/mysqld_tmp_file_case_insensitive_test.lower-test"
pid=10740 comm="mysqld" requested_mask="c" denied_mask="c" fsuid=0 ouid=0
Jan  1 23:13:44 scutech kernel: [3445518.180727] audit: type=1400
audit(1609555424.061:146): apparmor="DENIED" operation="mknod"
profile="/usr/sbin/mysqld" name="/disk1/data/error.log" pid=10740
comm="mysqld" requested_mask="c" denied_mask="c" fsuid=123 ouid=123
Jan  1 23:13:44 scutech kernel: [3445518.180731] audit: type=1400
audit(1609555424.061:147): apparmor="DENIED" operation="mknod"
profile="/usr/sbin/mysqld" name="/disk1/data/error.log" pid=10740
comm="mysqld" requested_mask="c" denied_mask="c" fsuid=123 ouid=123
```

此时可以使用如下代码将该配置文件状态改成compaining模式:

```
$ sudo aa-complain /usr/sbin/mysqld
Setting /usr/sbin/mysqld to complain mode.
```

但更安全的做法是相应地修改配置文件。MySQL的默认AppArmor配置文件是/etc/apparmor.d/usr.sbin.mysqld,除了该配置文件外,还有一个补充配置文件/etc/apparmor.d/local/usr.sbin.mysqld,该文件初始时为空,通常可以将补充的访问配置规则加入该配置文件,例如,可以在该文件中增加下面的内容:

```
/disk1/data/ r,
/disk1/data/** rwk,
```

其中,将读(r)的权限赋予目录/disk1/data/,将读(r)、写(w)、锁(k)的权限递归地赋予/disk1/data/目录下的所有内容。

然后使用如下代码将修改后的配置文件重新装载到Linux的内核中:

```
$ sudo service apparmor reload
```

重新启动MySQL后,新的规则即可生效。

1.7 启动排错

很多用户安装完MySQL数据库后,启动时都会遇到错误,通常遇到的第一个错误是MySQL服务无法启动,解决了这个问题后还可能遇到MySQL客户端连接不上的问题。本节将说明如何排除这两类常见的错误。

1.7.1 MySQL 服务无法启动

MySQL 服务启动失败最常见的原因有两类，分别是无法访问系统资源和系统参数设置错误。

1. 无法访问系统资源

MySQL 不能访问启动需要的资源，如文件、端口等，是造成 MySQL 无法启动的一个常见原因。可以使用 Linux 里最常用的命令 ls 来查看对应文件和目录的权限。另外一个更实用的方法就是进行测试，但由于 Linux 中用于启动 mysqld 进程的 mysql 用户通常是个伪用户，是不能登录的，因此可以用 sudo 命令将当前用户转换成 mysql 用户后进行测试，例如：

```
$ sudo -u mysql touch /var/lib/mysql/fileb
```

该命令可以用来测试 mysql 用户对/var/lib/mysql/目录的写权限，其中参数-u mysql 是指这条命令用 mysql 用户来执行，如果 mysql 用户对这个目录没有写权限，该命令会执行失败。遇到这种情况，通常通过修改对应文件或目录的权限，或属主来解决。

如果文件系统出了问题，从操作系统层面也不能访问资源，例如，有如下一个案例：

```
$ sudo cp ibdata1 /tmp
cp: error reading 'ibdata1': Input/output error
```

如果操作系统能正确地访问资源，并且 mysql 用户的权限也是正确的，mysqld 进程访问相应的文件和目录也仍然有可能被拒绝，如下面这个例子：

```
mysql> system sudo -u mysql touch  /home/mysql/data/a
mysql> create table t1 (id int primary key,n varchar(10)) data
directory='/home/mysql/data';
ERROR 1030 (HY000): Got error 168 from storage engine
```

测试证明，mysql 用户有/home/mysql/data 目录的访问权限，但创建文件还是提示失败，这时通常是因为 mysqld 进程的访问被 Linux 的 SELinux 或 AppArmor 阻止。从上面的命令可以看到，创建的表不在 mysql 的默认目录下，因此 SELinux 或 AppArmor 的 policy 里没有包含这个目录的访问权限，此时只要修改对应的 policy 即可，也可以通过终止 SELinux 或 AppArmor 来启动 mysqld 进程。

有时虽然对系统资源有访问权限，但此时可能系统资源已经被其他进程占用，从而出现故障，例如：

```
$ sudo mysqld --no-defaults --user mysql
2020-11-03T03:36:07.519419Z 0 [System] [MY-010116] [Server] /usr/sbin/mysqld (mysqld
8.0.19) starting as process 21171
2020-11-03T03:36:07.740347Z 1 [ERROR] [MY-012574] [InnoDB] Unable to lock ./ibdata1
error: 11
```

如上故障产生的原因是另外一个 mysqld 进程已经启动并占用了对应的文件。

2. 系统参数设置错误

MySQL 8.0 共有六百多个系统参数，这些系统参数之间还有依赖关系，因此因系统参数设置错误造成 MySQL 无法启动的原因也很常见。MySQL 启动时调用的系统参数放在参数文件中，在 mysqld 的帮助信息中可以查询到 MySQL 启动时读取参数文件的顺序，代码如下：

```
$ mysqld --verbose --help | grep "Default options "  -A 1
Default options are read from the following files in the given order:
/etc/my.cnf /etc/mysql/my.cnf ~/.my.cnf
```

可以看到，MySQL 启动时最优先读取的参数文件是/etc/my.cnf，其次是/etc/mysql/my.cnf，最后是 home(家)目录下的.my.cnf。知道了 MySQL 参数文件的读取顺序，就可以检查对应的参数文件，找出其中的错误，如果觉得参数文件的可读性不强，可以使用 mysqld 命令的选项--print-defaults 显示将要调用的系统参数，代码如下：

```
$ mysqld --print-defaults
/usr/sbin/mysqld would have been started with the following arguments:
...
```

mysqld--print-defaults 命令是对参数文件进行解析，然后显示 mysqld 在实际运行中调用的系统参数。注意该命令显示完系统参数后就退出，并不会真正运行 mysqld。mysqld--print-defaults 命令和 my_print_defaults mysqld 类似，只不过后者的显示方式是一行一个系统参数。

知道了 mysqld 启动时调用的系统参数就可以开始对可疑的系统参数进行调试，这里介绍一种使用排除法找出阻止 MySQL 启动的错误系统参数的方法，首先使用一个稳妥的方法启动 mysqld，代码如下：

```
$ sudo mysqld --no-defaults --console --log-error-verbosity=3 --user mysql &
```

该指令里没有可疑的系统参数，只包括了如下四个系统参数。

(1) 第一个系统参数是--no-defaults，该系统参数的作用是通知 mysqld 在启动时不要读任何参数文件，这样 mysqld 启动时就避免了参数文件里可能存在的错误系统参数的干扰。

(2) 第二个系统参数是--console，该系统参数仅仅用于 Windows 操作系统，Linux 下不需要，它会把错误信息输出到屏幕上，该系统参数带来的一个弊端是所有的信息都输出到屏幕上，让屏幕显得比较乱，但对于调试却很方便。

(3) 第三个系统参数是--log-error-verbosity＝3，该系统参数表示将显示详细的错误日志。

(4) 第四个系统参数是--user mysql，表示使用 mysql 用户调用 mysqld 进程。

使用这四个系统参数启动 mysqld，如果能够启动成功，说明参数文件中剩下的系统参数有错误，就用 mysqladmin shutdown 关闭 MySQL 服务，然后在这个命令行后加上一个可疑的系统参数，再重新启动 mysqld。如果不能启动成功，就证明新增加的系统参数有误；如果能够启动成功，就再次关闭 MySQL 服务，继续增加其他可疑的系统参数后再测试，就这样采用排除法逐步地找出错误的系统参数。

1.7.2 MySQL 客户端连接失败

刚刚安装的系统经常遇到 MySQL 客户端连接不上的情况，这里介绍如何排除这类故障。

1. 设置参数 log_error_verbosity 为 3

MySQL 8.0 使用参数 log_error_verbosity 控制日志记录的详细程度，该参数可以设置为 1、2 和 3，默认值为 2，设置为 3 时记录的信息最全，该参数可以采用如下命令联机设置：

```
mysql >  set global log_error_verbosity = 3;
```

这时错误日志中会记录用户登录失败的信息，而设置为 2 时不会记录这类信息，例如：

```
44 [Note] [MY - 010926] [Server] Access denied for user 'wrong_user'@'localhost' (using
password: YES)
```

2. 检查服务进程 mysqld

首先到服务器上检查 mysqld 进程是否还在运行，可以采用 mysqladmin ping 命令进行检查，代码如下：

```
$ sudo mysqladmin ping - uroot - p
Enter password:
mysqld is alive
```

也可以使用 ps 命令检查进程是否存在，指令如下：

```
$ ps - ef|grep mysqld|grep - v grep
mysql     18020     1  0 Dec24 ?          00:06:20 /usr/sbin/mysqld
```

3. 检查客户端和服务进程 mysqld 之间的通信

使用 telnet 在服务端进行网络连通的测试，代码如下：

```
$ telnet localhost 3306
```

如果出现 connected to localhost 表示本地能通，再到客户端的机器上把 localhost 换成 MySQL 服务器的 IP 地址进行测试。如果不通，通常有两种原因：OS 或网络的问题，如被防火墙隔绝；mysqld 自身根本没有侦听客户端的连接请求。

MySQL 的服务使用参数--skip-networking 时会跳过侦听客户端的网络连接，代码如下：

```
$ sudo mysqld -- no - defaults -- user mysql   -- skip - networking &
```

此时采用如下命令在服务端查找监听进程不会有任何返回记录：

```
$ sudo netstat - plunt|grep 3306
```

参数--bind-address 会限制监听的地址，如把该参数设置为 127.0.0.1 时启动 mysqld，将只接受来自本地的连接，可以把该参数设置为 *、0.0.0.0 或::，让其接受所有地址的连接。

4. 账号密码错误

当在客户端输入的密码错误时，会有如下提示：

```
$ sudo mysql -uroot -perrorpassword
mysql: [Warning] Using a password on the command line interface can be insecure.
ERROR 1045 (28000): Access denied for user 'root'@'localhost' (using password: YES)
```

在 MySQL 的错误日志里有如下记录：

```
7 [Note] [MY-010926] [Server] Access denied for user 'root'@'localhost' (using password: YES)
```

如果把参数--log-error-verbosity 设置成默认值 2 时是没有这个提示的，也就是说没有 note 类型的信息。

MySQL 中的一个账户由 User 和 Host 两部分组成，在 MySQL 中 mysql 数据库的 user 表中有两个字段：Host 和 User，这两个字段组成表的主键(Primary Key)列，主键列有唯一约束，唯一表示一个用户。如果出现类似下面的提示：

```
ERROR 1130 (HY000): Host '192.168.17.149' is not allowed to connect to this MySQL server
```

那么通常是 Host 字段指定的地址和客户端的地址不符，如 Host 字段是 localhost，而客户端和服务端却不在同一台服务器上。使用如下命令将 Host 字段改成通配符"%"，再刷新权限表即可：

```
mysql> update mysql.user set Host = '%' where user = 'root';
Query OK, 1 row affected (0.01 sec)
Rows matched: 1  Changed: 1  Warnings: 0
mysql> flush privileges;
Query OK, 0 rows affected (0.02 sec)
```

请注意，账户错误时提示的是 is not allowed to connect to this MySQL server，而密码错误时提示的是 Access denied for user。

1.7.3 MySQL 客户端到服务端通信出错

客户端和服务端之间的通信包大小受系统参数 max_allowed_packet 的控制，默认是 64MB，超过这个值时会被截断，例如：

```
mysql> \W
Show warnings enabled.
mysql>  select repeat('A', 10000000000);
+--------------------------+
| repeat('A', 10000000000) |
+--------------------------+
| NULL                     |
+--------------------------+
1 row in set, 1 warning (0.00 sec)

Warning (Code 1301): Result of repeat() was larger than max_allowed_packet (67108864)
 - truncated
```

客户端和服务端之间交互连接的超时时间受系统参数 wait_timeout 控制，默认为 28 800 秒，也就是 8 小时，超过该时间也没有通信将中断连接，例如，采用如下命令将该参数

设置成1秒后,连接很快会中断:

```
mysql > set wait_timeout = 1;
Query OK, 0 rows affected (0.00 sec)

mysql > select 1;
ERROR 2013 (HY000): Lost connection to MySQL server during query
```

1.8 实验

视频演示

1. 使用 RPM 包安装 MySQL 并初始化和启动

(1) 登录 MySQL 官方网站,下载 RPM 格式的 MySQL 安装包。

(2) 安装 MySQL。

(3) 初始化 MySQL。

(4) 修改 root@localhost 用户的初始密码。

(5) 安装 sakila 示例数据库。

视频演示

2. 使用排除法找出阻止 MySQL 启动的错误参数

使用排除法找出参数文件中的错误系统参数,并进行更正。

视频演示

3. SELinux 阻止 MySQL 启动的故障排除

因为 MySQL 的数据目录没有设置到默认的目录,SELinux 会阻止 mysqld 的启动。分析 mysqld 启动受阻的原因并进行排除。

第2章

账号和权限

MySQL 安装完成后的第一个工作通常是创建账号，本章介绍了账号、权限、访问控制、角色、代理用户、密码、登录等方面的内容。在开发测试环境下，很多人为了方便，登录时不愿输入密码，本章介绍了三种不输入密码登录 MySQL 的方法，并分析了它们的安全性；用户忘记密码的情况也会经常发生，本章还介绍了重置 root 密码的方法。

2.1 账号

2.1.1 账号的组成

MySQL 的账号由用户名和客户端主机名组成，账号信息保存在 mysql 系统数据库的 user 表中，通过如下代码可查看该表的主键如下：

```
mysql > select index_name,column_name  from information_schema.statistics where table_schema
= 'mysql' and table_name = 'user';
+------------+-------------+
| INDEX_NAME | COLUMN_NAME |
+------------+-------------+
| PRIMARY    | Host        |
| PRIMARY    | User        |
+------------+-------------+
2 rows in set (0.02 sec)
```

可以看到，表的主键有两个字段：一个是 User 字段代表用户名；另一个是 Host 字段代表客户端主机名。这两个字段唯一标识了一个账号，同样的用户名从不同主机名的客户端登录，表示不是同一个用户，如下面的用户 root 有三个账号分别对应三个主机名：

```
mysql > select host from mysql.user where user = 'root';
+-----------+
```

```
| host        |
+-------------+
| %           |
| 127.0.0.1   |
| localhost   |
+-------------+
3 rows in set (0.00 sec)
```

如上三个主机名中，%代表任意主机名；127.0.0.1和localhost都表示本机，但这两个主机名分别对应不同的账号。在Linux中，环境变量MYSQL_HOST指定默认的主机名，使用如下命令将当前主机名设定为localhost：

```
$ export MYSQL_HOST=localhost
```

查询当前的账号，相关命令及其运行结果如下：

```
$ mysql -e 'select current_user()'
+-----------------+
| current_user()  |
+-----------------+
| root@localhost  |
+-----------------+
```

把当前主机名设定为127.0.0.1，命令如下：

```
$ export MYSQL_HOST=127.0.0.1
```

查询当前的账号，相关命令及其运行结果如下：

```
$ mysql -e 'select current_user()'
+-----------------+
| current_user()  |
+-----------------+
| root@127.0.0.1  |
+-----------------+
```

从以上结果可以看到，root@localhost和root@127.0.0.1并不是同一个账号。

现在再创建一个非本地用户，该用户只能从client_a主机上进行登录，命令如下：

```
mysql> create user scutech@client_a identified by 'dingjia';
```

再在/etc/hosts文件中插入一个客户端的IP地址和对应的主机名，命令如下：

```
$ sudo bash -c 'echo "192.168.18.227 client_a">> /etc/hosts'
```

然后从客户端连接到MySQL，命令如下：

```
$ mysql -uscutech -pdingjia -h 192.168.18.229
```

登录成功后，检查客户端的主机，相关命令及其运行结果如下：

```
mysql> select host from information_schema.processlist
where id = connection_id();
+-----------------+
```

```
| host             |
+------------------+
| client_a: 55780  |
+------------------+
1 row in set (0.07 sec)
```

检查客户端的账号，相关命令及其运行结果如下：

```
mysql> select user(),current_user();
+-------------------+-------------------+
| user()            | current_user()    |
+-------------------+-------------------+
| scutech@client_a  | scutech@client_a  |
+-------------------+-------------------+
1 row in set (0.01 sec)
```

再修改 IP 地址的解析，将 client_a 改成 client_b，命令如下：

```
$ sudo sed -e 's/client_a/client_b/' -i /etc/hosts
```

重新启动 MySQL 服务端后，再次从客户端连接到 MySQL，命令如下：

```
$ mysql -uscutech -pdingjia -h 192.168.18.229
mysql: [Warning] Using a password on the command line interface can be insecure.
ERROR 1045 (28000): Access denied for user 'scutech'@'client_b' (using password: YES)
```

从如上提示中可以看到，客户端到 MySQL 的连接失败，因为该地址已经被解析成 client_b，而 client_b 在 MySQL 服务端里没有对应的账号。

远程连接时，MySQL 服务端会对 IP 进行反向解析，这个行为可以通过把系统参数 skip_name_resolve 设置为 true 来跳过。设置完成后再次从客户端连接 MySQL 到服务端，命令如下：

```
$ mysql -uscutech -pdingjia -h 192.168.18.229
mysql: [Warning] Using a password on the command line interface can be insecure.
ERROR 1045 (28000): Access denied for user 'scutech'@'192.168.18.227' (using password: YES)
```

可以看到，该 IP 地址没有被解析成主机名，而该 IP 在 MySQL 服务端里没有对应的账号，因此登录失败。

2.1.2 四个保留账号

MySQL 数据库在初始化时会创建四个保留账号，在 mysql. user 表里查询这四个账号的 host 和 user 字段如下：

```
mysql> select user,host,account_locked from mysql.user;
+-------------------+-----------+-----------------+
| user              | host      | account_locked  |
+-------------------+-----------+-----------------+
| mysql.infoschema  | localhost | Y               |
| mysql.session     | localhost | Y               |
```

```
| mysql.sys         | localhost  | Y              |
| root              | localhost  | N              |
+-------------------+------------+----------------+
4 rows in set (0.00 sec)
```

（1）'root'@'localhost'：用于管理的系统账号，它具有在MySQL数据库中执行所有操作的权限，也被称为超级用户。严格地讲，它并不是一个保留账号，因为它是这四个账号中唯一一个没有被锁住的账号。host对应的字段是localhost代表这个账号只能从本机登录，剩下的三个账号也是这样。

（2）'mysql.sys'@'localhost'：用于sys系统数据库的定义者（DEFINER）的账号，该账号是被锁住的，也就是不能用于客户端登录。

（3）'mysql.session'@'localhost'：用于内部插件登录MySQL的账号，该账号也是被锁住的。

（4）'mysql.infoschema'@'localhost'：给INFORMATION_SCHEMA系统数据库的定义者（DEFINER）的账号，该账号也是被锁住的。

2.1.3 创建账号

创建账号的基本语法如下：

```
CREATE USER [IF NOT EXISTS] account_name IDENTIFIED BY 'password' | IDENTIFIED WITH auth_plugin
BY 'password';
```

字段说明：

（1）account_name是用户的账号，由用户名和客户端主机名两部分组成，如账号scutech@localhost中，scutech是用户名，localhost是客户端主机名。当使用scutech用户从MySQL服务器的本机登录时使用此账号，但从其他机器登录时不能使用此账号。

（2）auth_plugin是账号的验证插件，在MySQL 8.0中默认的验证插件是caching_sha2_password，在MySQL 5.7中默认的验证插件是mysql_native_password。

（3）password是账号的密码明文，它经过验证插件加密后存储在mysql.user表的authentication_string字段中。

增加一个只能从本地登录的（用户名为scutech）账号的代码如下：

```
mysql > create user scutech@localhost identified by 'dingjia';
```

创建完成后将向mysql.user表中增加一条记录，查询该记录中相关字段的代码如下：

```
mysql > select user, host, authentication_string from mysql.user where user = 'scutech'\G
*************************** 1. row ****************************
                 user: scutech
                 host: localhost
authentication_string: *C784601FFE12A6CA353714520632A795C5B7A2CA
1 row in set (0.01 sec)
```

其中authentication_string是加密后的密码。

删除账号的代码如下：

```
mysql> drop user scutech@localhost;
```

2.1.4 修改密码

修改用户的密码可以使用如下命令：

```
mysql> alter user scutech@localhost identified by 'newpassword';
```

或者：

```
mysql> set password for scutech@localhost = 'newpassword';
```

修改当前用户密码的命令如下：

```
mysql> alter user user() identified by 'newpassword';
```

或者：

```
mysql> set password = 'newpassword';
```

也可以使用 mysqladmin 修改密码，命令如下：

```
$ mysqladmin -u scutech -p password 'newpassword'
```

使用如下命令可以强制用户的密码过期：

```
mysql> alter user scutech@localhost password expire;
```

用户仍然可以使用过期的密码登录，但登录后在修改密码前不能执行其他任何操作，相关命令及其运行结果如下：

```
mysql> select current_user();
+--------------------+
| current_user()     |
+--------------------+
| scutech@localhost  |
+--------------------+
1 row in set (0.01 sec)
mysql> select 1;
ERROR 1820 (HY000): You must reset your password using ALTER USER statement before executing
this statement.
```

即使执行一个简单的 select 操作都会被禁止，使用如下命令重新设置密码：

```
mysql> set password = 'dingjia';
Query OK, 0 rows affected (0.03 sec)
```

再次执行这个 select 操作，命令如下：

```
mysql> select 1;
+---+
| 1 |
+---+
```

```
| 1 |
+---+
1 row in set (0.00 sec)
```

可以使用 create user 或 alter user 给用户设定一个密码过期时间，例如：

```
mysql > create user scutech@localhost identified by 'please_change' password expire interval
30 day;
```

可以在配置文件中加入如下的系统参数，设置一个全局的默认用户密码过期时间：

```
[mysqld]
default_password_lifetime = 90
```

系统参数 default_password_lifetime 默认为 0，表示永远不过期。

从 MySQL 8.0.14 开始，MySQL 允许给一个账号设置两个密码，分别为主密码和辅助密码，这样可以方便应用程序的过渡，例如，当主库有多个副本时，主库的账号密码已修改，而复制到副本还需要一定的时间，这时账号可以用主密码在主库上登录，还可以用旧密码在副本上登录，例如：

```
mysql > alter user scutech@localhost identified by 'password_zhu' retain current password;
```

然后等待密码被复制到所有的副本，并且把所有的应用程序代码里登录到 scutech@localhost 用户的密码改成 password_zhu，都完成后再抛弃旧密码，命令如下：

```
mysql > alter user scutech@localhost discard old password;
```

2.2 权限

2.2.1 权限的分类

MySQL 的全部数据由若干数据库（database 或 schema）组成，数据库中包括对象，如表、视图、索引、存储过程等。权限从层次上也对应的分为如下三级。

(1) 全局级的管理权限。允许用户对 MySQL 进行管理操作，不局限于某个具体的数据库，管理权限包括关闭数据库、备份、管理防火墙等。如果在全局级定义了对数据库的操作权限，将对所有的数据库都起作用。

(2) 数据库级的权限。针对单个数据库及其里面的对象，如创建、增加、删除、修改、查询等权限。

(3) 对象级的权限。针对单个数据库里的具体对象的权限，这些对象包括表、视图、索引、存储过程等，权限包括创建、增加、删除、修改、查询、执行等。

MySQL 的管理权限可以分为静态权限和动态权限两类，静态权限是在 MySQL 代码编译时定义的；动态权限是在 MySQL 启动时定义的。动态权限还包括加载的插件定义的权限，动态权限存放在表 mysql.global_grants 中。使用 show privileges 可以显示 MySQL 服务支持的权限列表，显示的权限包括所有静态权限和所有当前注册的动态权限。代码如下：

```
mysql > show privileges\G
```

```
*************************** 1. row ***************************
Privilege: Alter
  Context: Tables
  Comment: To alter the table
*************************** 2. row ***************************
Privilege: Alter routine
  Context: Functions,Procedures
  Comment: To alter or drop stored functions/procedures
*************************** 3. row ***************************
Privilege: Create
  Context: Databases,Tables,Indexes
  Comment: To create new databases and tables
...
```

使用如下 show grants 命令可以查询某个用户拥有的权限：

```
mysql > show grants for scutech@localhost;
+-----------------------------------------------+
| Grants for scutech@localhost                  |
+-----------------------------------------------+
| GRANT USAGE ON *.* TO 'scutech'@'localhost'   |
+-----------------------------------------------+
1 row in set (0.00 sec)
```

可以发现，该用户只有 usage 权限，就是只能登录和查询 information_schema 数据库中的少量表，其他什么权限也没有。

运行如下命令将对这个用户从全局级(grant 语句中的 *.* 代表全局级)赋予所有(all)的权限：

```
mysql > grant all on *.* to  scutech@localhost;
Query OK, 0 rows affected (0.02 sec)
```

再用 show grants 查询这个用户拥有的权限，代码如下：

```
mysql > show grants for  scutech@localhost\G
*************************** 1. row ***************************
Grants for scutech@localhost:  GRANT SELECT, INSERT, UPDATE, DELETE, CREATE, DROP, RELOAD,
SHUTDOWN, PROCESS, FILE, REFERENCES, INDEX, ALTER, SHOW DATABASES, SUPER, CREATE TEMPORARY
TABLES, LOCK TABLES, EXECUTE, REPLICATION SLAVE, REPLICATION CLIENT, CREATE VIEW, SHOW VIEW,
CREATE ROUTINE, ALTER ROUTINE, CREATE USER, EVENT, TRIGGER, CREATE TABLESPACE, CREATE ROLE,
DROP ROLE ON *.* TO 'scutech'@'localhost'
*************************** 2. row ***************************
Grants for scutech@localhost:  GRANT APPLICATION_PASSWORD_ADMIN,AUDIT_ADMIN,BACKUP_ADMIN,
BINLOG_ADMIN,BINLOG_ENCRYPTION_ADMIN,CLONE_ADMIN,CONNECTION_ADMIN,ENCRYPTION_KEY_ADMIN,
GROUP_REPLICATION_ADMIN,INNODB_REDO_LOG_ARCHIVE,INNODB_REDO_LOG_ENABLE,PERSIST_RO_
VARIABLES_ADMIN,REPLICATION_APPLIER,REPLICATION_SLAVE_ADMIN,RESOURCE_GROUP_ADMIN,RESOURCE_
GROUP_USER,ROLE_ADMIN,SERVICE_CONNECTION_ADMIN,SESSION_VARIABLES_ADMIN,SET_USER_ID,SHOW_
ROUTINE,SYSTEM_USER,SYSTEM_VARIABLES_ADMIN,TABLE_ENCRYPTION_ADMIN,XA_RECOVER_ADMIN ON *.
* TO 'scutech'@'localhost'
2 rows in set (0.00 sec)
```

可以看到，在用 show grants 查询这个用户的权限时，发现有两条赋权语句，第一条语句赋予的是所有静态全局权限，第二条语句赋予的是所有动态权限。

2.2.2 权限表

MySQL 对用户权限的定义放在 mysql 系统数据库中的若干表中，正常情况下，用户不应该直接修改这些表，而是通过账号管理语句来修改这些表，账号管理语句包括创建账号(create user)、赋予(grant)权限和回收(revoke)权限等。下面分别介绍这些表。

1. 全局静态权限在表 mysql.user 中定义

用户拥有的全局静态权限在表 mysql.user 中定义，查看该表中以_priv 结尾的字段，相关代码及其运行结果如下：

```
mysql> select column_name,column_type from information_schema.columns where table_schema =
'mysql' and table_name = 'user' and column_name like '%_priv';
+-------------------------+---------------+
| COLUMN_NAME             | COLUMN_TYPE   |
+-------------------------+---------------+
| Alter_priv              | enum('N','Y') |
| Alter_routine_priv      | enum('N','Y') |
| Create_priv             | enum('N','Y') |
| Create_role_priv        | enum('N','Y') |
| Create_routine_priv     | enum('N','Y') |
| Create_tablespace_priv  | enum('N','Y') |
| Create_tmp_table_priv   | enum('N','Y') |
| Create_user_priv        | enum('N','Y') |
| Create_view_priv        | enum('N','Y') |
| Delete_priv             | enum('N','Y') |
| Drop_priv               | enum('N','Y') |
| Drop_role_priv          | enum('N','Y') |
| Event_priv              | enum('N','Y') |
| Execute_priv            | enum('N','Y') |
| File_priv               | enum('N','Y') |
| Grant_priv              | enum('N','Y') |
| Index_priv              | enum('N','Y') |
| Insert_priv             | enum('N','Y') |
| Lock_tables_priv        | enum('N','Y') |
| Process_priv            | enum('N','Y') |
| References_priv         | enum('N','Y') |
| Reload_priv             | enum('N','Y') |
| Repl_client_priv        | enum('N','Y') |
| Repl_slave_priv         | enum('N','Y') |
| Select_priv             | enum('N','Y') |
| Show_db_priv            | enum('N','Y') |
| Show_view_priv          | enum('N','Y') |
| Shutdown_priv           | enum('N','Y') |
| Super_priv              | enum('N','Y') |
| Trigger_priv            | enum('N','Y') |
| Update_priv             | enum('N','Y') |
+-------------------------+---------------+
```

```
31 rows in set (0.00 sec)
```

可以发现，这个表里有 31 个字段以_priv 结尾，它们代表用户对应的权限，从定义中也可以看到这些字段是枚举类型，只能有'N'或'Y'两种值，默认是'N'，也就是没有相应的权限。

赋予用户全局级 insert 权限的例子如下：

```
mysql > grant insert on * . * to scutech@localhost;
```

其中，*. * 代表全局级，也就是对所有对象生效，这样赋予权限后，用户在 mysql. user 表里对应的字段就被修改成 Y。代码如下：

```
mysql > select Insert_priv from user where user = 'scutech';
+-------------+
| Insert_priv |
+-------------+
| Y           |
+-------------+
1 row in set (0.00 sec)
```

2. 全局动态权限在表 global_grants 中定义

用户拥有的全局动态权限在表 global_grants 中定义，运行如下代码查询该表的 priv 字段可以得到如下动态权限的清单：

```
mysql > select distinct priv from mysql.global_grants ;
+----------------------------+
| priv                       |
+----------------------------+
| SYSTEM_USER                |
| BACKUP_ADMIN               |
| CLONE_ADMIN                |
| CONNECTION_ADMIN           |
| PERSIST_RO_VARIABLES_ADMIN |
| SESSION_VARIABLES_ADMIN    |
| SYSTEM_VARIABLES_ADMIN     |
| APPLICATION_PASSWORD_ADMIN |
| AUDIT_ADMIN                |
| BINLOG_ADMIN               |
| BINLOG_ENCRYPTION_ADMIN    |
| ENCRYPTION_KEY_ADMIN       |
| GROUP_REPLICATION_ADMIN    |
| INNODB_REDO_LOG_ARCHIVE    |
| INNODB_REDO_LOG_ENABLE     |
| REPLICATION_APPLIER        |
| REPLICATION_SLAVE_ADMIN    |
| RESOURCE_GROUP_ADMIN       |
| RESOURCE_GROUP_USER        |
| ROLE_ADMIN                 |
| SERVICE_CONNECTION_ADMIN   |
| SET_USER_ID                |
| SHOW_ROUTINE               |
```

```
| TABLE_ENCRYPTION_ADMIN        |
| XA_RECOVER_ADMIN              |
+-------------------------------+
25 rows in set (0.01 sec)
```

运行代码后可以看到，共有 25 项动态权限，当赋予一个用户某项动态权限后，会在这个表里相应地增加一条记录，例如下面的赋值语句：

```
mysql > grant audit_admin on * . * to  scutech@localhost;
```

可以发现，在表中增加了一条对应的记录：

```
mysql > select * from mysql.global_grants where user = 'scutech';
+---------+-----------+-------------+-------------------+
| USER    | HOST      | PRIV        | WITH_GRANT_OPTION |
+---------+-----------+-------------+-------------------+
| scutech | localhost | AUDIT_ADMIN | N                 |
+---------+-----------+-------------+-------------------+
1 row in set (0.00 sec)
```

其中，字段 WITH_GRANT_OPTION 表示是否有权把这项权限再赋予其他用户。

3. 数据库级别的权限在 db 表中定义

用户拥有的数据库权限在表 mysql. db 中定义，运行如下代码查看该表的主键：

```
mysql > select column_name from information_schema. statistics where table_schema = 'mysql' and
table_name = 'db' and index_name = 'PRIMARY';
+-------------+
| COLUMN_NAME |
+-------------+
| Host        |
| Db          |
| User        |
+-------------+
3 rows in set (0.00 sec)
```

可以发现，mysql. db 表的主键由三个字段决定，其中 Host 和 User 字段唯一表示了一个账号，Db 字段表示某个具体的数据库。

当赋予一个账号数据库级别的权限后，会在这个表里相应地增加一条记录，例如下面的赋值语句赋予了用户 sakila 数据库的 select 权限：

```
mysql > grant select on sakila. * to scutech@localhost;
```

运行如下代码即在表里面增加了如下一条对应的记录：

```
mysql > select * from mysql.db where user = 'scutech'\G
*************************** 1. row ***************************
                 Host:  localhost
                   Db:  sakila
                 User:  scutech
          Select_priv:  Y
          Insert_priv:  N
```

```
           Update_priv: N
           Delete_priv: N
           Create_priv: N
             Drop_priv: N
            Grant_priv: N
       References_priv: N
            Index_priv: N
            Alter_priv: N
 Create_tmp_table_priv: N
      Lock_tables_priv: N
      Create_view_priv: N
        Show_view_priv: N
   Create_routine_priv: N
    Alter_routine_priv: N
          Execute_priv: N
            Event_priv: N
          Trigger_priv: N
1 row in set (0.00 sec)
```

可以发现,除了这三个组成主键的字段外还有 19 个以_priv 结尾的字段分别代表对应的权限,新增的记录除了 Select_priv 的字段是 Y,其他的字段都是 N,代表用户 scutech@localhost 对 sakila 只有 select 的权限。

4. 其他权限表

除了前面介绍的表外,可以保存账号权限信息的表还有如下几种。

(1) 表级的权限在表 tables_priv 中定义。

(2) 字段级的权限在表 columns_priv 中定义。

(3) 存储过程和函数的权限在表 procs_priv 中定义。

(4) 代理账号的权限在表 proxies_priv 中定义。

收回权限使用 revoke 语句,如下面的语句收回了用户在 sakila 数据的 select 权限:

```
mysql > revoke select on sakila. * from scutech@localhost;
```

尽量不要直接用 SQL 语句修改这些权限表,如果出错可能会让你无法登录,如果一定要直接修改这些表,修改完成后要使用 FLUSH PRIVILEGES 把变更载入内存才能生效。要尽量使用账号管理语句,如 GRANT、REVOKE、SET PASSWORD 和 RENAME USER 来管理账号,这些语句在修改这些权限表时,会同时修改内存中的表。

对密码和全局级的权限变更在重新登录后才能生效;对数据库级的权限变更当用户下次使用 use 数据库名命令时才能生效;对角色、表、字段、存储过程的权限变更立即生效。

2.3 访问控制

MySQL 对客户端的访问进行两个阶段的控制,首先验证客户端的连接是否合法,如果验证通过,再检查用户对具体对象的操作是否有权限。

2.3.1 连接请求验证

MySQL 对客户端的连接请求进行两步验证，首先验证账号和凭证(如密码或密钥)是否正确，然后验证账号是否被锁住。

账号是否被锁是由 mysql 系统数据库中的 user 表里的 account_locked 字段决定的，这个字段必须是'N'才能验证通过，对这个字段的修改可以通过语句 CREATE USER 或 ALTER USER 进行。在实际工作中，如果某个用户离开比较长的时间，可以用 ALTER USER 语句将这个用户的账号锁住，等该用户回来后再解锁。

MySQL 的账号由用户名和客户端主机名组成，分别对应 mysql 系统数据库的 user 表中的 User 和 Host 字段，User 字段如果为空则可以匹配任何用户名，Host 字段如果没有指定，默认为'%'，可以匹配任意主机名，'%'也可以匹配主机名的一部分，允许连接的账号和 User、Host 字段的对应关系如表 2-1 所示。

表 2-1

User 字段的值	Host 字段的值	允许连接的账号
'scott'	'hr. scutech. net'	用户 scott 从主机 hr. scutech. net 上连接
''	'hr. scutech. net'	任何用户从主机 hr. scutech. net 上连接
''	'%'	任何用户从任意主机上连接
'scott'	'%. scutech. net'	scott 用户从域 scutech. net 上连接
'scott'	'192. 168. 87. 100'	scott 用户从 IP 地址为 192. 168. 87. 100 的机器上连接
'scott'	'192. 168. 87. %'	scott 用户从 C 类子网 192. 168. 87 上的任意机器上连接

当一个用户可以对应于多个账号时，优先对应其中最具体的账号，例如，从本机上登录的 scutech 用户可以对应 scutech@localhost 和 scutech@%两个账号，则优先对应 scutech@localhost 账号。很多时候你认为账号的权限不对，其实是因为对应了不是你想象中的账号，必要时可以使用如下命令查看当前的实际账号：

```
mysql > SELECT CURRENT_USER();
+--------------------+
| CURRENT_USER()     |
+--------------------+
| scutech@localhost  |
+--------------------+
1 row in set (0.00 sec)
```

确定了用户的账号后，MySQL 再调用 user 表里 plugin 字段指定的验证插件，结合 authentication_string 字段里的值验证对应账号的凭证，最常见的凭证是密码。

2.3.2 操作请求验证

连接成功后，就要验证用户对具体操作的对象是否有权限，对具体对象的验证过程如图 2.1 所示。

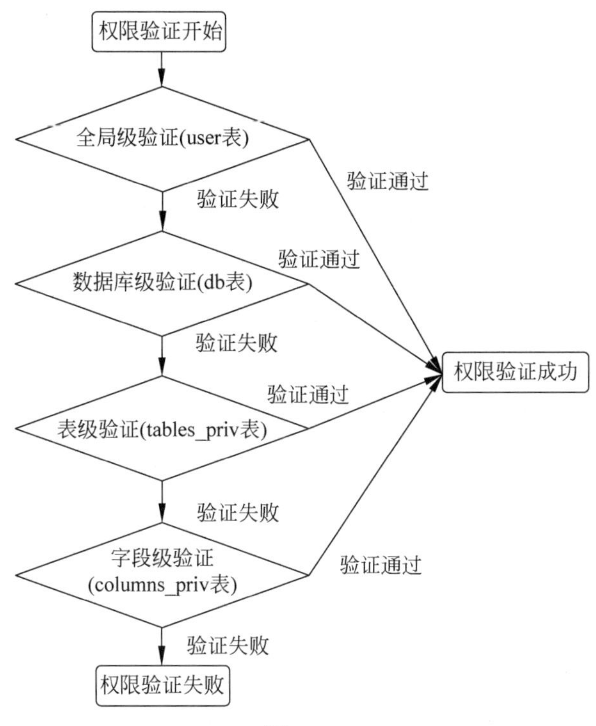

图　2-1

验证步骤如下。

（1）首先查询 user 表进行全局级验证，如果验证通过就不需要进行下一步的验证，例如，某个账号在 user 表对应记录的 Delete_priv 字段为'Y'，它就可以对所有数据库里的对象进行删除。

（2）如果全局级验证失败，再查询 db 表验证对某个数据库的权限，验证通过就结束。

（3）如果数据库级的验证失败，再查询 tables_priv 进行表级验证，验证通过就结束。

（4）如果表级的验证失败，再查询 columns_priv 进行字段级验证，验证通过就结束；验证失败就拒绝客户端的操作请求。

赋予字段级的权限和其他赋权语法稍微有些不同，字段名放在权限后面并置于括号中，例如，下面的语句将表 test.employees1 中 id 字段的查询权限赋予账号 scutech@localhost：

```
mysql> grant select (id) on test.employees1 to scutech@localhost;
```

字段级的权限是所有权限的最低一级，如下代码显示，即使没有其他任何权限，只有字段级别的权限，仍然可以执行相应的操作：

```
mysql> show grants;
+------------------------------------------------------------------+
| Grants for scutech@localhost                                     |
+------------------------------------------------------------------+
| GRANT USAGE ON *.* TO 'scutech'@'localhost'                      |
| GRANT SELECT ('id') ON 'test'.'employees1' TO 'scutech'@'localhost' |
+------------------------------------------------------------------+
```

```
2 rows in set (0.00 sec)
mysql > select id from test.employees1;
+----+
| id  |
+----+
| 1   |
+----+
1 row in set (0.00 sec)
```

2.3.3 部分权限回收

MySQL 数据库在对被操作的对象进行权限验证时，如果对高级别的对象权限验证得到通过，就不再对其下一级的对象进行权限的验证。这种验证方式虽然简单，但在需要对一大批对象中的个别对象进行排除时就很不方便，例如，如果需要把 100 个数据库中除某一个数据库之外的访问权限赋予某个用户，就需要进行 99 次赋权。从 MySQL 8.0.16 开始，这种情况得到了改善，MySQL 推出了一种部分权限回收(partial revokes)功能，可以将在高级别上赋予的权限在低级别上部分回收。

1. 部分权限回收功能

要使用部分权限回收功能需要将系统参数 partial_revokes 设置成 on，这个参数默认是 off，即默认不允许使用部分权限回收功能，在默认状态使用部分权限回收功能时会遇到下面的错误：

```
mysql > revoke select on mysql.* from scutech;
ERROR 1141 (42000):  There is no such grant defined for user 'scutech' on host '%'
```

可以使用如下命令将这个参数打开：

```
mysql > set persist partial_revokes = ON;
```

使用部分权限回收功能的一个典型应用场景是对用户赋予除系统数据库 mysql 之外的所有数据库的查询权限，相应命令如下：

```
mysql > grant select on *.* to scutech;
mysql > revoke select on mysql.* from scutech;
```

赋权完成后可以使用 show grants 命令进行检查，代码如下：

```
mysql > show grants for scutech;
+------------------------------------------------+
| Grants for scutech@%                           |
+------------------------------------------------+
| GRANT SELECT ON *.* TO 'scutech'@'%'           |
| REVOKE SELECT ON 'mysql'.* FROM 'scutech'@'%'  |
+------------------------------------------------+
2 rows in set (0.00 sec)
```

赋权完成后，在 mysql.user 表中对应记录的 User_attributes 字段会有 Restrictions 的

属性,如下所示:

```
mysql > select  User_attributes from mysql.user where user = 'scutech' and host = ' % ';
+-----------------------------------------------------------------+
| User_attributes                                                 |
+-----------------------------------------------------------------+
| {"Restrictions":  [{"Database":  "mysql", "Privileges":  ["SELECT"]}]} |
+-----------------------------------------------------------------+
1 row in set (0.00 sec)
```

2. 回收部分权限回收功能

回收部分权限回收功能可以通过再次赋予相应的权限完成,例如:

```
mysql > grant select on 'mysql'. * to scutech;
Query OK, 0 rows affected (0.01 sec)

mysql > show grants for scutech;
+--------------------------------------+
| Grants for scutech@ %                |
+--------------------------------------+
| GRANT SELECT ON * . * TO 'scutech'@' % ' |
+--------------------------------------+
1 row in set (0.00 sec)
```

也可以从高级别上回收权限,这样低级别的部分权限回收就没有必要存在了,例如:

```
mysql >  revoke select on * . * from scutech;
Query OK, 0 rows affected (0.01 sec)

mysql > show grants for scutech;
+-------------------------------------+
| Grants for scutech@ %               |
+-------------------------------------+
| GRANT USAGE ON * . * TO 'scutech'@' % ' |
+-------------------------------------+
1 row in set (0.00 sec)
```

2.4 角色

角色在其他数据库上已经存在很长时间了。从 MySQL 8.0 开始,MySQL 终于有了这个功能。这个功能最重要的优点是只定义一次包含“权限集”的角色,然后将其分配给每个用户,避免浪费时间单独声明它们。

2.4.1 使用角色

角色名称类似于用户账户,由两部分组成,分别为名称和主机,如果省略主机部分,则默认为%,表示任何主机。创建角色的命令和创建用户类似,例如,可以使用如下代码创建三个角色：

```
mysql > create role sakila_dev, sakila_read, sakila_write;
```

如下语句将 sakila 数据库的所有特权授予 sakila_dev 角色：

```
mysql > grant all on sakila. * TO sakila_dev;
```

如下语句将 sakila 数据库的 select 权限授予 sakila_read 角色：

```
mysql > grant select on sakila. * TO sakila_read;
```

如下语句将 sakila 数据库的 insert、update 和 delete 权限授予 sakila_write 角色：

```
mysql > grant insert, update, delete on sakila. * to sakila_write;
```

可以使用如下 grant 语句将创建好的角色通过赋权给用户，例如：

```
mysql > grant sakila_dev to sakila_dev1;
mysql > grant sakila_read to sakila_read1;
mysql > grant sakila_write to sakila_write1;
```

角色和用户的对照关系可以在 mysql. role_edges 表里查询到，例如，当完成了前面三个赋权语句后查询该表会有下面的记录：

```
mysql > select * from mysql. role_edges;
+-----------+--------------+----------+---------------+---------------+
| FROM_HOST | FROM_USER    | TO_HOST  | TO_USER       | WITH_ADMIN_OPTION |
+-----------+--------------+----------+---------------+---------------+
| %         | sakila_dev   | %        | sakila_dev1   | N             |
| %         | sakila_read  | %        | sakila_read1  | N             |
| %         | sakila_write | %        | sakila_write1 | N             |
+-----------+--------------+----------+---------------+---------------+
```

使用 revoke 命令可以回收角色，回收角色的命令和回收权限的命令类似，代码如下：

```
mysql > revoke role from user;
```

2.4.2 激活角色

仅仅把角色赋予用户并不能使用户马上使用相应的权限，例如，使用 show grants 命令检查用户 sakila_dev1 的权限，代码如下：

```
mysql > show grants;
+------------------------------------------+
| Grants for sakila_dev1@ %                |
+------------------------------------------+
| GRANT USAGE ON * . * TO 'sakila_dev1'@'%' |
| GRANT 'sakila_dev'@'%' TO 'sakila_dev1'@'%' |
+------------------------------------------+
2 rows in set (0.00 sec)
```

可以发现，这个用户已经被赋予了 sakila_dev 角色，但此时这个角色并没有被激活，可以使用 select current_role()命令检查当前已经激活的角色，代码如下：

```
mysql > select current_role();
+-----------------+
| current_role()  |
+-----------------+
| NONE            |
+-----------------+
1 row in set (0.00 sec)
```

可以发现，并没有处于激活状态的角色，当该用户要使用 sakila 数据库时，发现没有权限，代码如下：

```
mysql > use sakila;
ERROR 1044 (42000):  Access denied for user 'sakila_dev1'@'%' to database 'sakila'
```

因为角色 sakila_dev 没有被激活，可以使用 set role 命令将角色激活，代码如下：

```
mysql > set role sakila_dev;
```

再使用 select current_role()命令检查，会发现 sakila_dev 角色已经被激活了，代码如下：

```
mysql > select current_role();
+-------------------+
| CURRENT_ROLE()    |
+-------------------+
| 'sakila_dev'@'%'  |
+-------------------+
1 row in set (0.00 sec)
```

再使用 sakila 数据库就有权限了，代码如下：

```
mysql > use sakila;
Database changed
```

再用 show grants 命令检查当前用户的权限，发现多了一条语句，即被激活的角色 sakila_dev 拥有的权限，代码如下：

```
mysql > show grants;
+-----------------------------------------------------------+
| Grants for sakila_dev1@%                                  |
+-----------------------------------------------------------+
| GRANT USAGE ON *.* TO 'sakila_dev1'@'%'                   |
| GRANT ALL PRIVILEGES ON 'sakila'.* TO 'sakila_dev1'@'%'   |
| GRANT 'sakila_dev'@'%' TO 'sakila_dev1'@'%'               |
+-----------------------------------------------------------+
```

除了可以使用 set role 语句激活角色外，还可以使用 set default role 指定用户一登录就处于激活状态的角色，例如，如下语句指定了 sakila_dev 角色对于 sakila_dev1 用户默认是激活的：

```
mysql > set default role sakila_dev to sakila_dev1;
```

还可以使用如下命令将 sakila_dev1 用户所有的角色设置成默认激活：

```
mysql > set default role all to sakila_dev1;
```

2.4.3 强制角色

可以使用系统参数 mandatory_roles 指定对整个数据库起作用的角色，使用如下命令将 sakila_read 角色赋予所有登录到这个数据库的用户：

```
mysql > set persist mandatory_roles = 'sakila_read';
```

也可以修改 MySQL 的参数配置文件，在[mysqld]组里加上如下内容：

```
mandatory_roles = 'sakila_read'
```

请注意，这样配置的角色默认也是没有被激活的，还需要使用 set role 命令将角色激活。另外，MySQL 中还有一个系统参数 activate_all_roles_on_login，如果被设置为 true，所有用户拥有的角色在登录时都会被自动激活，这个系统参数的默认值是 false，默认只有用户的默认角色是激活的。

系统参数 mandatory_roles 的设置对在不同环境下使用 MySQL 很有帮助，例如，可以设置所有的用户在测试环境的 MySQL 数据库中有写的权限，在生产环境的 MySQL 数据库中只有读的权限。

2.5 代理用户

MySQL 的登录验证插件可以使一个用户在登录时被映射成另一个用户，第一个用户称为代理用户，第二个用户称为被代理用户。

2.5.1 参数设置

运行如下代码，可以查看 MySQL 中与代理相关的参数如下：

```
mysql > show variables like " % proxy % ";
+-----------------------------------+-------+
| Variable_name                     | Value |
+-----------------------------------+-------+
| check_proxy_users                 | OFF   |
| mysql_native_password_proxy_users | OFF   |
| proxy_user                        |       |
| sha256_password_proxy_users       | OFF   |
+-----------------------------------+-------+
4 rows in set (0.03 sec)
```

系统参数 check_proxy_users 需要设置为 ON 才会在登录时进行代理关系的检查，如果使用 sha256_password 验证插件，需要把系统参数 sha256_password_proxy_users 设置为 ON，MySQL 5.7 的默认验证插件是 mysql_native_password，如果要使用这个验证插件，需要把系统参数 mysql_native_password_proxy_users 设置为 ON。

2.5.2　配置用户代理

使用如下命令创建一个代理用户 client：

```
mysql> create user client identified with sha256_password by 'dingjia';
```

使用的验证插件是 sha256_password，需要对应地把系统参数 sha256_password_proxy_users 设置为 ON。再创建一个被代理用户 manager，代码如下：

```
mysql>  create user manager identified by 'dingjia';
```

将用户 manager 的代理权限赋予 client 的命令如下：

```
mysql> grant proxy on manager to client;
```

将赋权的结果保存在 mysql.proxies_priv 表中，代码如下：

```
mysql> select host,User,Proxied_host,Proxied_user from mysql.proxies_priv;
+-----------+--------+--------------+--------------+
| host      | User   | Proxied_host | Proxied_user |
+-----------+--------+--------------+--------------+
| localhost | root   |              |              |
| %         | client | %            | manager      |
+-----------+--------+--------------+--------------+
2 rows in set (0.00 sec)
```

使用 client 用户登录后，使用如下命令检查：

```
mysql> select user(),current_user();
+------------------+----------------+
| user()           | current_user() |
+------------------+----------------+
| client@localhost | manager@%      |
+------------------+----------------+
1 row in set (0.00 sec)
```

可以发现，client 用户已经被 manager 用户代理了，此时执行命令使用的权限将是 manager 用户的权限。

2.5.3　禁止被代理用户登录

为了安全起见，通常应该禁止被代理用户登录，一个简单的方法是将这个用户账户锁住，例如，可以使用如下命令将 manager 用户的账户锁住：

```
mysql> alter user manager account lock;
```

另外一个简单的方法是给这个用户设置一个复杂的密码，然后不告诉任何人。

还有一个方法是将这个用户的验证插件改成不能登录的插件，如 mysql_no_login，该插件默认是没有载入的，需要使用如下命令载入该插件：

```
mysql > install plugin mysql_no_login soname 'mysql_no_login.so';
```

或者在 mysql 的参数配置文件中加入如下配置，让 mysql 在启动时就载入这个插件：

```
[mysqld]
plugin - load - add = mysql_no_login.so
```

然后使用如下命令将 manager 用户的登录验证插件改成 mysql_no_login：

```
mysql > alter user manager identified with mysql_no_login;
```

此时，该用户自己就无法登录了，只能被映射到代理用户。

2.6 无密码登录

每次登录都要求输入密码是个有点烦琐的事情，而且在脚本里输入密码很不方便，因此有时需要无密码登录，本节介绍三种不需要输入密码登录 MySQL 的方法。

2.6.1 在参数配置文件中保存账户和密码

MySQL 的参数配置文件里的参数是按对应的程序名进行分组的，组名放在[]中。如[mysqld] 和 [mysql] 组分别对应 mysqld 服务和 mysql 客户端程序。[client] 参数组对应所有的客户端程序，包括 mysql、mysqldump 和 mysqladmin 等，不包括 mysqld 程序。如果把账号和密码放在[mysql] 组里，使用 mysql 程序登录时就不需要输入账号和密码；如果放到[client]组里，所有的客户端程序登录时都不需要输入账号和密码。例如，在参数配置文件里增加如下内容：

```
[client]
user = root
password = dingjia
```

有了这两个系统参数，在客户端登录时可以不用输入账号和密码，仅仅调用如下 mysql 程序：

```
$ mysql
```

就相当于：

```
$ mysql  - uroot  - pdingjia
```

本书中的很多例子都采用这种方法，但这样做存在的风险显而易见，即密码是以明文存放在配置文件中的。

2.6.2　使用 mysql_config_editor 工具保存登录凭证

在脚本里登录 MySQL 时，无法采用交互的方式进行输入，MySQL 提供了一个工具 mysql_config_editor 可以解决这个问题，它可以生成被扰乱后的登录凭证，注意不是加密后的凭证，这里的加密不是绝对安全的。该工具会在用户的家目录下生成一个名字为 .mylogin.cnf 的配置文件。这个配置文件分成若干组，每个组对应一个登录凭证，称为登录路径(Login Path)，每个登录路径包括五个参数：host、user、password、port 和 socket。默认的登录路径为 client。

如下命令为使用 mysql_config_editor 的 set 选项生成一个 root 用户的登录路径：

```
$ mysql_config_editor set --user=root --password
Enter password: (输入密码)
```

如下命令为使用 mysql_config_editor 的 print 选项查看生成的登录路径：

```
$ mysql_config_editor print
[client]
user = root
password = *****
```

这里只列出五个选项中的两个，没有列出的选项采用默认值。

再使用 mysql_config_editor 的 set 选项生成一个 scutech 用户的登录路径，这里指定登录路径名为 scutech，因为默认的登录路径名 client 已经被占用了，代码如下：

```
$ mysql_config_editor set --login-path=scutech --user=scutech --password
Enter password: (输入密码)
```

再使用 mysql_config_editor 的 print 选项查看生成的登录路径，代码如下：

```
$ mysql_config_editor print --all
[client]
user = root
password = *****
[scutech]
user = scutech
password = *****
```

登录客户端时指定与用户对应的登录路径即可登录，不用输入用户名和密码，如下命令使用 scutech 用户的对应路径登录：

```
$  mysql --login-path=scutech -e 'select user()'
mysql> select user();
+-------------------+
| user()            |
+-------------------+
| scutech@localhost |
+-------------------+
1 row in set (0.00 sec)
```

如果需要 root 用户进行登录,可以不输入--login-path 选项,因为 root 用户对应的登录路径是默认的 client。

2.6.3 使用 auth_socket 验证插件无密码登录

1. auth_socket 的特点

MySQL 中登录验证插件 auth_socket 对登录用户的验证也不需要密码,该插件有如下三个特点。

(1) 这种验证方式不要求输入密码,即使输入了密码也不验证。这个特点让很多人觉得很不安全,实际上仔细研究一下这种方式,发现还是相当安全的,因为它还有另外两个限制。

(2) 只能用 socket 的连接方式登录,这就保证了只能本地登录,而且操作系统只能是 Linux 或 UNIX,用户在使用这种登录方式时已经通过了操作系统的安全验证。

(3) 操作系统的用户名必须和 MySQL 数据库的用户名或密码一致,例如,要登录 MySQL 的 root 用户,必须用操作系统的 root 用户,或者将 MySQL 用户的密码设置成 root。

2. auth_socket 的使用

验证插件 auth_socket 默认没有载入,使用如下代码可以加载这个插件:

```
mysql> install plugin auth_socket soname 'auth_socket.so';
```

如下命令将账号'scutech'@'localhost'的验证插件改成 auth_socket,密码设置为 dingjia:

```
mysql> alter user 'scutech'@'localhost' identified with auth_socket as 'dingjia';
```

检查系统表 mysql.user 中的对应记录,代码如下:

```
mysql> select user,host,plugin,authentication_string from mysql.user  where user = 'scutech';
+---------+-----------+-------------+-----------------------+
| user    | host      | plugin      | authentication_string |
+---------+-----------+-------------+-----------------------+
| scutech | localhost | auth_socket | dingjia               |
+---------+-----------+-------------+-----------------------+
```

可以发现,authentication_string 字段里保存的是明文,因为这个字段并不是用于保存密码的。

当操作系统的用户为 scutech 时,无须密码即可登录,代码如下:

```
$ id
uid=1000(scutech) gid=1000(scutech) groups=1000(scutech),127(mysql)
$ mysql -uscutech
Welcome to the MySQL monitor.  Commands end with ; or \g.
Your MySQL connection id is 19
...
```

当操作系统的用户为 root 时,登录失败,代码如下:

```
# id
uid = 0(root) gid = 0(root) groups = 0(root)
# mysql  - uscutech
ERROR 1698 (28000):  Access denied for user 'scutech'@'localhost'
```

如果要在系统的用户为 root 时也可以登录，可以使用如下代码将密码改成 root：

```
mysql > alter user 'scutech'@'localhost' identified with auth_socket as 'root';
Query OK, 0 rows affected (0.09 sec)
```

再次登录即可成功，代码如下：

```
# id
uid = 0(root) gid = 0(root) groups = 0(root)
# mysql  - uscutech
Welcome to the MySQL monitor.  Commands end with ;  or \g.
Your MySQL connection id is 21
...
```

这样这个 MySQL 账号就可以在当操作系统用户是 scutech 或 root 时进行登录了，其他的操作系统用户不能登录。

3. auth_socket 的适用场景

auth_socket 很适合在系统投产前进行安装调试时使用。在系统投产前经常需要同时使用操作系统的 root 用户和 MySQL 的 root 用户，无须输入密码，很方便。当系统投产后，操作系统的 root 用户和 MySQL 的 root 用户就不能随便使用了，这时就可以换成其他的验证方式。

2.7 重置 root 用户密码

在实际工作中常常出现用户忘记密码的情况，如果一般的用户忘记了密码可以使用 root 用户登录，再使用 alter user 的命令重置用户的密码，但如果忘记了 root 用户的密码就比较麻烦，本节介绍两种重置 root 用户密码的方法。

2.7.1 使用系统参数 skip-grant-tables 启动 mysqld

在启动 mysqld 时，如果加上系统参数 skip-grant-tables 将会跳过读取系统数据库 mysql 的授权表，此时任何用户都可以不用密码登录 MySQL，并且可以访问或修改 MySQL 中的任何数据。通常使用 skip-grant-tables 系统参数启动 mysqld 时也会加上系统参数 skip-networking 禁止远程连接，但从 MySQL 8.0 开始使用系统参数 skip-grant-tables 时 MySQL 会自动加上系统参数 skip-networking。

可以在命令行加上该系统参数，如 mysqld_safe --skip-grant-tables &，或者修改启动的参数文件，如/etc/my.cnf 或者/etc/mysql/mysql.cnf，在[mysqld]下面加上 skip-grant-tables 后再重新启动 msyqld。启动完成后即可不用密码登录，登录后使用如下 SQL 语句修改 mysql.user 表里的字段 authentication_string：

```
mysql > update mysql.user set authentication_string = password('dingjia') where user = 'root';
```

修改完成后需要使用如下命令将权限表重新载入：

```
mysql > flush privileges;
```

也可以使用 mysqladmin flush-privileges 或 mysqladmin reload 载入权限表。

但在 MySQL 8.0 版本中，这种方式已经不可行，因为其已移除了 password()函数。可以使用 alter user 命令修改密码，但注意要载入权限表后才能用 alter user 命令，如下为修改过程：

```
mysql > alter user root identified by 'dingjia';
ERROR 1290 (HY000): The MySQL server is running with the -- skip - grant - tables option so it
cannot execute this statement
mysql > flush privileges;
Query OK, 0 rows affected (0.07 sec)
mysql > alter user root@localhost identified by 'dingjia';
Query OK, 0 rows affected (0.02 sec)、
```

密码修改完成后要把参数文件里的系统参数 skip-grant-tables 去掉，重新启动 mysqld 即可。

2.7.2 使用系统参数 init-file 修改 root 用户的密码

采用跳过载入授权表的方法启动 mysqld 会让 MySQL 处于非常不安全的状态，因为此时任何用户均可修改数据库里的任何数据。但采用系统参数 init-file 修改 root 用户的密码就安全很多，而且只需要重启一次 MySQL 实例。

先关闭实例，再创建一个 SQL 文件，如 chpw.sql，文件里包含修改 root 密码的语句如下：

```
alter user 'root'@'localhost' identified by 'dingjia';
```

然后使用系统参数 init-file 启动实例，该系统参数指向修改 root 密码的 SQL 文件，例如：

```
$ mysqld_safe -- init - file = /tmp/chpw.sql &
```

也可以在参数配置文件里加入如下一行命令再启动实例：

```
init - file = /tmp/chpw.sql
```

MySQL 实例启动时会自动执行这个 SQL 语句，启动完成后密码即修改完毕。

2.8 实验

用户名、主机名和账号的对应关系

视频演示

(1) 分别设置 MYSQL_HOST 为 Localhost 和 127.0.0.1 时，登录 mysql，查看其账号的对应关系。

(2) 观察同一个客户端 IP 地址通过/etc/host 文件解析成不同主机名时，其账号的对应关系。

第3章

日　志

MySQL 的四类日志分别从不同的角度记录着 MySQL 内部的运行信息，出现故障时首先应查询错误日志；通用查询日志记录着 MySQL 执行的 SQL；查询慢查询日志可以找到需要优化的 SQL；二进制日志记录着数据库的变化，可以用于复制和时间点恢复。本章将介绍这四类日志。

3.1　错误日志

MySQL 的错误日志记录着 mysqld 进程运行中的重要信息，包括 mysqld 启动和关闭时间，还包括在启动和关闭期间以及运行期间发生的错误、警告和说明等诊断消息。例如，如果 mysqld 发现一个表需要自动检查或修复，它会将一条消息写入错误日志。

3.1.1　系统参数 log_error_services

MySQL 记录错误日志的功能是由组件(component)完成的。MySQL 有一套基于组件的基础架构用于扩展 MySQL 的功能，组件之间通过服务(service)进行交互，这和插件(plugin)不同，插件之间是通过 API(application programming interface，应用程序接口)调用进行交互的。

MySQL 记录错误日志的功能由两类组件完成：一类是过滤(filter)组件；另一类是写(sink 或 writer)组件。系统参数 log_error_services 控制激活哪些组件用于错误日志的记录，运行如下代码可得到默认值：

```
mysql> show variables like 'log_error_services';
+--------------------+----------------------------------------+
| Variable_name      | Value                                  |
+--------------------+----------------------------------------+
```

```
| log_error_services        | log_filter_internal;  log_sink_internal           |
+---------------------+------------------------------------------+
```

其中,log_filter_internal 是过滤组件;log_sink_internal 是写组件。internal 表示这两个组件是内置组件,即无须载入。非内置组件使用前要先载入,载入后在系统表 mysql.component 中可以查到。如下代码使用 install component 载入 log_sink_syseventlog 组件:

```
mysql > install component 'file: //component_log_sink_syseventlog';
```

载入完成后,运行如下命令可以在 mysql.component 视图中可以查询到刚刚载入的组件:

```
mysql > select component_id,component_urn  from mysql.component;
+--------------+--------------------------------------+
| component_id | component_urn                        |
+--------------+--------------------------------------+
|            1 | file: //component_log_sink_syseventlog |
+--------------+--------------------------------------+
1 row in set (0.00 sec)
```

系统参数 log_error_services 不能仅指定一个日志过滤组件,而不指定写组件,例如:

```
mysql > set global log_error_services  =  'log_filter_internal';
ERROR 1231 (42000): Variable 'log_error_services' can't be set to the value of 'log_filter_
internal'
```

系统参数 log_error_services 可以指定多个日志写组件,例如:

```
mysql > set global log_error_services  =  'log_filter_internal; log_sink_internal; log_sink_
syseventlog';
```

这样设置的系统参数 log_error_services 会通过两个组件完成错误日志的写入,除了使用组件 log_sink_internal 写入默认的错误日志,还会使用组件 log_sink_syseventlog 将错误日志记录到操作系统的系统日志中。

系统参数 log_error_services 中记录的组件顺序也很重要,因为执行组件过程是从左向右进行的,例如:

```
log_filter_internal; log_sink_1; log_sink_2
```

如上日志事件先传递到内置的过滤组件(log_filter_internal),然后传递到第一个写组件(log_sink_1),再传递到第二个写组件(log_sink_2),但如果更换顺序,例如:

```
log_sink_1; log_filter_internal; log_sink_2
```

在这种情况下,日志事件先传递到第一个写组件,再传递到内置的过滤组件,然后传递到第二个写组件,第一个写组件接收未过滤的事件,第二个写组件接收过滤后的事件。如果希望一个日志包含所有日志事件的信息,而另一个日志仅包含日志事件子集的信息,则可以通过这种方式配置错误日志记录。

3.1.2 log_filter_internal 组件

log_filter_internal 组件是 MySQL 内置的日志过滤组件，它基于优先级和错误代码对日志事件进行过滤，过滤规则由系统参数 log_error_verbosity 和 log_error_suppression_list 进行控制。

1. 系统参数 log_error_verbosity

MySQL 日志事件的优先级分为三级，分别是 ERROR、WARNING 和 INFORMATION，系统参数 log_error_verbosity 的值和日志事件优先级的对应关系见表 3.1。

表 3.1

系统参数 log_error_verbosity	日志事件的优先级
1	ERROR
2	ERROR、WARNING
3	ERROR、WARNING、INFORMATION

log_error_verbosity 默认设置是 2，为了获得更详细的信息，可以将该参数设置为 3，设置为 1 时记录的信息很少，通常不推荐。MySQL 日志事件的优先级分为四级，除了表中列出的三级外，还有 SYSTEM 级别的信息，此类信息会一直记录到错误日志中，不受系统参数 log_error_verbosity 的控制，SYSTEM 级别的信息包括 MySQL 的启动、关闭，以及对系统影响大的设置的变更。

2. 系统参数 log_error_suppression_list

系统参数 log_error_suppression_list 可以用来屏蔽某个具体的错误事件，当一个错误事件经常发生但又不重要时，可以将这个错误事件的错误号加入该参数中进行屏蔽，这样，这个错误事件就不会记录到错误日志中，例如，如下的设置将不会记录错误号为 MY-010926 的错误事件：

```
mysql > set global log_error_suppression_list = 'MY - 010926';
```

进行了如上设置后，错误日志中将不会出现类似下面的信息：

```
2021 - 01 - 17T04: 12: 36.457722 - 03: 30 17 [Note] [MY - 010926] [Server] Access denied for
user 'no_such_user'@'localhost' (using password: YES)
```

log_error_suppression_list 参数只能屏蔽 WARNING 和 INFORMATION 类型的错误事件，不能屏蔽 ERROR 和 SYSTEM 类型的信息。

3.1.3 log_sink_internal 组件

log_sink_internal 组件是 MySQL 内置的日志写入组件，它会把错误日志记录到系统参数 log_error 指定的文件和 performance_schema.error_log 表中，系统参数 log_error 的默认值如下：

```
mysql > show variables like 'log_error';
+---------------+-------------------------+
| Variable_name | Value                   |
+---------------+-------------------------+
| log_error     | /var/log/mysql/error.log |
+---------------+-------------------------+
1 row in set (0.00 sec)
```

如下是 log_sink_internal 组件记录信息的一个例子：

```
2021-01-17T01:40:26.676173Z 0 [Note] [MY-010304] [Server] Skipping generation of SSL certificates as certificate files are present in data directory.
2021-01-17T01:40:26.681883Z 0 [Warning] [MY-010068] [Server] CA certificate ca.pem is self signed.
2021-01-17T01:40:26.682157Z 0 [System] [MY-013602] [Server] Channel mysql_main configured to support TLS. Encrypted connections are now supported for this channel.
2021-01-17T01:40:26.682459Z 0 [Note] [MY-010308] [Server] Skipping generation of RSA key pair through --sha256_password_auto_generate_rsa_keys as key files are present in data directory.
```

log_sink_internal 组件记录信息的格式如下：

```
time thread [priority] [err_code] [subsystem] msg
```

其中，方括号是记录的格式，并不意味着可以省略这些项。

3.1.4 log_filter_dragnet 组件

log_filter_dragnet 组件对错误事件的过滤基于用户定义的规则，该组件不是内置组件，因此使用前要先载入，如下代码是使用该组件的例子：

```
mysql > install component 'file://component_log_filter_dragnet';
Query OK, 0 rows affected (0.06 sec)

mysql > set global log_error_services = 'log_filter_dragnet;
log_sink_internal';
Query OK, 0 rows affected (0.00 sec)
```

用户设置过滤规则是通过设置系统参数 dragnet.log_error_filter_rules 来进行的，运行如下代码可得到该参数的默认值：

```
mysql > show variables like 'dragnet%'\G
*************************** 1. row ***************************
Variable_name: dragnet.log_error_filter_rules
        Value: IF prio >= INFORMATION THEN drop. IF EXISTS source_line THEN unset source_line.
1 row in set (0.01 sec)
```

过滤规则的设置以 IF 开始，以圆点结束，基本语法格式如下：

```
IF condition THEN action
[ELSEIF condition THEN action] ...
```

```
[ELSE action]
...
```

具体设置方法参见 https://dev.mysql.com/doc/refman/8.0/en/error-log-rule-based-filtering.html。

3.1.5 log_sink_json 组件

log_sink_json 组件会以 JSON 格式记录错误日志，该组件也不是内置组件，如下代码为载入和使用 log_sink_json 组件的例子：

```
mysql> install component 'file://component_log_sink_json';
Query OK, 0 rows affected (0.00 sec)

mysql> set persist log_error_services = 'log_filter_internal; log_sink_json';
Query OK, 0 rows affected (0.00 sec)
```

log_sink_json 组件记录信息的位置也是由系统参数 log_error 决定的，该组件会在文件名后面加上"NN.json"的后缀，NN 从 00 开始编号，下一次日志刷新时会变为 01。如下代码为该组件记录信息的一个例子：

```
$ cat /var/log/mysql/error.log.00.json
...
{ "log_type" : 1, "prio" : 0, "err_code" : 10931, "msg" : "/usr/sbin/mysqld: ready for
connections. Version: '8.0.22'  socket:
'/var/run/mysqld/mysqld.sock'  port: 3306  MySQL Community Server - GPL.", "time" : "2021-
01-17T08:38:03.416301Z", "ts" : 1610872683416, "err_symbol" : "ER_SERVER_STARTUP_MSG",
"SQL_state" : "HY000", "subsystem" : "Server", "label" : "System" }
```

3.1.6 log_sink_syseventlog 组件

log_sink_syseventlog 组件会将错误日志记录到操作系统的系统日志中，Windows 操作系统记录到事件日志，UNIX 和 Linux 操作系统记录到 syslog 中。该组件也不是内置组件，如下代码为载入和使用 log_sink_syseventlog 组件的例子：

```
mysql> install component 'file://component_log_sink_syseventlog';
Query OK, 0 rows affected, 1 warning (0.00 sec)

mysql> set global log_error_services = 'log_filter_internal;
log_sink_syseventlog';
Query OK, 0 rows affected, 1 warning (0.00 sec)
```

在 Linux 操作系统中激活这个组件后会增加三个系统参数，代码如下：

```
mysql> show variables like 'syseventlog%';
+-------------------------+--------+
| Variable_name           | Value  |
+-------------------------+--------+
```

```
| syseventlog.facility          | daemon  |
| syseventlog.include_pid       | ON      |
| syseventlog.tag               |         |
+-------------------------------+---------+
3 rows in set (0.01 sec)
```

（1）syseventlog.facility：指定记录在系统日志中的MySQL错误日志的设施(facility)，默认为daemon。

（2）syseventlog.include_pid：指定是否记录mysqld的进程号。

（3）syseventlog.tag：用于为记录的MySQL的错误日志加上一个标签，当在同一个系统日志中记录多个MySQL的错误日志时，可以给不同的MySQL指定不同的标签进行区别。

log_sink_syseventlog组件记录的错误日志的格式是由所在操作系统的系统日志格式决定的，如下代码是在Ubuntu平台上记录的错误日志的例子：

```
Jan 17 05:53:18 scutech mysqld[19857]: Dumping buffer pool(s) to /var/lib/mysql/ib_buffer
_pool
Jan 17 05:53:18 scutech mysqld[19857]: Buffer pool(s) dump completed at 210117  5:53:18
Jan 17 05:53:18 scutech mysqld[19857]: Log background threads are being closed...
Jan 17 05:53:19 scutech mysqld[19857]: Shutdown completed; log sequence number 151947099
Jan 17 05:53:19 scutech mysqld[19857]: Removed temporary tablespace data file: "ibtmp1"
Jan 17 05:53:19 scutech mysqld[19857]: Shutting down plugin 'MEMORY'
Jan 17 05:53:19 scutech mysqld[19857]: Shutting down plugin 'CSV'
```

3.1.7 系统参数log_timestamps

系统参数log_timestamps用于控制写入错误日志、通用查询日志和慢速查询日志中的时间戳格式。MySQL在错误事件到达任何日志写入组件之前应用该参数，因此所有日志写入组件时都会受其影响，但performance_schema.error_log表中记录的日志信息不受该参数的控制。这个参数可以设置成UTC或SYSTEM，默认值是UTC，如下代码为将该参数的值设置成SYSTEM：

```
mysql> show variables like 'log_timestamps';
+----------------+-------+
| Variable_name  | Value |
+----------------+-------+
| log_timestamps | UTC   |
+----------------+-------+
1 row in set (0.03 sec)

mysql> set global log_timestamps = 'SYSTEM';
Query OK, 0 rows affected (0.00 sec)
```

下面是错误日志里记录的两条登录失败的信息：第一条信息是当log_timestamps为UTC时记录的信息，时间格式是YYYY-MM-DDThh:mm:ss.uuuuuu，结尾有个字母Z，代表祖鲁时间(zulu time)，也就是格林威治时间；第二条信息是log_timestamps为

SYSTEM 时记录的信息，在时间结尾有“+08：00”，代表东八区，也就是北京时间。

```
2021-01-17T01:48:22.450646Z 11 [Note] [MY-010926] [Server] Access denied for user 'kl'@
'localhost' (using password: YES)
2021-01-17T09:49:04.234165+08:00 13 [Note] [MY-010926] [Server] Access denied for
user 'ds'@'localhost' (using password: YES)
```

错误日志中默认记录的是 UTC 时间，对于中国用户来说，加 8 小时就是当地时间了。

3.1.8 备份错误日志

FLUSH LOGS 这个 SQL 命令会关闭和重新打开 MySQL 正在使用的所有日志文件，该命令和 mysqladmin flush-logs 等价，同时它还相当于如下 6 个 SQL 命令：

```
FLUSH BINARY LOGS
FLUSH ENGINE LOGS
FLUSH ERROR LOGS
FLUSH GENERAL LOGS
FLUSH RELAY LOGS
FLUSH SLOW LOGS
```

因此可以使用类似下面的脚本进行错误日志的备份：

```
$ mv /var/log/mysql/error.log /var/log/mysql/error.log.err-old
$ mysqladmin flush-logs
$ mv /var/log/mysql/error.log.err-old backup-directory
```

3.1.9 错误日志记录到数据库中

错误日志写入组件 log_sink_internal 和 log_sink_json，把错误信息记录到系统参数 log_error 指定的文件中的同时，还会同时记录到 performance_schema 数据库的 error_log 表中。这样做有两个明显的好处；一是只要将该表的 select 权限赋予用户，即使没有权限访问传统错误日志文件的用户也可以查询错误日志的记录；二是可以使用 SQL 语句访问错误日志，这种方式比查询传统的文件灵活得多。当然错误日志记录到文件中也是非常必要的，因为很多时候需要在 MySQL 无法启动时查看错误日志进行排错。

MySQL 会为 error_log 表在内存中分配一段固定大小的空间，因此它只包括最近的记录，超过其大小的记录会被抛弃。通过如下代码可以查询与该表相关的 4 个状态参数：

```
mysql> show status like 'error_log_%';
+----------------------------+--------------------+
| Variable_name              | Value              |
+----------------------------+--------------------+
| Error_log_buffered_bytes   | 134032             |
| Error_log_buffered_events  | 335                |
| Error_log_expired_events   | 0                  |
| Error_log_latest_write     | 1610876058836067   |
+----------------------------+--------------------+
```

```
4 rows in set (0.00 sec)
```

(1) Error_log_buffered_bytes：表示 error_log 表的大小。

(2) Error_log_buffered_events：表示 error_log 表记录的事件个数。

(3) Error_log_expired_events：表示被抛弃的事件个数。

(4) Error_log_latest_write：表示最后一次记录到这个表的时间戳，这里的时间戳是个整数，是从 1970-01-01 到现在的毫秒数(millisecond)，可以使用 from_unixtime 函数将其转换成当前的时间，代码如下：

```
mysql> select from_unixtime(1610876058836067/1000000.0) from_time;
+----------------------------+
| from_time                  |
+----------------------------+
| 2021-01-17 17:34:18.8360   |
+----------------------------+
1 row in set (0.01 sec)
```

如下是查看 error_log 表记录信息的例子：

```
mysql>  select * from performance_schema.error_log limit 4 \G
*************************** 1. row ***************************
    LOGGED: 2020-12-31 11:50:02.614050
 THREAD_ID: 0
      PRIO: Warning
ERROR_CODE: MY-011070
 SUBSYSTEM: Server
      DATA: 'Disabling symbolic links using --skip-symbolic-links (or equivalent) is the
default. Consider not using this option as it' is deprecated and will be removed in a future
release.
*************************** 2. row ***************************
    LOGGED: 2020-12-31 11:50:02.622128
 THREAD_ID: 0
      PRIO: System
ERROR_CODE: MY-010116
 SUBSYSTEM: Server
      DATA: /usr/sbin/mysqld (mysqld 8.0.22) starting as process 5854
*************************** 3. row ***************************
    LOGGED: 2020-12-31 11:50:02.690152
 THREAD_ID: 1
      PRIO: System
ERROR_CODE: MY-011012
 SUBSYSTEM: Server
      DATA: Starting upgrade of data directory.
*************************** 4. row ***************************
    LOGGED: 2020-12-31 11:50:02.691017
 THREAD_ID: 1
      PRIO: System
ERROR_CODE: MY-013576
 SUBSYSTEM: InnoDB
      DATA: InnoDB initialization has started.
```

```
4 rows in set (0.00 sec)
```

error_log 表中包括如下一些字段：

（1）LOGGED：错误事件的时间戳，精确到微秒。

（2）THREAD_ID：MySQL 的线程 id，对于后台线程，该值为 0。

（3）PRIO：错误事件的优先级。

（4）ERROR_CODE：错误代码。

（5）SUBSYSTEM：MySQL 的子系统。

（6）DATA：错误事件的文本描述，该字段的记录格式由写组件决定，log_sink_internal 组件记录的是传统格式，而 log_sink_json 组件记录的是 JSON 格式。

除 DATA 字段之外的 5 个字段上都各有一个 hash 索引，可以提高针对这些字段的查询速度。

3.2 通用查询日志

通用查询日志记录着 mysqld 正在做什么，当客户端连接或断开连接时，服务器将信息写入此日志中，并记录从客户端接收的每个 SQL 语句。查看通用查询日志可以知道客户端发送给 mysqld 的确切内容。

3.2.1 系统参数说明

1. log_output

log_output 参数用于指定慢查询日志和通用查询日志记录的位置。

- --log_output=FILE 时，错误日志和通用查询日志记录到文件，这是默认值。
- --log_output=TABLE 时，错误日志和通用查询日志记录到数据库的表中。
- --log_output=TABLE,FILE 时，错误日志和通用查询日志记录到数据库的表中同时记录到文件。
- --log_output=NONE 时，不记录错误日志和通用查询日志。

2. general_log

general_log 参数可以设定为 on(1)或 off(0)，分别表示打开和关闭通用查询日志。该参数的默认值是 off，它可以动态设置。

3. general_log_file

general_log_file 参数用于指定通用查询日志的文件名，默认文件名是< hostname >. log。

4. sql_log_off

sql_log_off 参数用于指定当通用查询日志激活时，当前的会话是否记录通用查询日志，当设置为 off 时进行记录，off 也是默认值，设置为 on 时不记录。

3.2.2 记录内容

通用查询日志会根据系统参数 log_output 的设置将信息记录到文件和数据库的表中。

1. 通用查询日志文件中的记录

如下为通用查询日志文件中记录的一个用户登录时产生的信息：

```
2021-01-18T02:15:28.331010Z   10 Connect  root@localhost on using Socket
2021-01-18T02:15:28.331610Z   10 Query   select @@version_comment limit 1
```

第一行记录的是用户登录的信息。第一项是时间；第二项 10 是 MySQL 中的会话号；第三项 Connect 表示用户登录；最后一项包括账号(root@localhost)和登录时使用的协议：Socket。每个登录信息中都会有 using 连接类型的信息，连接类型包括 TCP/IP、SSL/TLS、Socket、Named Pipe 和 Shared Memory 等。

第二行记录着用户登录时还自动执行了一个查询语句，查询 MySQL 的版本。

2. 通用查询日志表中的记录

当 log_output 设置的值中包括 TABLE 时，通用查询日志和慢查询日志还可以记录到 mysql 数据库的 general_log 表和 slow_log 表中，这样查询起来更方便，例如，可以只查询满足特定条件的日志记录，如下是 general_log 表的定义语句：

```
mysql> show create table  mysql.general_log \G
*************************** 1. row ***************************
       Table: general_log
Create Table: CREATE TABLE 'general_log' (
  'event_time' timestamp(6) NOT NULL DEFAULT CURRENT_TIMESTAMP(6) ON UPDATE CURRENT_TIMESTAMP
(6),
  'user_host' mediumtext NOT NULL,
  'thread_id' bigint unsigned NOT NULL,
  'server_id' int unsigned NOT NULL,
  'command_type' varchar(64) NOT NULL,
  'argument' mediumblob NOT NULL
) ENGINE = CSV DEFAULT CHARSET = utf8 COMMENT = 'General log'
1 row in set (0.00 sec)
```

可以看到，general_log 表使用的是 CSV 引擎，使数据可以很方便地导入 windows access 中，但需要注意的是，这种引擎是没有索引的，查询的速度不一定很快。

如下代码为查询线程号是 2249 执行的 SQL，并将查询的结果按 SQL 执行时间由后到先输出：

```
mysql> select event_time,user_host,thread_id, server_id,command_type,convert(argument using
utf8mb4)
from mysql.general_log
where thread_id  =  2249 order by event_time desc\G
*************************** 1. row ***************************
                    event_time: 2021-08-09 10:06:24.896843
                     user_host: root[root] @ localhost []
```

```
                      thread_id: 2249
                      server_id: 3940373588
                   command_type: Query
convert(argument using utf8mb4): select * from tab_a
*************************** 2. row ***************************
                     event_time: 2021-08-09 10:06:15.592542
                      user_host: root[root] @ localhost []
                      thread_id: 2249
                      server_id: 3940373588
                   command_type: Query
convert(argument using utf8mb4): show tables
*************************** 3. row ***************************
                     event_time: 2021-08-09 10:00:57.131268
                      user_host: root[root] @ localhost []
                      thread_id: 2249
                      server_id: 3940373588
                   command_type: Query
convert(argument using utf8mb4): select count(*) from actor
```

需要注意的是,因为字段 argument 的类型是 mediumblob,所以需要使用 convert 或 cast 函数将其转换成可以显示的字符集,如 utf8。

3.3 慢查询日志

慢查询日志记录 MySQL 中运行时间长的 SQL 语句,这些语句通常是需要进行优化的对象。

3.3.1 配置参数

1. long_query_time

long_query_time 参数规定了响应时间超过该参数值的 SQL 语句被定义为慢 SQL,状态变量 slow_queries 记录了慢查询 SQL 语句的个数。这些 SQL 语句将被记录到慢查询日志中。该参数的默认值是 10,单位是秒,它可以设置成小数,精确到微秒(microseconds),也就是百万分之一秒。

2. slow_query_log

slow_query_log 参数决定是否激活慢查询日志,默认值是 off。

3. log_output

log_output 参数在通用查询日志里已经介绍过,此处不再赘述。

4. slow_query_log_file

slow_query_log_file 参数用于指定慢速查询日志文件的路径和文件名,默认位置在数据目录中,默认文件名是<hostname>-slow.log。

5. log_queries_not_using_indexes

log_queries_not_using_indexes 参数为 true 时,会记录所有做全表扫描和索引扫描的

SQL 语句，无论这些 SQL 语句的实际执行时间有多长，该参数默认值是 false。

6. min_examined_row_limit

min_examined_row_limit 参数规定只有检查的行数超过该参数规定值的 SQL 语句才会被记录，该参数默认是 0，也就是没有限制。该参数可以和另一个参数 log_queries_not_using_indexes 结合使用，当 log_queries_not_using_indexes 设置为 true 时，min_examined_row_limit 可以设置成一个合理的正整数，不记录访问一些小表的 SQL 语句。

7. log_throttle_queries_not_using_indexes

当参数 log_queries_not_using_indexes 为 true 时，该参数限制每分钟记录到错误日志中 SQL 语句的数量，默认值是 0，也就是不限制。

8. log_slow_admin_statements

log_slow_admin_statements 参数控制是否记录管理类的慢 SQL，默认值是 off，当被设置成 on 时，管理类的慢 SQL 如 alter table 或 optimize table 将被记录到慢查询日志中。

9. log_slow_slave_statements

log_slow_slave_statements 参数用于控制是否记录复制中的慢 SQL，默认值是 off，当被设置成 on 时，复制中的慢 SQL 被记录到慢查询日志中，但该设置只有采用语句模式进行复制时才起作用，采用行模式进行复制时不起作用。

10. log_short_format

log_short_format 参数被激活时，记录到慢查询日志里的信息变得简短，其默认值是 off。

11. log_slow_extra

log_slow_extra 参数是从 MySQL 8.0.14 版本开始才有的，当它为 true 时，将记录与执行的 SQL 语句相关的额外信息，如 Handler_%状态参数。该参数的默认值为 off。通常建议将参数值设置为 on，除非解析慢查询日志的工具不支持。

12. 记录内容

如下是设置 long_query_time = 0 时记录的一条 SQL 语句的默认内容：

```
# Time: 2021-01-22T16:22:21.177507+08:00
# User@Host: root[root] @ localhost []  Id:   115
# Query_time: 0.000781  Lock_time: 0.000361 Rows_sent: 1  Rows_examined: 1
SET timestamp=1611303741;
SELECT s_quantity, s_data, s_dist_01 FROM stock WHERE s_i_id = 48241 AND s_w_id = 3;
```

其中，第一行是该 SQL 语句执行的时间；第二行是执行该 SQL 语句的账号和会话 ID；第三行是该 SQL 语句执行过程中的一些统计信息，这里的执行时间（Query_time）和锁时间（Lock_time）都很小，如果这两个时间较长就需要关注。发送到客户端的行数（Rows_sent）和检查的行数（Rows_examined）都是 1，如果检查的行数和发送的行数比值比较大，如检查了 1000 行才发送 1 行，那么该 SQL 语句的执行效率就比较低，一个常见的解决办法是加索引；第四行是时间标签，这里的时间是从 1970 年 1 月 1 日到执行点的毫秒数；最后一行是 SQL 语句的原文。

当 log_slow_extra 参数值设置为 on 时，执行同样的 SQL 语句，记录的信息如下：

```
# Time: 2021-01-22T17:13:08.765664+08:00
# User@Host: root[root] @ localhost []  Id:   117
# Query_time: 0.000558  Lock_time: 0.000330 Rows_sent: 1  Rows_examined: 1 Thread_id: 117
Errno: 0 Killed: 0 Bytes_received: 0 Bytes_sent: 265 Read_first: 0 Read_last: 0 Read_key: 1
Read_next: 0 Read_prev: 0 Read_rnd: 0 Read_rnd_next: 0 Sort_merge_passes: 0 Sort_range_count:
0 Sort_rows: 0 Sort_scan_count: 0 Created_tmp_disk_tables: 0 Created_tmp_tables: 0 Start:
2021-01-22T17:13:08.765106+08:00 End: 2021-01-22T17:13:08.765664+08:00
SET timestamp=1611306788;
SELECT s_quantity, s_data, s_dist_01 FROM stock WHERE s_i_id = 48241 AND s_w_id = 3;
```

可以看到，这时多了很多状态参数，其中 Read_key 值是 1，其他状态参数值都是 0，说明该 SQL 语句根据索引只读取了一行记录，证明该 SQL 语句的执行效率是很高的，已经没有优化的空间了。

3.3.2　使用 mysqldumpslow 解析慢查询日志

当慢查询日志里的记录较多时，看起来比较困难。可以使用 MySQL 自带的慢查询日志分析工具 mysqldumpslow（在 Windows 操作系统中是 mysqldumpslow.pl）对慢查询日志进行汇总分析。

mysqldumpslow 工具会使用摘要函数对 SQL 语句进行摘要分析，MySQL 8.0 中新增加了如下两个摘要函数：

（1）statement_digest_text()：返回 SQL 语句对应的摘要文本。

（2）statement_digest()：计算 SQL 语句的摘要哈希值。

摘要函数会将 SQL 语句中能替换成变量的数字、字符、字符串替换成统一的占位符，还会合并其中多余的空格，如对于如下三条语句，使用函数 statement_digest_text()生成的摘要文本是一样的：

```
mysql> select statement_digest_text('SELECT s_quantity FROM stock WHERE s_i_id = 48241 AND s_
w_id = 3')  digest_text;
+------------------------------------------------------------------+
| digest_text                                                      |
+------------------------------------------------------------------+
| SELECT 's_quantity' FROM 'stock' WHERE 's_i_id' = ? AND 's_w_id' = ?  |
+------------------------------------------------------------------+
1 row in set (0.00 sec)

mysql> select statement_digest_text('SELECT s_quantity FROM stock WHERE s_i_id = 3177 AND s_w
_id = 10') digest_text;
+------------------------------------------------------------------+
| digest_text                                                      |
+------------------------------------------------------------------+
| SELECT 's_quantity' FROM 'stock' WHERE 's_i_id' = ? AND 's_w_id' = ?  |
+------------------------------------------------------------------+
1 row in set (0.00 sec)
```

```
mysql> select statement_digest_text('SELECT s_quantity FROM stock WHERE s_i_id = 81375 AND s_
w_id = 3') digest_text;
+-------------------------------------------------------------------+
| digest_text                                                       |
+-------------------------------------------------------------------+
| SELECT 's_quantity' FROM 'stock' WHERE 's_i_id' = ? AND 's_w_id' = ? |
+-------------------------------------------------------------------+
1 row in set (0.00 sec)
```

对这三个语句，使用函数statement_digest()生成的摘要哈希值也是一样的：

```
mysql> select statement_digest('SELECT s_quantity FROM stock WHERE s_i_id = 48241 AND s_w_id
= 3')  digest_hash;
+-------------------------------------------------------------------+
| digest_hash                                                       |
+-------------------------------------------------------------------+
| 644bcc2d9735094a731c1661c080318f4f3b490209482d1183408672eed76ac6  |
+-------------------------------------------------------------------+
1 row in set (0.00 sec)

mysql> select statement_digest('SELECT s_quantity FROM stock WHERE s_i_id = 3177 AND s_w_id
= 10') digest_hash;
+-------------------------------------------------------------------+
| digest_hash                                                       |
+-------------------------------------------------------------------+
| 644bcc2d9735094a731c1661c080318f4f3b490209482d1183408672eed76ac6  |
+-------------------------------------------------------------------+
1 row in set (0.00 sec)

mysql> select statement_digest('SELECT s_quantity FROM stock WHERE s_i_id = 81375 AND s_w_id
= 3') digest_hash;
+-------------------------------------------------------------------+
| digest_hash                                                       |
+-------------------------------------------------------------------+
| 644bcc2d9735094a731c1661c080318f4f3b490209482d1183408672eed76ac6  |
+-------------------------------------------------------------------+
1 row in set (0.00 sec)
```

慢查询日志工具mysqldumpslow会对摘要哈希值相同的SQL语句合并进行处理，其有以下常用的选项：

- -a：不对SQL语句进行摘要转换。
- -g：对SQL语句进行模式匹配，只输出匹配对的SQL语句，语法和grep命令一样。
- -t：指定返回的SQL语句个数。
- -r：逆序输出SQL语句。
- -s：指定排序规则，默认的排序规则是at，即根据平均执行时间排序，共有7种排序规则，如表3.2所示。

表 3.2

选项	说　明
c	根据 SQL 被执行的次数(count)排序
l	根据合计锁时间排序
al	根据平均锁时间排序
r	根据合计向客户端发送的行数排序
ar	根据平均向客户端发送的行数排序
t	根据合计执行时间排序
at	根据平均执行时间排序

如下是使用 mysqldumpslow 工具解析慢 SQL 日志的一个例子。

首先清空慢查询日志，类似的命令如下：

```
$ sudo rm /disk1/data/redhat7-slow.log
$ mysqladmin flush-logs
```

然后在 mysql 客户端中执行如下 SQL 语句：

```
set global slow_query_log = on;
set long_query_time = 0;
Use tpcc1000;
SELECT s_quantity, s_data, s_dist_01 FROM stock WHERE s_i_id = 48241 AND s_w_id = 3;
SELECT s_quantity, s_data, s_dist_01 FROM stock WHERE s_i_id = 3177 AND s_w_id = 10;
SELECT s_quantity, s_data, s_dist_01 FROM stock WHERE s_i_id = 81375 AND s_w_id = 3;
SELECT count(c_id) FROM customer WHERE c_w_id = 2 AND c_d_id = 4 AND c_last = 'ABLEESECALLY';
SELECT count(c_id) FROM customer WHERE c_w_id = 10 AND c_d_id = 8 AND c_last = 'OUGHTPRIESE';
SELECT count(c_id) FROM customer WHERE c_w_id = 6 AND c_d_id = 6 AND c_last = 'ANTIPRESOUGHT';
SELECT count(c_id) FROM customer WHERE c_w_id = 8 AND c_d_id = 10 AND c_last = 'ESEOUGHTEING';
```

最后使用 mysqldumpslow 工具对慢查询日志进行分析，输出执行次数最多的两条 SQL 语句，代码如下：

```
$ mysqldumpslow -s c -t 2 /disk1/data/redhat7-slow.log

Reading mysql slow query log from /disk1/data/redhat7-slow.log
Count: 4  Time=0.00s (0s)  Lock=0.00s (0s)  Rows=1.0 (4), root[root]@localhost
  SELECT count(c_id) FROM customer WHERE c_w_id = N AND c_d_id = N AND c_last = 'S'

Count: 3  Time=0.00s (0s)  Lock=0.00s (0s)  Rows=1.0 (3), root[root]@localhost
  SELECT s_quantity, s_data, s_dist_01 FROM stock WHERE s_i_id = N AND s_w_id = N
```

可以看到，对这些 SQL 语句进行摘要后，其中的数字被 N 代替，字符串被 S 代替，这样，四条对 customer 表进行查询的 SQL 语句并成了一条 SQL 语句，三条对 stock 进行查询的 SQL 语句合并成了另一条 SQL 语句。输出的内容还包括执行时间、锁和行数，括号里是合计值。

3.3.3 使用 pt-query-digest 解析慢查询日志

pt-query-digest 是 Percona 公司开发的一个功能强大的 MySQL 的 SQL 语句分析工具，它不仅可以分析慢查询日志、二进制日志、通用查询日志、show processlist 命令的输出结果，还可以分析 tcpdump 抓取的 MySQL 的网络流量包。它对要分析的 SQL 语句进行摘要分组，对分组的结果进行统计，计算出各组 SQL 语句的执行时间、次数、占比等信息；它还可以将分析结果保存到 MySQL 的表中，用于以后的对比和趋势分析。

1. 语法和选项

pt-query-digest 的基本语法格式如下：

```
pt-query-digest [OPTIONS] [FILES] [DSN]
```

其中，FILES 可以是慢查询日志、二进制日志、通用查询日志，也可以是 tcpdump 抓取的 MySQL 的网络流量包。

DSN(DATA SOURCE NAME)是指一组用逗号分隔的“键＝值”的方式描述连接到 MySQL 的连接参数，例如：

```
h=host1,P=3306,u=scott
```

表示 MySQL 的主机是 host1，端口是 3306，连接用户是 scott。

pt-query-digest 工具的重要选项如下。

(1) --ask-pass：当连接到 MySQL 时，提示输入密码。

(2) --continue-on-error：决定当遇到一个错误时是否继续，默认是 yes。但并不是永远继续，如果遇到 100 个错误就停止，因为很可能是输入文件的错误或自身的 Bug。

(3) --create-review-table：当使用--review 参数把分析结果输出到表中时，如果没有表就自动创建，默认值是 yes。

(4) --create-history-table：当使用--history 参数把分析结果输出到表中时，如果没有表就自动创建，默认值是 yes。

(5) --defaults-file：指定 MySQL 的参数文件名，这里必须是一个带绝对路径的文件名。

(6) --explain：该参数的类型是 DSN，表示根据 DSN 连接到 MySQL 服务，生成 SQL 语句的执行计划并将执行计划输出到报告中。

(7) --filter：该参数是一串 perl 代码或是一个包含 perl 的文件名，可使用该参数对要分析的文件进行过滤后再分析，将不符合这段 perl 代码的事件全部抛弃。

(8) --history：将分析结果保存到表中，以后再使用--history 时，如果存在相同的 SQL 语句，也会记录到数据表中，但 SQL 语句所在的时间区间不同。

(9) --limit：使用百分比或 SQL 语句的数量来限制输出结果，如果设定一个百分比，则输出达到这个百分比时停止；如果是一个数字，则输出的分析 SQL 达到这个数字时停止；如果两种类型都设置了，则先达到其中任何一个值时就停止。默认值是 95%：20，表示输出到 95%或输出了默认最慢的 20 个 SQL 语句时停止。

(10) --max-line-length：表示裁剪输出行的长度，默认是 74，为 0 不进行裁剪。

(11) --order-by：该参数用于指定输出的排序方式，格式是 attribute：aggregate，attribute 是属性，包括执行时间（Query_time）、检查行数（Rows_examined）和发送行数（Rows_sent）等，aggregate 是汇总方式，包括总计（sum）、最大（max）、最小（min）和计数（cnt 即 count）等，默认的排序方式是 Query_time：sum，也就是按照 SQL 语句的总执行时间进行排序。但当--type 为 genlog 时，也就是解析通用查询日志时，默认的排序规则是 Query_time：cnt，即按照 SQL 语句的执行次数进行排序，因为通用查询日志里没有记录 SQL 语句的执行时间。

(12) --output：该参数用于指定分析结果的输出类型，值可以是 report（标准分析报告）、slowlog（慢查询日志）、json（JSON 格式的报告）、json-anon（没有 SQL 语句例子的 JSON 格式的报告）和 secure-slowlog（没有 SQL 语句例子的慢查询日志），为便于阅读，默认值是 report。

(13) --review：将分析过的 SQL 语句保存到表中，具有同一个摘要哈希值的 SQL 语句只保留一条记录。以后再次使用--review 时，如果存在相同的 SQL 语句将不会记录到数据表中。

(14) --since：该参数用于指定只分析从什么时间开始的 SQL 语句，该参数的格式可以是指定的某个 yyyy-mm-dd [hh：mm：ss]格式的时间点，也可以是简单的一个时间值：s(秒)、h(小时)、m(分钟)、d(天)，如 12h 就表示从 12 小时前开始统计。

(15) --type：该参数用于指明输入文件的类型，可以是通用查询日志（genlog）、二进制日志（binlog）、慢查询日志（slowlog）、抓取的网络流量包（tcpdump）和 SQL 语句文本（rawlog）等。

(16) --until：该参数用于指定被分析的 SQL 语句的截止时间，可以配合--since 使用，只分析一段时间内的 SQL 语句。

2. 用法示例

1) 分析慢查询日志

分析慢查询日志 slow.log 的代码如下：

```
$ pt-query-digest slow.log
```

2) 分析 processlist 的输出

分析在 host1 主机上采用 show processlist 输出的 SQL 语句如下：

```
$ pt-query-digest --processlist h=host1
```

3) 分析网络流量包

从 3306 端口抓取 1000 个网络流量包输出到文件 mysql.tcp.txt 的代码如下：

```
$ tcpdump -s 65535 -x -nn -q -tttt -i any -c 1000 port 3306 > mysql.tcp.txt
```

分析前面抓取到的网络流量包的代码如下：

```
$ pt-query-digest --type tcpdump mysql.tcp.txt
```

4）保存分析过的SQL语句到数据库的表中

将分析过的SQL语句保存到数据库表中的代码如下(不输出分析报告)：

```
$ pt-query-digest --review h=192.168.17.103 --no-report redhat7-slow.log
```

默认的保存表是percona_schema.query_review，通过运行如下代码可以得到该表的结构：

```
mysql> desc query_review ;
+--------------+-----------------+------+-----+---------+-------+
| Field        | Type            | Null | Key | Default | Extra |
+--------------+-----------------+------+-----+---------+-------+
| checksum     | bigint unsigned | NO   | PRI | NULL    |       |
| fingerprint  | text            | NO   |     | NULL    |       |
| sample       | text            | NO   |     | NULL    |       |
| first_seen   | datetime        | YES  |     | NULL    |       |
| last_seen    | datetime        | YES  |     | NULL    |       |
| reviewed_by  | varchar(20)     | YES  |     | NULL    |       |
| reviewed_on  | datetime        | YES  |     | NULL    |       |
| comments     | text            | YES  |     | NULL    |       |
+--------------+-----------------+------+-----+---------+-------+
8 rows in set (0.01 sec)
```

5）保存分析结果到数据库的表中

将分析后的结果保存到数据库表中的代码如下(不输出分析报告)：

```
$ pt-query-digest --history h=192.168.17.103 --no-report redhat7-slow.log
```

默认的保存表是percona_schema.query_history。

--review和--history有如下两个不同之处：

(1) --review只保存分析过的SQL语句，--history不仅保存分析过的SQL语句，还保存分析的结果，通过比较表的结构会发现，percona_schema.query_history记录的字段比percona_schema.query_review要多得多。

(2) 以后分析时，--review对相同的SQL语句不进行保存，但--history还会保存，从这两个表的主键也可以看出这一点，表percona_schema.query_review的主键只有一个字段checksum，而表percona_schema.query_history的主键有三个字段，除checksum外，还有两个时间的字段表示时间段，因此可以用percona_schema.query_history表来分析相同checksum的SQL语句在不同时间段的执行效率。

3. 解析输出

使用工具pt-query-digest对慢查询日志进行解析后，输出的结果可以分为三部分，分别为①汇总的统计结果；②按摘要哈希值对SQL语句进行分组后的统计结果；③每组SQL语句的统计结果。

下面是对pt-query-digest的一个输出结果进行说明的例子。

(1) 汇总的统计结果(中文部分是添加的说明)。

```
# 执行时所用的用户时间、系统时间、物理内存占用大小、虚拟内存占用大小
```

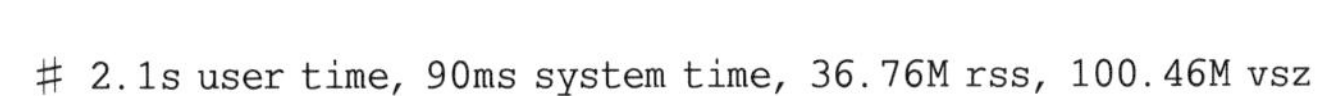

```
# 2.1s user time, 90ms system time, 36.76M rss, 100.46M vsz
# 执行时间
# Current date: Sat Jan 30 05: 10: 28 2021
# 运行时的主机名
# Hostname: scutech
# 被分析的文件名
# Files: redhat7 - slow.log
# 语句总数量,唯一的语句数量,每秒执行的 SQL 语句数量,并发数
# Overall: 9.73k total, 40 unique, 157 QPS, 1.51x concurrency ____________
# 日志记录的时间范围
# Time range: 2021 - 01 - 30T07: 14: 42 to 2021 - 01 - 30T07: 15: 44
# 属性,            总计,       最小,       最大,       平均,      95%百分位,标准,     中位数
# Attribute        total       min         max         avg          95%      stddev     median
# =========       =======     =======     =======     =======     =======    =======    =======
# SQL 语句执行时间
# Exec time         93s        32us        755ms       10ms         51ms       38ms       332us
# SQL 语句执行时锁占用时间
# Lock time         10s         0          726ms       978us        2ms        15ms       84us
# SQL 语句发送到客户端的行数
# Rows sent         7.53k       0          209         0.79         0.99       6.55       0.99
# SQL 语句扫描的行数
# Rows examine      9.63k       0          210        1.01         0.99       6.71       0.99
# SQL 语句查询的字符数
# Query size        983.63k     6          586        103.48       202.40     65.85      80.10
```

（2）按摘要哈希值对 SQL 语句进行分组后的统计结果。

```
# Profile
# Rank Query ID                  Response      time    Calls    R/Call     V/M      Item
# ==== ==========  ============  =====  ======  =====  ===============
#    1 0x813031B8BBC3B329        33.7160      36.1%    369     0.0914     0.16     COMMIT
#    2 0x5E61FF668A8E8456        12.8970      13.8%    2185    0.0059     0.06     SELECT stock
…
```

其中，每行代表一组摘要哈希值一样的 SQL 语句；每列的含义如下。

- Rank：SQL 语句的排名，默认按总的查询时间由长到短排列。
- Query ID：SQL 语句的摘要哈希值。
- Response：每组 SQL 语句总的响应时间。
- time：每组 SQL 语句的用时在总时间中的占比。
- Calls：执行次数。
- R/Call：平均每次执行的响应时间。
- V/M：响应时间的方差均值(variance-to-mean)。
- Item：抽象 SQL 语句。

（3）每组 SQL 语句的统计结果(中文部分是添加的说明)。

```
# 排名第 2 的 SQL,每秒执行的 SQL 语句数量,并发数,SQL ID,at byte 表示这组 SQL 语句中最慢的
SQL 语句在日志中的位置,可以使用 tail  - c 206881 redhat7 - slow.log|head 进行查询
# Query 2: 218.50 QPS, 1.29x concurrency, ID 0x5E61FF668A8E8456 at byte 206881
# This item is included in the report because it matches -- limit.
```

```
# 响应时间的方差均值
# Scores: V/M = 0.06
# 日志记录的时间范围
# Time range: 2021-01-30T07:15:34 to 2021-01-30T07:15:44
# 属性,在本次分析中的占比,总计,最小,最大,平均,95%百分位,标准,中位数
# Attribute    pct   total    min     max     avg     95%     stddev  median
# ============ === ======= ======= ======= ======= ======= ======= =======
# 执行次数
# Count         22    2185
# 执行时间
# Exec time     13     13s   128us   469ms     6ms    34ms    20ms   247us
# 锁占用时间
# Lock time      8   782ms    53us     6ms   358us     2ms   726us    76us
# 发送到客户端的行数
# Rows sent     28   2.13k       1       1       1       1       0       1
# 扫描的行数
# Rows examine   2     267       0       1    0.12    0.99    0.32       0
# 查询的字符数
# Query size    17 174.95k      80      83   81.99   80.10    0.16   80.10
# String:
# 数据库
# Databases    tpcc1000
# 主机名
# Hosts        localhost
# 用户
# Users        root
# 执行时间分布,可以看到大部分SQL语句是在100μs内完成的
# Query_time distribution
#   1us
#  10us
# 100us  ####################################################
#   1ms  #####################
#  10ms  ###########
# 100ms  #
#    1s
#  10s+
# Tables
# 该SQL语句涉及的表,可把下面的SHOW TABLE STATUS和SHOW CREATE TABLE命令直接粘贴到MySQL中执行。
#    SHOW TABLE STATUS FROM 'tpcc1000' LIKE 'stock'\G
#    SHOW CREATE TABLE 'tpcc1000'.'stock'\G
# 如下语句可以生成SQL语句的执行计划,该SQL语句是这组SQL中最慢的SQL语句。
# EXPLAIN /*!50100 PARTITIONS*/
SELECT count(*) FROM stock WHERE s_w_id = 3 AND s_i_id = 44865 AND s_quantity < 12\G
```

3.4 二进制日志

3.4.1 简介

二进制日志(binlog)记录着数据库的变更(如创建表或更改表数据),还记录每个变更

花费时间的信息。记录的单位是事件(event),请注意是事件不是事务(transaction),一个事务可以包含多个事件。

从MySQL 8.0开始,二进制日志默认是激活的,启用二进制日志会使MySQL的性能稍微降低,但二进制日志的两个重要用途带来的好处通常超过了这一较小的性能下降。二进制日志有如下两个重要用途:

(1) 用于复制,将主服务器上的二进制日志发送到从服务器上,并将这些事件应用到从服务器上,实现主服务器和从服务器的数据同步。

(2) 用于数据恢复,恢复全量备份后,再应用二进制日志中的事件使数据库前滚,这种恢复方式称为时间点恢复或增量恢复。

二进制日志和通用查询日志都记录MySQL中执行的语句,但这两种日志有如下两个不同之处:

(1) 二进制日志只记录数据库的变更,不记录只查询数据的语句,而通用查询日志记录所有执行的语句。

(2) 这两种日志记录语句的顺序不同。mysqld按照接收语句的顺序将语句写入通用查询日志,这不一定与语句的执行顺序相同。而二进制日志中的语句是在执行语句之后,但在释放任何锁之前写入,这就保证了语句是按照提交的顺序记录的。

在MySQL 5.7中,启用二进制日志记录时必须指定服务器ID(server_id),否则服务器将无法启动。在MySQL 8.0中,服务器ID默认设置为1。启用二进制日志记录时,可以使用此默认ID启动服务器,但如果未使用系统参数server_id显式指定服务器ID,则会发出一条信息性消息。对于复制拓扑中使用的服务器,必须为每个服务器指定唯一的非零服务器ID。

二进制日志有如下三种不同的格式:

(1) 语句模式(STATEMENT):记录每条会修改数据的SQL语句,这样记录产生的日志量比较小,但记录的非确定(nondeterministic)语句在恢复或复制时可能会出错,如包含now()或uuid()函数的语句,指定这种模式需要设置系统参数--binlog-format=STATEMENT。

(2) 行模式(ROW):不记录SQL语句,仅记录哪条数据已被修改及如何被修改。当单个SQL语句的修改记录数较多时,产生的日志量也相应增大,但可以避免非确定语句在恢复或复制时出错的情况发生,这种方式是MySQL 8.0中的默认记录模式,也可以通过设置系统参数--binlog-format=ROW来指定行模式。

(3) 混合模式(MIX):默认采用语句模式记录二进制日志,但对于非确定语句采用行模式进行记录时,指定这种模式需要设置系统参数--binlog-format=MIXED。

需要说明的是,使用系统参数--binlog-format指定记录格式时会有例外的情况。当修改系统数据库mysql的授权表时,如果使用SQL语句直接修改这些表,则按照系统参数--binlog-format指定记录格式进行记录,这些语句包括insert、update、delete、replace、do、load data、select和truncate;如果使用授权语句间接修改这些表时,无论系统参数--binlog-format指定何种格式均按照语句模式记录这些变更,这些授权语句包括grant、revoke、set password、rename user、create、alter和drop。另外create table… select语句是数据定义和修改的结合体,该语句的create table部分按照语句模式进行记录,而select部分按照系统参数--binlog-format 指定的记录格式进行记录。

3.4.2　系统参数

与二进制日志相关的主要系统参数如下。

1. log_bin

参数 log_bin 用于决定是否激活二进制日志，当该参数值为 on 时，激活二进制日志，这在 MySQL 8.0 里是默认值；当该参数值为 off 时，二进制日志不激活。该参数还可以用于指定二进制日志的路径和文件名的前缀，没有指定时，二进制日志的默认路径在 datadir 目录下，前缀是 binlog。

如果希望当前会话中执行的命令不写入二进制日志，可以把当前会话的参数 sql_log_bin 设置为 0 或 off，例如：

```
mysql > set session sql_log_bin = 0;
```

2. log_bin_index

参数 log_bin_index 用于指定二进制索引文件的路径与名称，默认路径是在 datadir 目录下，默认文件名是 binlog.index。

3. binlog_do_db

参数 binlog_do_db 表示只记录指定数据库的二进制日志。在配置多个数据库时，即使中间用逗号分开，也不能把多个数据库名写在一行里，如 binlog_do_db＝db1，db2，db3，MySQL 会把“db1，db2，db3”当成一个数据库，因为数据库名里可以包括逗号。因此要分成三行写，每行只能写一个数据库，例如：

```
binlog_do_db = db1
binlog_do_db = db2
binlog_do_db = db3
```

4. binlog_ignore_db

参数 binlog_ignore_db 用于指定记录二进制日志时忽略的数据库。指定多个数据库时的写法与 binlog_do_db 一样。

5. max_binlog_cache_size

参数 max_binlog_cache_size 用于指定单个事务能使用的内存最大值，当超过这个值时，会出错，并提示 Multi-statement transaction required more than 'max_binlog_cache_size' bytes of storage，该参数的最小值是 4096，其最大值同时也是默认值为 16EiB。

6. max_binlog_stmt_cache_size

参数 max_binlog_stmt_cache_size 用于指定内存里用于存放的非事务语句的空间大小，当单个事务里的非事务语句超过该值时会出错，该参数的最小值是 4096，最大值同时也是默认值为 16EiB。

7. binlog_cache_size

参数 binlog_cache_size 用于指定每个客户端在事务中使用的二进制日志的缓存大小，如果需要保存的数据超过该值，将在硬盘上创建临时表保存数据，因此可以通过状态变量 binlog_cache_use 和 binlog_cache_disk_use 来帮助确定该参数是否合适，该参数的默认值是 32768，也就是 32K，如果事务较大，可以增加该参数。事务结束时，缓存会被释放，硬盘上的临时表会被截断（truncate）。这个参数 binlog_cache_size 指定的是事务的缓存，参数 binlog_stmt_cache_size 指定的是语句的缓存。

8. binlog_stmt_cache_size

在事务里有非事务语句时，会在 binlog statement 缓存里暂存相关的 SQL 语句，该参数决定客户端独立分配的缓存大小。

9. binlog_error_action

参数 binlog_error_action 有两个默认值，分别是 abort_server 和 ignore_error，默认值为 abort_server 时，表示当二进制日志出现无法写入、无法刷新，或无法同步的情况时，MySQL 主动关闭实例。当该参数值为 ignore_error 时，表示当二进制日志出现无法写入、无法刷新，或无法同步时，关闭 binlog 模式，但 MySQL 实例继续对外提供服务。要恢复 binlog 功能，需要重新开启 log_bin 参数，这需要重新启动实例。该参数在较早的版本里对应于 binlogging_impossible_mode。

10. max_binlog_size

参数 max_binlog_size 规定单个二进制日志文件的最大值，其最大值和默认值都是 1GB。但该参数并不能完全保证二进制日志不会超过这个值，当二进制日志接近最大值并遇到一个比较大的事务时，为了保证事务的完整性，不会进行日志切换，只能将该事务的所有变更都记录进当前的二进制日志文件，直到事务结束，因此二进制日志文件偶尔会超过这个值的设定。

11. sync_binlog

参数 sync_binlog 用于规定将二进制日志刷新到磁盘的频率。当该参数值为 0，事务提交时，MySQL 仅仅将二进制日志缓存中的数据写入二进制日志文件，并不执行 fsync 之类的磁盘同步指令，因此数据可能仍然在操作系统的文件缓存中，何时刷新到磁盘由操作系统决定，这种设置对 MySQL 的性能是最好的，但如果出现机器掉电或操作系统崩溃的情况，二进制文件里的记录可能不全；当该参数值为 1 时，也是默认设置，事务提交时同时刷新二进制日志文件到磁盘，这样即使出现机器掉电或操作系统崩溃的情况，二进制文件里的记录也是完整的，但这种设置对系统的性能有一定程度的负面影响；当该参数为 N 时（N 不为 0 或 1），每进行 N 次事务的提交会刷新一次磁盘，N 越大对 MySQL 的性能影响越小，但丢失数据的可能性也越大。

12. binlog_rows_query_log_events

参数 binlog_rows_query_log_events 决定对于行模式的二进制日志是否记录对应的 SQL，默认值是 off，当参数值设置为 on 时会记录对应的 SQL，此时用 show binlog events 命令在 mysql 客户端里可以查看到对应的 SQL 语句，如图 3.1 所示。

```
mysql> show binlog events in  'binlog.000017';
+---------------+-----+----------------+-----------+-------------+-------------------------------------------------------------------+
| Log_name      | Pos | Event_type     | Server_id | End_log_pos | Info                                                              |
+---------------+-----+----------------+-----------+-------------+-------------------------------------------------------------------+
| binlog.000017 |   4 | Format_desc    |         1 |         125 | Server ver: 8.0.22, Binlog ver: 4                                 |
| binlog.000017 | 125 | Previous_gtids |         1 |         196 | 472444b5-45b9-11eb-bfa1-fa163e368ff4:1-6500                       |
| binlog.000017 | 196 | Gtid           |         1 |         275 | SET @@SESSION.GTID_NEXT= '472444b5-45b9-11eb-bfa1-fa163e368ff4:6501' |
| binlog.000017 | 275 | Query          |         1 |         354 | BEGIN                                                             |
| binlog.000017 | 354 | Rows_query     |         1 |         439 | # insert into actor(first_name,last_name) values('san','zhang')   |
| binlog.000017 | 439 | Table_map      |         1 |         505 | table_id: 107 (sakila.actor)                                      |
| binlog.000017 | 505 | Write_rows     |         1 |         557 | table_id: 107 flags: STMT_END_F                                   |
| binlog.000017 | 557 | Xid            |         1 |         588 | COMMIT /* xid=8837 */                                             |
| binlog.000017 | 588 | Gtid           |         1 |         667 | SET @@SESSION.GTID_NEXT= '472444b5-45b9-11eb-bfa1-fa163e368ff4:6502' |
```

图 3.1

也可以用mysqlbinlog工具的选项--verbose（两次）来解析二进制日志，在解析结果中可以查看到执行的SQL语句，命令mysqlbinlog -vv或mysqlbinlog --verbose --verbose及其输出结果如下：

```
# mysqlbinlog -vv /disk1/data/binlog.000017
…
BEGIN
/*!*/;
# at 354
#210126 17:53:00 server id 1  end_log_pos 439 CRC32 0xc6284f63 Rows_query
# insert into actor(first_name,last_name) values('san','zhang')
# at 439
…
```

13. binlog_expire_logs_seconds和expire_logs_days

参数binlog_expire_logs_seconds用于指定以秒为单位的二进制日志失效的时间，参数expire_logs_days用于指定以天为单位的二进制日志失效的时间，超过这个时间点，二进制日志在刷新日志或MySQL启动时将会被自动删除。推荐使用binlog_expire_logs_seconds，expire_logs_days将在以后的版本中被废弃，binlog_expire_logs_seconds的默认值是2592000，即30天。如果MySQL启动时，binlog_expire_logs_seconds和expire_logs_days都没有设置，则使用binlog_expire_logs_seconds的默认值，即30天；如果两个参数都设置了，则采用binlog_expire_logs_seconds的设置，忽略expire_logs_days的设置，并产生一个警告信息。

如下代码为设置二进制日志失效的时间为一个星期：

```
mysql> set global binlog_expire_logs_seconds = 7 * 24 * 60 * 60;
```

3.4.3 管理二进制日志

1. 查询二进制日志

MySQL的二进制日志包括一组二进制日志文件和一个索引文件，索引文件里记录着二进制日志文件名，如下是在Linux操作系统里检查二进制日志的一个例子：

```
$ ls -l binlog.*
-rw-r-----. 1 mysql mysql    7642178 Jan 21 19:22 binlog.000001
-rw-r-----. 1 mysql mysql        179 Jan 21 19:23 binlog.000002
```

```
-rw-r-----. 1 mysql mysql 1075217757 Jan 21 19:35 binlog.000003
-rw-r-----. 1 mysql mysql 1073754878 Jan 21 19:41 binlog.000004
-rw-r-----. 1 mysql mysql   80599841 Jan 21 19:41 binlog.000005
-rw-r-----. 1 mysql mysql         80 Jan 21 19:41 binlog.index
$ cat binlog.index
./binlog.000001
./binlog.000002
./binlog.000003
./binlog.000004
./binlog.000005
```

也可以在 mysql 客户端里使用 show binary logs 或 show master logs 进行检查，如下代码为使用 show master logs 进行检查：

```
mysql> show master logs;
+----------------+------------+-----------+
| Log_name       | File_size  | Encrypted |
+----------------+------------+-----------+
| binlog.000001  |    7642178 | No        |
| binlog.000002  |        179 | No        |
| binlog.000003  | 1075217757 | No        |
| binlog.000004  | 1073754878 | No        |
| binlog.000005  |   80599841 | No        |
+----------------+------------+-----------+
5 rows in set (0.01 sec)
```

使用 show master status 查询当前日志状态的代码如下：

```
mysql> show master status \G
*************************** 1. row ***************************
             File: binlog.000005
         Position: 80599841
     Binlog_Do_DB:
 Binlog_Ignore_DB:
Executed_Gtid_Set:
1 row in set (0.00 sec)
```

在 mysql 客户端里查询的二进制文件的大小(File_size)和在当前日志状态里的位置(Position)就是从操作系统里查询的文件的字节数，该原则在解析二进制文件时一样适用。

2. 删除二进制日志文件

二进制日志文件不能从操作系统层面直接删除，否则可能导致 binlog.index 里的记录和磁盘上的文件不一致，进而导致 show master logs 之类的命令失败。应该使用相应的 SQL 命令进行删除，删除二进制日志文件需要获取 binlog_admin 的权限，可以使用 reset master 命令删除所有的二进制日志文件，也可以使用 purge 命令删除部分二进制日志文件，使用 purge 命令删除二进制日志文件的语法格式如下：

```
PURGE { BINARY | MASTER } LOGS {
    TO 'log_name'
  | BEFORE datetime_expr
```

```
}
```

删除 binlog.000018 之前的二进制日志文件的代码如下：

```
mysql> purge binary logs to 'binlog.000018';
```

删除'2021-04-02 18：0：0'时间点之前的二进制日志文件的代码如下：

```
mysql> purge binary logs before '2021-04-02 18:0:0';
```

3. 移动二进制日志文件

MySQL 的所有二进制日志文件通常位于同一个目录下，但有时需要将全部或部分二进制日志文件移动到其他目录中，例如，对二进制日志文件进行归档时，或二进制日志文件所在的硬盘空间不够时。MySQL 工具集里的 mysqlbinlogmove 工具可以将二进制日志文件移动到一个新位置，同时正确更新相应的索引文件。

把指定目录的二进制文件全部移动到一个新目录的代码如下：

```
$ sudo mysqlbinlogmove --binlog-dir=/server/data /new/binlog_dir
#
# Moving bin-log files...
# - binlog.000001
# - binlog.000002
# - binlog.000003
# - binlog.000004
# - binlog.000005
# - binlog.000006
#
#...done.
#
```

从服务器上将两天前修改的二进制文件移动到一个归档目录的代码如下，--log-type=all 表示包括二进制日志和中继日志两种文件：

```
$ sudo mysqlbinlogmove --server=user:pass@localhost:3306
--log-type=all  --modified-before=2 /archive/slave/binlog_dir
#
# Applying modified date filter to bin-log files...
#
# Moving bin-log files...
# - slave-bin.000001
# - slave-bin.000002
# - slave-bin.000003
#
# Flushing binary logs...
#
#
# Applying modified date filter to relay-log files...
#
# Moving relay-log files...
# - slave-relay-bin.000001
```

```
# - slave-relay-bin.000002
# - slave-relay-bin.000003
# - slave-relay-bin.000004
# - slave-relay-bin.000005
#
# Flushing relay logs...
#
# ...done.
```

mysqlbinlogmove 工具会在默认位置查找二进制日志文件，但如果没有找到，则会出现下面的错误提示：

```
# WARNING: No bin-log files found to move.
#
# ...done.
#
```

此时可以通过指定参数--bin-log-index 和--bin-log-basename 来定位二进制日志文件，代码如下：

```
$ sudo mysqlbinlogmove --binlog-dir=/disk1/data
--bin-log-index=/disk1/data/binlog.index --bin-log-basename=binlog /tmp
#
# Moving bin-log files...
# - binlog.000018
# - binlog.000019
#
# ...done.
#
```

移动完成后再查询索引文件，发现里面包括不同路径的二进制日志文件，代码如下：

```
$ sudo cat binlog.index
/tmp/binlog.000018
/tmp/binlog.000019
./binlog.000001
./binlog.000002
```

此时在 mysql 客户端里查询的结果如下：

```
mysql> show binary logs;
+---------------+-----------+-----------+
| Log_name      | File_size | Encrypted |
+---------------+-----------+-----------+
| binlog.000018 |       240 | No        |
| binlog.000019 |       240 | No        |
| binlog.000001 |       240 | No        |
| binlog.000002 |       196 | No        |
+---------------+-----------+-----------+
4 rows in set (0.00 sec)
```

3.4.4 mysqlbinlog工具

数据库中的变更事件以二进制的格式记录到二进制日志中,可以使用 mysqlbinlog 工具将其解析成文本格式。

1. 解读 mysqlbinlog 工具的输出

mysqlbinlog 工具的调用方法如下:

```
mysqlbinlog [options] binlog_file …
```

使用 mysqlbinlog 工具解析 binlog.000002 二进制文件的代码如下:

```
$ mysqlbinlog binlog.000002
```

下面对 binlog.000002 的解析输出进行说明。

首先,每个事件的开头会有类似下面的信息:

```
# at 275
#210127 15:32:00 server id 1  end_log_pos 354 CRC32 0xc8739458 Query thread_id=14 exec_
time=0 error_code=0
```

其中:at 275 是该事件在二进制文件中的位置;210127 是日期 2021 年 1 月 27 日;15:32:00 是时间;server id 是 1;CRC32 0xc8739458 是采用 CRC32 算法的二进制日志事件的校验值;thread_id=14 是指连接的会话号;exec_time=0 指该事件的执行时间;error_code=0 是该事件的返回值,这里表示执行成功。

接下来是事件的主体,也就是 BINLOG 语句,这种语句是 MySQL 内部使用的语句,格式如下:

```
BINLOG 'str'
```

其中,str 是采用 Base64 编码的字符串,如下是 BINLOG 语句的一个例子:

```
BINLOG '
ESQRYBMBAAAAQgAAAOACAAAAAFsAAAAAAAMABnNha2lsYQAFYWN0b3IABAIPDxEFtACOAAAAAQGA
AgP8/wBaZ8NC
ESQRYB8BAAAAQQAAACEDAAAAAFsAAAAAAAEAAgAE//8AzQACc2kCbGlgD+dDAM0AAnNpBXpoYW5n
YBEkEfgQ9WM=
'/*!*/;
```

其中,最后的/*!*/;表示事件的结束。

BINLOG 语句的内容是为机器准备的,不是人类可以阅读的,可以使用选项--base64-output=decode-rows 抑制 BINLOG 语句的输出,并使用选项--verbose 输出与 BINLOG 语句对应的伪 SQL 语句,相应代码如下:

```
$ mysqlbinlog --base64-output=decode-rows --verbose binlog.000002
```

对前面的 BINLOG 语句进行解析后输出了如下以三个以#开头的伪 SQL 语句:

```
### UPDATE 'sakila'.'actor'
```

```
###  WHERE
###    @1 = 205
###    @2 = 'si'
###    @3 = 'li'
###    @4 - 1611654979
###  SET
###    @1 = 205
###    @2 = 'si'
###    @3 = 'zhang'
###    @4 = 1611736081
```

指定两次选项--verbose 可以输出更详细的信息，包括数据类型、元数据信息等，采用选项--verbose --verbose 或-vv，例如：

```
$ mysqlbinlog --base64-output=decode-rows -vv binlog.000002
```

现在生成的伪 SQL 语句比之前的伪 SQL 信息更加丰富：

```
###  UPDATE 'sakila'.'actor'
###  WHERE
###    @1 = 205 /* SHORTINT meta = 0 nullable = 0 is_null = 0 */
###    @2 = 'si' /* VARSTRING(180) meta = 180 nullable = 0 is_null = 0 */
###    @3 = 'li' /* VARSTRING(180) meta = 180 nullable = 0 is_null = 0 */
###    @4 = 1611654979 /* TIMESTAMP(0) meta = 0 nullable = 0 is_null = 0 */
###  SET
###    @1 = 205 /* SHORTINT meta = 0 nullable = 0 is_null = 0 */
###    @2 = 'si' /* VARSTRING(180) meta = 180 nullable = 0 is_null = 0 */
###    @3 = 'zhang' /* VARSTRING(180) meta = 180 nullable = 0 is_null = 0 */
###    @4 = 1611736081 /* TIMESTAMP(0) meta = 0 nullable = 0 is_null = 0 */
```

需要注意的是，对 Binlog 语句的解析有可能不成功，如下面这个例子：

```
$ mysqlbinlog --start-position=420 --base64-output=decode-rows --verbose
binlog.000002
...
# at 420
#210127 15:32:00 server id 1  end_log_pos 472 CRC32 0xd0900530 Delete_rows: table id 91
flags: STMT_END_F
### Row event for unknown table #91
# at 472
#210127 15:32:00 server id 1  end_log_pos 503 CRC32 0x43b3cb20 Xid = 116
...
```

从事件头可以看出，这是一个对表 id 为 91 的表进行删除的语句，在位置 420～472 本应该输出解析后的 SQL 语句，但因为没有解析成功，所以在三个 # 后没有相应的输出。因此如果需要知道二进制日志中执行的具体 SQL 语句，还是应该把参数 binlog_rows_query_log_events 设置为 on，这样在写二进制日志时记录 SQL 语句就变得更可靠。

2. 重放二进制日志

mysqlbinlog 解析的结果可以通过管道定向到 mysql 客户端再次执行，将变更应用到 MySQL 服务，或者称为重放到 MySQL 服务，这种方法常用于时间点恢复，如下面这个

例子:

```
$ mysqlbinlog binlog.000001 | mysql -u root -p
```

binlog.000001中记录的数据库的变更被mysqlbinlog解析后通过mysql客户端再次在MySQL服务中重新执行。如果二进制日志里包括BLOB类型的数据,调用mysql时需要使用--binary-mode选项。还可以将mysqlbinlog的解析结果输出到一个文本文件,然后对这个文本进行编辑,如去掉一些不需要的变更,再通过mysql客户端执行这个编辑后的文本文件。这里用如下代码说明这个过程:

```
$ mysqlbinlog binlog.000001 > tmpfile
$ vi tmpfile
$ mysql -u root -p < tmpfile
```

也可以使用--start-datetime和--stop-datetime指定解析的时间段,或者使用--start-position和--stop-position指定解析二进制日志的开始和结束位置,这样可以选择部分内容重放到MySQL服务。

例如,如果要跳过一个不希望执行的SQL语句,运行如下代码先查询出该SQL语句在二进制日志中的位置:

```
$ mysqlbinlog -v binlog.000016|grep -B7 'DELETE FROM 'sakila'.'tab_c''
# at 939
#210705  3:06:22 server id 143  end_log_pos 992 CRC32 0xd372fd33
    Delete_rows: table id 322 flags: STMT_END_F

BINLOG '
VpriYBOPAAAAOQAAAKsDAAAAAEIBAAAAAAAEABnNha2lsYQAFdGFiX2MAAf4C/igBAgP8/wBUYiJZ
VpriYCCPAAAANQAAAOADAAAAAEIBAAAAAAAEAAgAB/wAEYWFhYQAEMTExMQAEMTExMTP9ctM=
'/*!*/;
### DELETE FROM 'sakila'.'tab_c'
### WHERE
###   @1='aaaa'
### DELETE FROM 'sakila'.'tab_c'
### WHERE
###   @1='1111'
### DELETE FROM 'sakila'.'tab_c'
```

可以发现,这个需要跳过的SQL语句的开始位置是939,结束位置是992。另外一个查询需要跳过的SQL语句在二进制日志位置的方法是使用show binlog events in binlog_filename命令。把二进制日志重放到939位置的命令如下:

```
$ mysqlbinlog --stop-position=939 binlog.000016|mysql
```

再从992位置继续开始重放二进制日志的命令如下:

```
$ mysqlbinlog --start-position=992 binlog.000016|mysql
```

这样,就跳过了不希望执行的SQL语句。注意在使用mysqlbinlog工具对二进制日志里的事件进行重放时,--base64-output只能设置成auto,这也是默认设置,设置成decoderows 或其他值可能会丢失事件。

当把多个二进制日志重放到 MySQL 服务时，要在同一个连接中进行重放，下面的做法是错误的：

```
$   mysqlbinlog binlog.000001 | mysql -u root -p
$   mysqlbinlog binlog.000002 | mysql -u root -p
```

正确的做法是：

```
$   mysqlbinlog binlog.000001 binlog.000002 | mysql -u root -p
```

因为不同的二进制日志中可以用到同一个临时表，如果中间连接释放了，MySQL 会删除临时表，后面的连接执行可能会出错。

从 MySQL 8.0.12 开始，可以通过管道将二进制日志定向到 mysqlbinlog 的输入进行解析，例如：

```
$   gzip -cd binlog.000001.gz | mysqlbinlog - | mysql -uroot  -p
```

这种做法减少了临时文件的生成，并提高了执行的效率。

3. 备份二进制日志

mysqlbinlog 不仅可以将二进制日志解析成文本格式，还可以将二进制日志保存成二进制格式，该功能可以用于二进制日志的备份，这样可以不登录到 MySQL 所在的操作系统而直接备份二进制日志，实现这种备份方式需要增加如下两个选项。

(1) --read-from-remote-server(或-R)：该选项用于告诉 mysqlbinlog 工具连接到远端的 MySQL 服务。

(2) --raw：该选项用于告诉 mysqlbinlog 工具将读取到的二进制日志保存成二进制格式而不是文本格式。

连接到远端 MySQL 服务的选项包括--host、--user 和—password，这些选项和 mysql 客户端的使用方法一样。例如，将远端 MySQL 的二进制日志全部备份到本地的代码如下：

```
$ mysqlbinlog --read-from-remote-server --host=192.168.17.103 --raw
--user=root --to-last-log binlog.000001
```

mysqlbinlog 工具还可以动态地备份远端 MySQL 服务即时产生的二进制日志，这时需要增加一个选项--stop-never，例如：

```
$ mysqlbinlog --read-from-remote-server --host=192.168.17.103 --raw
--user=root --stop-never binlog.000001
```

使用这种备份方式需要使用选项--connection-server-id 指定一个服务 ID，如果没有显示指定服务 ID，则默认为 1。该备份方式和复制类似，但不会自动重新连接，如果连接中断，需要手工启动 mysqlbinlog 再次进行备份。

3.4.5 “篡改”二进制日志

二进制日志是以事件为单位记录 MySQL 数据库的所有变更信息，了解二进制日志的结构可以帮助我们解析二进制日志，甚至对它做一些修改，或者说是“篡改”，实现类似事务回滚的功能，如恢复误删除的记录或者还原被修改的记录等。

1. 二进制日志的结构

从 MySQL 5.0 到 MySQL 8.0，二进制日志采用的是 v4 版本，其中包括事件的类型有如下 36 类：

```
enum Log_event_type {
  UNKNOWN_EVENT = 0,
  START_EVENT_V3 = 1,
  QUERY_EVENT = 2,
  STOP_EVENT = 3,
  ROTATE_EVENT = 4,
  INTVAR_EVENT = 5,
  LOAD_EVENT = 6,
  SLAVE_EVENT = 7,
  CREATE_FILE_EVENT = 8,
  APPEND_BLOCK_EVENT = 9,
  EXEC_LOAD_EVENT = 10,
  DELETE_FILE_EVENT = 11,
  NEW_LOAD_EVENT = 12,
  RAND_EVENT = 13,
  USER_VAR_EVENT = 14,
  FORMAT_DESCRIPTION_EVENT = 15,
  XID_EVENT = 16,
  BEGIN_LOAD_QUERY_EVENT = 17,
  EXECUTE_LOAD_QUERY_EVENT = 18,
  TABLE_MAP_EVENT = 19,
  PRE_GA_WRITE_ROWS_EVENT = 20,
  PRE_GA_UPDATE_ROWS_EVENT = 21,
  PRE_GA_DELETE_ROWS_EVENT = 22,
  WRITE_ROWS_EVENT = 23,
  UPDATE_ROWS_EVENT = 24,
  DELETE_ROWS_EVENT = 25,
  INCIDENT_EVENT = 26,
  HEARTBEAT_LOG_EVENT = 27,
  IGNORABLE_LOG_EVENT = 28,
  ROWS_QUERY_LOG_EVENT = 29,
  WRITE_ROWS_EVENT = 30,
  UPDATE_ROWS_EVENT = 31,
  DELETE_ROWS_EVENT = 32,
  GTID_LOG_EVENT = 33,
  ANONYMOUS_GTID_LOG_EVENT = 34,
  PREVIOUS_GTIDS_LOG_EVENT = 35,
  ENUM_END_EVENT
  /* end marker */
};
```

每个二进制日志文件都是以 Format Description 类型的事件开始，以 Rotate 类型的事件结束，如下是一个二进制日志的例子：

```
mysql>  show binlog events in 'scut.000023';
+---------+---+------------+--------+----------+------------------------------+
| Log_name |Pos | Event_type   | Server_id | End_log_pos | Info                          |
+---------+---+------------+--------+----------+------------------------------+
| scut.000023 |   4| Format_desc   |   1024   |      123 | Server ver: 8.0.22, Binlog ver: 4   |
| scut.000023 | 123 | Previous_gtids |   1024   |      154 |                               |
| scut.000023 | 154 | Anonymous_Gtid |   1024   |      219 | SET @@SESSION.GTID_NEXT= 'ANONYMOUS' |
| scut.000023 | 219 | Query         |   1024   |      291 | BEGIN                         |
| scut.000023 | 291 | Rows_query    |   1024   |      330 | # delete from tt1             |
| scut.000023 | 330 | Table_map     |   1024   |      378 | table_id: 111 (test.tt1)      |
| scut.000023 | 378 | Delete_rows   |   1024   |      434 | table_id: 111 flags: STMT_END_F |
| scut.000023 | 434 | Xid           |   1024   |      465 | COMMIT /* xid=216 */          |
| scut.000023 | 465 | Rotate        |   1024   |      507 | scut.000024; pos=4            |
+---------+---+------------+--------+----------+------------------------------+
9 rows in set (0.00 sec)
```

每列的说明如下。

- Log_name：当前事件所在的二进制日志文件名。
- Pos：当前事件的开始位置，每个事件都占用固定的字节大小，结束位置(End_log_pos)减去 Pos，就是该事件占用的字节数。
- Event_type：事件的类型。
- Server_id：产生该事件的 MySQL Server_id。
- End_log_pos：下一个事件的开始位置。
- Info：当前事件的描述信息。

每行的事件类型(Event_type)说明如下。

- Format_desc：二进制日志文件的第一个事件。在 Info 列可以看到，其标明了 MySQL Server 的版本是 8.0.22，二进制日志版本是 4。
- Previous_gtids：表示之前的二进制日志文件中，已经执行过的 GTID。需要开启 GTID 选项，该事件才会有值。
- Anonymous_Gtid：没有开启 GTID 选项时，每个事务开始的事件。
- Query：向二进制日志发送一个语句，这里是事务的开始语句 begin。
- Rows_query：记录 SQL，该事件只有当参数 binlog_rows_query_log_events 为 TRUE 的情况下才会产生，该参数的默认值为 FALSE。
- Table_map：记录将要被修改的表的结构。
- Delete_rows：从表中删除一个记录。
- Xid：事务提交的时候写入事务 ID。
- Rotate：每个 Binlog 文件的结束事件。在 Info 列中记录了下一个二进制日志文件的名称是 scut.000024。

根据官方文档，每个事件的数据结构如下：

```
+=====================================+
| event  | timestamp         0 : 4    |
| header +----------------------------+
|        | type_code         4 : 1    |
```

```
|         +-------------------------------+
|         | server_id         5 : 4       |
|         +-------------------------------+
|         | event_length      9 : 4       |
|         +-------------------------------+
|         | next_position    13 : 4       |
|         +-------------------------------+
|         | flags            17 : 2       |
|         +-------------------------------+
|         | extra_headers    19 : x-19    |
+=========================================+
| event   | fixed part        x : y       |
| data    +-------------------------------+
|         | variable part                 |
+=========================================+
```

2. 恢复误删除的记录

了解了二进制日志的结构，就可以通过修改二进制日志里的数据实现一些特殊的功能。例如，二进制日志里的删除行（DELETE_ROWS_EVENT）事件和写行（WRITE_ROWS_EVENT）事件的数据结构是完全一样的，只是删除行的事件类型（type_code）是32，写行的事件类型是30。因此只要把32改成30，就可以把删除的记录重新插入。例如，前面二进制日志文件（scut.000023）删除行事件的位置是378，从事件的结构里可以看到事件类型在事件的第5个字节，因此只要把第383（378+5=383）字节改成30即可把删除的记录恢复，可以使用二进制编辑工具来进行修改，也可以使用如下的Python小程序来实现这个功能：

```
$ cat splitTran.py
#! /usr/bin/python3
import sys

if len(sys.argv) != 3:
    print ('Please run splitTrans.py inputBinlog changedBinlog.')
    sys.exit()

inputBinlog = open(sys.argv[1],"rb")
changedBinlog = open(sys.argv[2],"wb")

changedBinlog.write(inputBinlog.read(429))    # read from the head of  input binlog file to the first insert, then write into the changed binlog file.
firstInsert = inputBinlog.tell()
inputBinlog.seek(567,0)  # locate to the xid event
changedBinlog.write(inputBinlog.read(31))   # read from 567 to 598, write xid event, into the changed binlog file.
inputBinlog.seek(154,0)  # locate to the Anonymous_Gtid, Query events.
changedBinlog.write(inputBinlog.read(137))   # read from 154 to 291, write Anonymous_Gtid, Query events into changed binlog file.
inputBinlog.seek(firstInsert)
while True:
      line = inputBinlog.readline()
      if not line:
```

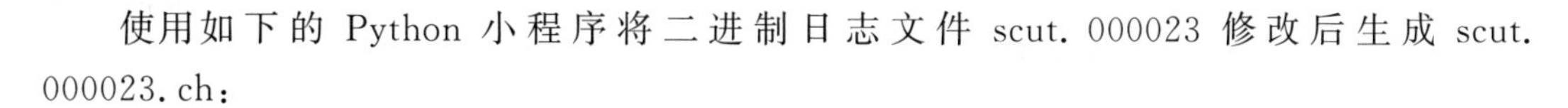

```
            break
        changedBinlog.write(line)

    inputBinlog.close()
    changedBinlog.close()
```

使用如下的 Python 小程序将二进制日志文件 scut.000023 修改后生成 scut.000023.ch：

```
$ ./ModiBinlog.py scut.000023 scut.000023.ch
```

然后分别将这两个二进制日志文件重放到 MySQL，看看执行的效果。

(1) 第一次查询表 test.tt1，发现没有记录，代码如下：

```
$ mysql  -e "select * from test.tt1";
```

(2) 将修改后的二进制日志文件 scut.000023_ch 重放到 MySQL，代码如下：

```
$ mysqlbinlog ./scut.000023_ch | mysql
```

(3) 第二次查询表 test.tt1，发现新增了一个记录，代码如下：

```
$ mysql  -e "select * from test.tt1";
+---------------------+
| col1                |
+---------------------+
| aaaaaaaaaaaaaaaaaaa |
+---------------------+
```

(4) 将原来的二进制日志文件 scut.000023 重放到 MySQL，代码如下：

```
$ mysqlbinlog ./scut.000023 |mysql
```

(5) 第三次查询表 test.tt1，发现记录又被删除了，代码如下：

```
$ mysql  -e "select * from test.tt1";
```

(6) 第二次将修改后的二进制日志文件 scut.000023_ch 重放到 MySQL，代码如下：

```
$ mysqlbinlog ./scut.000023_ch |mysql
```

(7) 第四次查询表 test.tt1，发现记录又被还原了，代码如下：

```
$ mysql  -e "select * from test.tt1";
+---------------------+
| col1                |
+---------------------+
| aaaaaaaaaaaaaaaaaaa |
+---------------------+
```

通过如上实验可以看到，修改二进制日志可以恢复被删除的记录。实际上还可以实现其他的一些功能，例如，二进制日志记录的更新事件(UPDATE_ROWS_EVENT)同时记录了更新前和更新后的值，因此可以利用这个特点将更新的值再改回去。

3.4.6 二进制日志中的大事务

由于行模式的二进制日志对每个记录的变更都记录一条日志，因此一个简单的SQL语句可能会在二进制日志中产生一个巨大的事务，这样的大事务经常是产生麻烦的根源。下面分析一下如何找出二进制日志中的大事务和如何切割大事务。

1. 找出二进制日志中的大事务

使用mysqlbinlog解析后的二进制日志，每个事务都是以BEGIN开头，BEGIN后的第二行是事务开始的地址，如下是使用grep命令找出事务的代码前三行：

```
$ mysqlbinlog  binlog.000004| grep "^BEGIN" -A 2|head -n 12
BEGIN
/*!*/;
# at 365
--
BEGIN
/*!*/;
# at 2549
--
BEGIN
/*!*/;
# at 4925
--
```

然后取出其中的事务的开始位置，代码如下：

```
$ mysqlbinlog  binlog.000004| grep "^BEGIN" -A 2| grep -E '^# at' | awk '{print $3}'|head
365
2549
4925
7857
10143
15473
29100
41840
44091
46146
```

再将相邻的两个位置相减可以算出每个事务的大小，代码如下：

```
$ mysqlbinlog binlog.000004| grep "^BEGIN" -A 2 | grep -E '^# at' | awk '{print $3}' | awk
'NR==1 {at=$1} NR>1 {print ($1-at); at=$1}' | head
2184
2376
2932
2286
5330
13627
12740
```

```
2251
2055
12560
```

最后将这些事务按从大到小排序,取最大的 10 个事务的值,代码如下:

```
$ mysqlbinlog binlog.000004| grep "^BEGIN" -A 2 | grep -E '^# at' | awk '{print $3}' | awk
'NR==1 {at=$1} NR>1 {print ($1-at); at=$1}' | sort -n -r | head
14656
14612
14603
14568
14566
14557
14554
14552
14541
14541
```

2. 切割 Binlog 中的大事务

分析了二进制日志的结构后会发现,如果在同一个事务的不同事件之间插入 xid、Anonymous_Gtid、Query 三个事件,可以把一个事务切割成两个事务。图 3.2 是一个包含两个 insert 语句的二进制日志文件的解析结果。

```
mysql> show binlog events in 'binlog.000010';
+---------------+-----+----------------+-----------+-------------+---------------------------------------------+
| Log_name      | Pos | Event_type     | Server_id | End_log_pos | Info                                        |
+---------------+-----+----------------+-----------+-------------+---------------------------------------------+
| binlog.000010 |   4 | Format_desc    |         1 |         125 | Server ver: 8.0.22, Binlog ver: 4           |
| binlog.000010 | 125 | Previous_gtids |         1 |         196 | 472444b5-45b9-11eb-bfa1-fa163e368ff4:1-8039 |
| binlog.000010 | 196 | Anonymous_Gtid |         1 |         275 | SET @@SESSION.GTID_NEXT= 'ANONYMOUS'        |
| binlog.000010 | 275 | Query          |         1 |         354 | BEGIN                                       |
| binlog.000010 | 354 | Table_map      |         1 |         420 | table_id: 99 (sakila.actor)                 |
| binlog.000010 | 420 | Write_rows     |         1 |         472 | table_id: 99 flags: STMT_END_F              |
| binlog.000010 | 472 | Table_map      |         1 |         538 | table_id: 99 (sakila.actor)                 |
| binlog.000010 | 538 | Write_rows     |         1 |         586 | table_id: 99 flags: STMT_END_F              |
| binlog.000010 | 586 | Xid            |         1 |         617 | COMMIT /* xid=1428 */                       |
| binlog.000010 | 617 | Rotate         |         1 |         661 | binlog.000011;pos=4                         |
+---------------+-----+----------------+-----------+-------------+---------------------------------------------+
10 rows in set (0.00 sec)
```

图 3.2

这两个 insert 语句位于同一个事务中,可以在第二个 insert 语句开始之前进行切割,位置是 472,如下是完成该切割的 Python 小程序:

```
$ cat splitTran.py
#! /usr/bin/python3
import sys

if len(sys.argv) != 3:
    print ('Please run splitTrans.py inputBinlog changedBinlog.')
    sys.exit()

inputBinlog = open(sys.argv[1],"rb")
changedBinlog = open(sys.argv[2],"wb")

changedBinlog.write(inputBinlog.read(472))     #从二进制日志文件的开头一直读取到第二个
                                               #insert 语句之前的位置
```

```
firstInsert = inputBinlog.tell()          # 将读取的内容插入新的二进制日志文件中
inputBinlog.seek(586,0)                   # 定位到 xid 事件的开始位置
changedBinlog.write(inputBinlog.read(31))    # 从 586 读取到 617,也就是读取整个 xid 事件,插
                                             # 入新的二进制日志文件中
inputBinlog.seek(196,0)                   # 定位到 Anonymous_Gtid、Query 事件
changedBinlog.write(inputBinlog.read(79))    # 读取 Anonymous_Gtid、Query 事件并插入到新的二
                                             # 进制日志文件中
inputBinlog.seek(firstInsert)
while True:
      line = inputBinlog.readline()
      if not line:
          break
      changedBinlog.write(line)

inputBinlog.close()
changedBinlog.close()
```

使用如下的 Python 小程序读取二进制日志文件 binlog.000010，生成新的二进制日志文件 binlog.000010.ch：

```
$ ./splitTran.py binlog.000010 binlog.000010.ch
```

将新的二进制日志文件重放到 MySQL，代码如下：

```
$ mysqlbinlog binlog.000010.ch |mysql
```

图 3.3 是通过解析当前二进制日志文件查看执行的效果。

```
mysql> show binlog events in 'binlog.000013';
+---------------+-----+----------------+-----------+-------------+---------------------------------------------------+
| Log_name      | Pos | Event_type     | Server_id | End_log_pos | Info                                              |
+---------------+-----+----------------+-----------+-------------+---------------------------------------------------+
| binlog.000013 |   4 | Format_desc    |         1 |         125 | Server ver: 8.0.22, Binlog ver: 4                 |
| binlog.000013 | 125 | Previous_gtids |         1 |         196 | 472444b5-45b9-11eb-bfa1-fa163e368ff4:1-8039       |
| binlog.000013 | 196 | Anonymous_Gtid |         1 |         282 | SET @@SESSION.GTID_NEXT= 'ANONYMOUS'              |
| binlog.000013 | 282 | Query          |         1 |         359 | BEGIN                                             |
| binlog.000013 | 359 | Table_map      |         1 |         425 | table_id: 99 (sakila.actor)                       |
| binlog.000013 | 425 | Write_rows     |         1 |         477 | table_id: 99 flags: STMT_END_F                    |
| binlog.000013 | 477 | Xid            |         1 |         508 | COMMIT /* xid=1576 */                             |
| binlog.000013 | 508 | Anonymous_Gtid |         1 |         594 | SET @@SESSION.GTID_NEXT= 'ANONYMOUS'              |
| binlog.000013 | 594 | Query          |         1 |         671 | BEGIN                                             |
| binlog.000013 | 671 | Table_map      |         1 |         737 | table_id: 99 (sakila.actor)                       |
| binlog.000013 | 737 | Write_rows     |         1 |         785 | table_id: 99 flags: STMT_END_F                    |
| binlog.000013 | 785 | Xid            |         1 |         816 | COMMIT /* xid=1584 */                             |
+---------------+-----+----------------+-----------+-------------+---------------------------------------------------+
12 rows in set (0.01 sec)
```

图　3.3

可以发现，这两个 insert 语句已经被分割到两个事务中了。

3.5　实验

视频演示

1. 错误日志

(1) 将系统参数 log_error_verbosity 分别设置成 2 和 3，观察错误日志记录信息的详细程度。

（2）将系统参数 log_timestamps 分别设置为 SYSTEM 和 UTC，观察时间格式的区别。

2. 通用日志

视频演示

（1）通过设置系统参数 general_log 为 on，启动通用日志。

（2）运行 mysqldump，同时从通用日志中观察 mysqldump 实际执行的命令。

3. 慢查询日志

视频演示

（1）通过设置系统参数 slow_query_log 为 on，启动慢查询日志。

（2）设置系统参数 log_slow_extra 为 on，观察慢查询日志里记录内容的区别。

4. 二进制日志

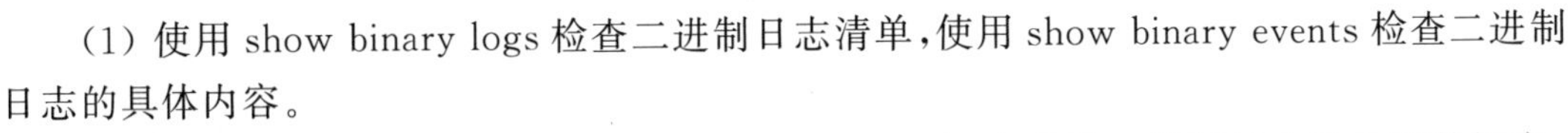
（1）使用 show binary logs 检查二进制日志清单，使用 show binary events 检查二进制日志的具体内容。

视频演示

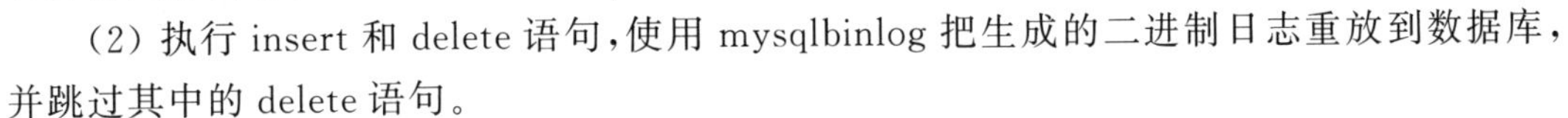
（2）执行 insert 和 delete 语句，使用 mysqlbinlog 把生成的二进制日志重放到数据库，并跳过其中的 delete 语句。

第4章

安　全

本章主要介绍如何提高 MySQL 的安全性，包括账号和密码的安全性、连接加密、数据加密和审计等。

4.1　密码验证组件

当使用 YUM 方式安装 MySQL 时，会自动安装密码验证组件，已经安装的组件可以在 mysql.component 表中查到，例如：

```
mysql> select * from mysql.component;
+--------------+--------------------+--------------------------------------+
| component_id | component_group_id | component_urn                        |
+--------------+--------------------+--------------------------------------+
|            1 |                  1 | file://component_validate_password   |
+--------------+--------------------+--------------------------------------+
1 row in set (0.00 sec)
```

如果没有安装密码验证组件，可以使用如下命令安装：

```
mysql> install component 'file://component_validate_password';
```

使用如下命令进行卸载：

```
mysql> uninstall component 'file://component_validate_password';
```

安装该组件后，会新增 7 个以 validate_password. 开头的系统变量用以配置密码的复杂度，运行如下代码后，可以看到这些系统变量的默认值：

```
mysql> show variables like 'validate%';
+--------------------------------------+--------+
| Variable_name                        | Value  |
```

```
+-----------------------------------+--------+
| validate_password.check_user_name       | ON     |
| validate_password.dictionary_file       |        |
| validate_password.length                | 8      |
| validate_password.mixed_case_count      | 1      |
| validate_password.number_count          | 1      |
| validate_password.policy                | MEDIUM |
| validate_password.special_char_count    | 1      |
+-----------------------------------+--------+
7 rows in set (0.01 sec)
```

validate_password. policy 决定设置密码时哪些以 validate_password. 开头的系统变量会被检查,它有如下三个值。

(1) 0 或 LOW:只检查密码的长度是否符合 validate_password. length 的规定。

(2) 1 或 MEDIUM:检查密码是否符合 validate_password. length、validate_password. number_count、validate_password. mixed_case_count 和 validate_password. special_char_count。

(3) 2 或 STRONG:在 1 的基础上还要检查密码是否与 validate_password. dictionary_file 指定的字典文件里的单词匹配,匹配的规则是密码中 4～100 位的字符子串(不区分大小写)和字典文件里的单词进行比较,如果能找到匹配的字符串则密码设置失败。

validate_password. check_user_name 如果设置成 ON,则拒绝密码和用户名相同(不管正序或倒序)。是否检查这个变量不受 validate_password. policy 的控制。

这 7 个以 validate_password. 开头的系统变量可以动态地进行设置,例如:

```
mysql> set global validate_password.policy = 2;
mysql> set global validate_password.length = 16;
```

4.2　连接控制插件

连接控制插件可以在用户连续多次输入错误密码后强制一个时延,输入错误密码的次数越多,时延越长,这可以有效地阻止暴力破解密码。该插件的安装方法如下:

```
mysql> install plugin connection_control soname 'connection_control.so';
```

安装该插件后,运行如下代码,会增加下面的系统变量:

```
mysql> show variables like 'connection_control%';
+-------------------------------------------------+------------+
| Variable_name                                   | Value      |
+-------------------------------------------------+------------+
| connection_control_failed_connections_threshold | 3          |
| connection_control_max_connection_delay         | 2147483647 |
| connection_control_min_connection_delay         | 1000       |
+-------------------------------------------------+------------+
3 rows in set (0.00 sec)
```

(1) connection_control_failed_connections_threshold 默认设置为 3,表示连续 3 次输

入错误密码后启动了时延机制。

（2）connection_control_min_connection_delay 默认设置为1000毫秒，也就是1秒，当用户第4次输错密码时，时延1秒，第5次时延2秒，以此类推。

（3）connection_control_max_connection_delay 设置最长时延时间，该变量不能设置的比 connection_control_min_connection_delay 小。

当前时延由状态参数 Connection_control_delay_generated 记录，代码如下：

```
mysql> show status like 'Connection_control%';
+------------------------------------+-------+
| Variable_name                      | Value |
+------------------------------------+-------+
| Connection_control_delay_generated | 4     |
+------------------------------------+-------+
1 row in set (0.01 sec)
```

可以安装 connection_control_failed_login_attempts 插件，记录更加详细的信息，安装方法如下：

```
mysql> install plugin connection_control_failed_login_attempts soname 'connection_control.so';
```

查询 information_schema 数据库的 connection_control_failed_login_attempts 表，可以得到密码输入失败的账号和次数。相关代码及其运行结果如下：

```
mysql> select * from information_schema.connection_control_failed_login_attempts;
+---------------+-----------------+
| USERHOST      | FAILED_ATTEMPTS |
+---------------+-----------------+
| 'scutech'@'%' |               7 |
+---------------+-----------------+
1 row in set (0.01 sec)
```

从查询结果中可以看到，用户 scutech 输入错误密码7次，因此当前时延是7减3，得到的延迟时间是4秒。

4.3 连接加密

4.3.1 证书与加密

MySQL 的客户端和服务端之间的通信通常是加密的，如果不加密，通信信息可能会被截取或篡改。通信加密和证书密切相关，在 MySQL 的数据目录下通常有8个以.pem 结尾的证书文件，具体说明如下。

（1）ca.pem：自己签名的 CA(certificate authority)证书。

（2）ca-key.pem：CA 证书的私钥。

（3）server-cert.pem：服务端证书。

（4）server-key.pem：服务端证书的私钥。

（5）client-cert.pem：客户端证书。

（6）client-key.pem：客户端证书的私钥。

（7）private_key.pem：公钥/私钥对中的私钥。

（8）public_key.pem：公钥/私钥对中的公钥。

如上8个证书中，前面6个用于客户端和服务端之间的通信加密，通常在数据库初始化时自动创建，也可以使用MySQL自带的工具mysql_ssl_rsa_setup调用openssl生成证书。最后两个证书用于当账号的验证插件是sha256_password或caching_sha2_password时的密码交换。

可以使用openssl查看这些证书，例如：

```
$ openssl x509 -text -in ca.pem
Certificate:
    Data:
        Version: 3 (0x2)
        Serial Number: 1 (0x1)
        Signature Algorithm: sha256WithRSAEncryption
        Issuer: CN = MySQL_Server_8.0.22_Auto_Generated_CA_Certificate
        Validity
            Not Before: Mar 16 11:57:43 2021 GMT
            Not After : Mar 14 11:57:43 2031 GMT
        Subject: CN = MySQL_Server_8.0.22_Auto_Generated_CA_Certificate
        Subject Public Key Info:
            Public Key Algorithm: rsaEncryption
                RSA Public-Key: (2048 bit)
...
```

MySQL客户端和服务端连接加密的过程如下：

（1）客户端发起一个到服务端的连接。

（2）服务端将数字证书传输到客户端。

（3）客户端根据数字证书验证服务端的身份。

（4）客户端使用服务端的数字证书将加密后的会话密钥传输到服务端（非对称加密）。

（5）服务端使用自己私钥解密客户端传过来的会话密钥。

（6）在剩下的会话里，客户端和服务端的通信将使用会话密钥进行加密（对称加密）。

如下是一个客户端连接的例子：

```
$ mysql --protocol=tcp
Welcome to the MySQL monitor.  Commands end with ; or \g.
Your MySQL connection id is 14
...
mysql> \s
--------------
mysql  Ver 8.0.22 for Linux on x86_64 (MySQL Community Server - GPL)

Connection id:      14
Current database:
Current user:       root@localhost
SSL:         Cipher in use is TLS_AES_256_GCM_SHA384
...
```

可以看到，客户端到服务端的会话是加密的，加密算法是 TLS_AES_256_GCM_SHA384。检查如下视图也可以查看连接加密的信息：

```
mysql> select * from sys.session_ssl_status\G
*************************** 1. row ***************************
          thread_id: 53
        ssl_version: TLSv1.3
         ssl_cipher: TLS_AES_256_GCM_SHA384
ssl_sessions_reused: 0
1 row in set (0.01 sec)
```

可以看到，ssl 的版本是 TLSv1.3，是目前最强壮的 ssl 版本。SSL(secure sockets layer，安全套接字层)和 TLS(transport layer security，传输层安全协议)都基于 OpenSSL API。这两种加密方法很多时候都称为 SSL，实际上 TLS 比 SSL 要强壮得多。TLS 不仅可以加密通信，还能验证身份，验证的标准是 X509。它使用 CA 签发证书，证书包括一个公钥和一个私钥，采用非对称加密，用公钥加密的信息只能被私钥解密。

4.3.2 服务端设置

服务端的 ssl 默认是激活的，由选项--ssl 控制，该选项的默认值是 ON。如果要知道正在运行的 MySQL 实例是否支持 ssl，可以查询系统参数 have_ssl，代码及其运行结果如下：

```
mysql> show variables like 'have_ssl';
+---------------+-------+
| Variable_name | Value |
+---------------+-------+
| have_ssl      | YES   |
+---------------+-------+
```

当 MySQL 启动时将--ssl 设置成 0 或 OFF，系统变量 have_ssl 是 DISABLED，则不支持 ssl。

服务端可通过设置 require_secure_transport＝ON 来强制客户端使用安全的连接，当该变量被设置成 ON 时，不安全的连接将会被禁止，例如：

```
$ mysql --protocol=tcp --ssl-mode=disable
ERROR 3159 (HY000): Connections using insecure transport are prohibited while --require_secure_transport=ON.
```

4.3.3 客户端设置

客户端对于加密连接的控制由选项--ssl-mode 控制，该选项有以下取值。

(1) PREFERRED：默认选项，如果可能就建立一个安全连接，如果不可能就建立一个不安全的连接。

(2) DISABLED：建立一个不安全的连接。

(3) REQUIRED：如果可能就建立一个安全连接，如果不可能就失败。

(4) VERIFY_CA：在 REQUIRED 的基础上，还要验证服务端的证书是由受信任的 CA 发行的。

(5) VERIFY_IDENTITY：在 VERIFY_CA 的基础上，还要验证服务端的证书和主机名是否匹配。

当--ssl-mode 设置为 VERIFY_CA 时，服务端的证书应是 CA 签发的，例如：

```
$ mysql --protocol=tcp --ssl-mode=VERIFY_CA --ssl-ca=ca.pem
Welcome to the MySQL monitor.  Commands end with ; or \g.
Your MySQL connection id is 25
```

也可以使用 openssl 指定 CA 证书来验证服务端的证书，例如：

```
$ openssl verify -CAfile ca.pem server-cert.pem
server-cert.pem: OK
```

如果使用了错误的 CA 证书，则验证不能通过，例如：

```
$ mysql -h 192.168.17.149 --ssl-mode=VERIFY_CA --ssl-ca=ca.pem
ERROR 2026 (HY000): SSL connection error: error: 14090086: SSL routines: ssl3_get_server_
certificate: certificate verify failed
```

在 MySQL 的错误日志中有类似下面的提示：

```
2021-05-05T12:36:02.535339Z 70 [Note] [MY-010914] [Server] Bad handshake
```

使用 openssl 指定错误的 CA 证书来验证服务端的证书也会失败，例如：

```
$ openssl verify -CAfile /home/scutech/mysql-sandboxes/3310/sandboxdata/ca.pem server-
cert.pem
CN = MySQL_Server_8.0.22_Auto_Generated_Server_Certificate
error 7 at 0 depth lookup: certificate signature failure
error server-cert.pem: verification failed
...
```

当--ssl-mode 设置为 VERIFY_IDENTITY 时，服务端的证书应是第三方 CA 签发的，MySQL 自己签发的证书不能用于验证主机名。

4.3.4 账号设置

可以在 create user 或 alter user 时使用 require 子句里指定账号的安全连接方式，有以下几个选项。

(1) NONE：这是默认方式，没有安全连接的要求。

(2) SSL：必须使用安全连接。

(3) X509：必须使用安全连接，而且客户端要有数字证书。

(4) ISSUER 'issuer'：必须使用安全连接，而且指定客户端数字证书的签发 CA。对于 MySQL 自己签发的证书，CA 是 MySQL_Server_version_Auto_Generated_CA_Certificate，其中的 version 是 MySQL 的版本号。

（5）SUBJECT 'subject'：必须使用安全连接，而且指定客户端数字证书的主题。对于 MySQL 自己签发的客户端证书，主题是 MySQL_Server_version_Auto_Generated_Client_Certificate，其中的 version 是 MySQL 的版本号。

（6）CIPHER 'cipher'：必须使用安全连接，而且指定加密算法。

例如，下面的语句要求用户 scutech 的证书发行者是 MySQL_Server_8.0.22_Auto_Generated_CA_Certificate。这也是 MySQL 自己的 CA 签发的证书。

```
mysql>  alter user scutech require issuer '/CN=MySQL_Server_8.0.22_Auto_Generated_CA_Certificate';
Query OK, 0 rows affected (0.02 sec)
```

这时指定使用 MySQL 自己的 CA 签发的证书可以连接成功，代码如下：

```
$ mysql --protocol=tcp --ssl-cert=client-cert.pem --ssl-key=client-key.pem -uscutech
Welcome to the MySQL monitor.  Commands end with ; or \g.
Your MySQL connection id is 49
```

把证书的发行者改成 scutech 的代码如下：

```
mysql>  alter user scutech require issuer 'scutech';
```

再次使用 MySQL 自己的 CA 签发的证书连接会失败，代码如下：

```
$ mysql --protocol=tcp --ssl-cert=client-cert.pem --ssl-key=client-key.pem -uscutech
ERROR 1045 (28000): Access denied for user 'scutech'@'localhost' (using password: YES)
```

在 MySQL 的错误日志里有如下的错误提示：

```
2021-05-06T11:19:28.010803Z 50 [Note] [MY-010290] [Server] X.509 issuer mismatch: should be 'scutech' but is '/CN=MySQL_Server_8.0.22_Auto_Generated_CA_Certificate'
2021-05-06T11:19:28.011187Z 50 [Note] [MY-010926] [Server] Access denied for user 'scutech'@'localhost' (using password: YES)
```

4.4 数据加密

4.4.1 密钥

InnoDB 使用两层加密密钥体系结构，由主密钥和表空间密钥组成。主密钥保存在钥匙环(keyring)中，表空间密钥被主密钥加密后存储在表空间头部。表空间密钥用于加密和解密表空间里的数据。

为了提高安全性，应定期更换主密钥，主密钥可以使用如下命令进行更换：

```
mysql> alter instance rotate innodb master key;
```

更换主密钥将重新加密表空间密钥，但不重新加密表空间里的数据。

4.4.2 钥匙环

MySQL 使用钥匙环管理主密钥。钥匙环功能采用 MySQL 插件的形式实现，MySQL 支持多种插件用于实现钥匙环，但免费的 MySQL 社区版只支持 keyring_file 这一种插件。

加载 keyring_file 插件需要在 MySQL 的配置文件里增加两个系统参数：一个是 early-plugin-load，设置为 keyring_file，这样 keyring_file 组件会在内置的插件和引擎初始化之前加载；另一个是 keyring_file_data 指向保存安全信息的文件，例如：

```
[mysqld]
early-plugin-load=keyring_file.so
keyring_file_data=/home/scutech/mysql-sandboxes/3310/sandboxdata/mysql-keyring/keyring
```

运行如下命令可以在 information_schema.plugins 视图里看到加载的插件：

```
mysql> select plugin_name, plugin_status
     from information_schema.plugins
     where plugin_name like 'keyring%';
+--------------+---------------+
| plugin_name  | plugin_status |
+--------------+---------------+
| keyring_file | ACTIVE        |
+--------------+---------------+
1 row in set (0.00 sec)
```

4.4.3 加密 InnoDB 表空间

成功加载钥匙环插件后，对于每个表对应一个文件的表空间，可以使用 alter table 语句将 encryption 属性设置成 Y 进行加密，例如：

```
mysql> alter table actor encryption='Y';
Query OK, 200 rows affected (0.21 sec)
Records: 200  Duplicates: 0  Warnings: 0
```

从如下 SQL 执行的提示中可以看到已经对该表中的所有记录进行了加密。如果钥匙环插件没有加载，运行前面的加密代码时就会出错：

```
mysql>  alter table actor encryption='Y';
ERROR 3185 (HY000): Can't find master key from keyring, please check in the server log if a
keyring plugin is loaded and initialized successfully.
```

通用表空间和系统表空间也使用类似的语句进行加密，当然也可以在创建表空间时将 encryption 属性设置成 Y。

如果把系统变量 default_table_encryption 设置为 on，例如：

```
mysql> set global default_table_encryption=on;
```

则新建的表空间默认是加密状态。数据和索引在保存到加密表空间的过程中会进行加密，读取时进行解密,数据缓存到内存中时是解密的。

4.4.4 加密重做日志表空间

将系统变量 innodb_redo_log_encrypt 设置为 on,启动重做日志表空间的加密,该参数可以联机进行设置,例如:

```
mysql> set @@global.innodb_redo_log_encrypt = on;
```

设置后会把新的重做日志进行加密,旧的重做日志不受影响。密钥保存在第一个重做日志文件的头部。

4.4.5 加密撤销表空间

将系统变量 innodb_redo_log_encrypt 设置为 on 启动撤销表空间的加密,该参数也可以联机进行设置,例如:

```
mysql> set @@global.innodb_undo_log_encrypt = on;
```

设置后会把新的撤销日志进行加密,旧的撤销日志不受影响。密钥保存在撤销表空间文件的头部。

4.4.6 二进制日志和中继日志的加密

从 MySQL 8.0.14 开始,可以对二进制日志和中继日志进行加密,方法是把参数 binlog_encryption 设置为 on,例如:

```
mysql> set @@global.binlog_encryption = on;
```

该参数设置为 on 后,将自动滚动一次二进制日志和中继日志,新的日志文件将采用新的设置,老的日志文件不受影响。日志文件是否加密可以使用 show binary logs 进行查看,例如:

```
mysql> show binary logs;
+--------------------+-----------+-----------+
| Log_name           | File_size | Encrypted |
+--------------------+-----------+-----------+
| scutech-bin.000001 |       179 | No        |
| scutech-bin.000002 |      4787 | No        |
| scutech-bin.000003 |      3139 | No        |
| scutech-bin.000004 |      1059 | No        |
| scutech-bin.000005 |       494 | No        |
| scutech-bin.000006 |       748 | Yes       |
+--------------------+-----------+-----------+
6 rows in set (0.00 sec)
```

从以上运行结果可以看出,最后一个日志文件是加密的,之前的文件保持不加密的状态。

加密后的二进制日志文件无法使用 mysqlbinlog 工具直接查看，会遇到下面的错误提示：

```
ERROR: Reading encrypted log files directly is not supported.
```

这时若使用 mysqlbinlog 工具，需要配置--read-from-remote-server 连接到正在运行的 MySQL 实例进行查看。

4.5 审计插件

Oracle 的 MySQL 社区版不带审计插件(audit plugin)，要想使用审计功能，通常只能用企业版，但这需要付费。另外一种方法是把遵守 GPL(GNU public license，GUN 公共许可)协议的审计插件迁移到 MySQL 上，这里介绍把 MariaDB 的审计插件迁移到 MySQL 的方法。

4.5.1 迁移 MariaDB 的审计插件

与 MySQL 5.7 对应的 MariaDB 版本是 10.1，登录其官网下载 Linux 的通用版本，解压后约 1.3GB，如下命令列出了下载的 MariaDB 安装包：

```
$ ll -h mariadb-10.1.46-linux-x86_64.tar
-rw-rw-r-- 1 scutech scutech 1.3G Aug 19 18:19 mariadb-10.1.46-linux-x86_64.tar
```

展开后找到需要的审计插件，命令如下：

```
./mariadb-10.1.46-linux-x86_64/lib/plugin/server_audit.so
```

将如上以 so 结尾的文件复制到 MySQL 的插件目录，如/usr/lib/mysql/plugin/，使用如下命令加载：

```
mysql> install plugin server_audit soname 'server_audit.so';
```

4.5.2 配置 MariaDB 的审计插件

运行如下代码，可得到如下与审计功能相关的参数：

```
mysql> show variables like '%audit%';
+-------------------------------+-----------------------+
| Variable_name                 | Value                 |
+-------------------------------+-----------------------+
| server_audit_events           |                       |
| server_audit_excl_users       |                       |
| server_audit_file_path        | server_audit.log      |
| server_audit_file_rotate_now  | OFF                   |
| server_audit_file_rotate_size | 1000000               |
```

```
| server_audit_file_rotations          | 9                        |
| server_audit_incl_users              |                          |
| server_audit_loc_info                |                          |
| server_audit_logging                 | OFF                      |
| server_audit_mode                    | 1                        |
| server_audit_output_type             | file                     |
| server_audit_query_log_limit         | 1024                     |
| server_audit_syslog_facility         | LOG_USER                 |
| server_audit_syslog_ident            | mysql - server_auditing  |
| server_audit_syslog_info             |                          |
| server_audit_syslog_priority         | LOG_INFO                 |
+--------------------------------------+--------------------------+
16 rows in set (0.00 sec)
```

配置 MariaDB 审计插件的参数说明参见 https://mariadb.com/kb/en/mariadb-audit-plugin-configuration/。

如上参数值可以用 set 语句进行设置，例如：

```
set global server_audit_logging = on;
set global server_audit_events = 'connect,query';
```

为了使相关参数在重新启动后也能生效，可以在 MySQL 的配置文件中添加如下相应的设置：

```
[server]
...
server_audit_logging = on
server_audit_events = connect,query
...
```

server_audit_logging 参数默认为 off，把该参数设置为 on 才能启动审计功能。server_audit_events 决定记录的事件，这里记录 connect 和 query，也就是记录用户的连接和查询语句。

4.5.3 审计记录

如下代码可以强制进行审计文件的切换：

```
mysql> set global server_audit_file_rotate_now = on;
```

此时，生成了一个新的审计文件，旧的审计文件名后加上点和数字表示文件的序号，如下所示：

```
$ ll server_au*
-rw-r----- 1 mysql mysql  26163 Aug 20 11:11 server_audit.log
-rw-r----- 1 mysql mysql 326651 Aug 20 11:09 server_audit.log.1
```

与审计记录文件相关的系统参数如下。

- server_audit_file_rotate_size：决定每个审计记录文件的大小，达到该阈值时自动进行审计记录文件的切换。

- server_audit_file_rotations：决定审计记录文件的数量，达到该阈值时会覆盖第一个审计记录文件，默认为 9。
- server_audit_output_type：设置为 file 时，记录成文件，默认目录是 MySQL 的 datadir 目录，默认文件名是 server_audit.log；设置为 syslog 时，审计记录会通过标准 <syslog.h> API 发送给本地的 syslogd daemon。

审计记录文件的格式如下：

```
[timestamp],[serverhost],[username],[host],[connectionid],[queryid],[operation],
[database],[object],[retcode]
```

一个对应的例子如下：

```
20200820
11:04:04,infohost,superuser,localhost,23,4759,QUERY,test_db,'select count(*) from test_
storage',0
```

4.6 实验

连接控制插件

实验连接控制插件对多次使用错误密码登录的延时控制，观察随着输入错误密码的次数增多，延长时间的变化。

视频演示

第二部分

监　控

第5章

information_schema数据库

MySQL 自带三个系统数据库用于监控 MySQL，这三个数据库分别从不同的角度观察 MySQL 数据库。

（1）information_schema 数据库主要保存 MySQL 的静态元数据。

（2）performance_schema 数据库主要保存与 MySQL 性能相关的动态数据。

（3）sys 数据库是建立在前两个数据库基础上的数据库，它包括一些视图、存储过程和函数，用于方便 DBA（database administrator，数据库管理员）、开发人员和操作人员的日常工作。

本章介绍 information_schema 数据库的相关知识，第 6、7 章将分别介绍 performance_schema 数据库和 sys 数据库。

5.1 数据组成

系统数据库 information_schema 中保存着 MySQL 的静态元数据。它和系统数据库 performance_schema 不同，performance_schema 主要保存动态元数据，而 information_schema 主要保存静态元数据，也有少量的动态元数据。information_schema 实际上是一个虚拟数据库，它并不包含任何实际的数据，它是唯一一个在操作系统的文件系统上没有对应目录的数据库，因此该数据库中的表只能被查询，不能被修改。如下 SQL 语句可查询出 information_schema 中表类型和数量：

```
mysql > select table_type, count( * ) from information_schema. tables where table_schema =
'information_schema' group by table_type;
+-------------+----------+
| TABLE_TYPE  | count( * ) |
+-------------+----------+
| SYSTEM VIEW |     79   |
+-------------+----------+
```

```
1 row in set (0.01 sec)
```

可以看到，在 MySQL 8.0 中 information_schema 数据库共有 79 个表，这些表都是同一个类型 SYSTEM VIEW，严格地说它们都是视图，并不是表，information_schema 中并没有 BASE TABLE 类型的表。

5.1.1 静态元数据

information_schema 中的静态元数据来源于数据字典，但为了保护 MySQL 的内部数据和对 MySQL 后续版本的兼容，用户并不能直接查询数据字典，例如，表 mysql.schemata 保存着数据库的信息，运行如下代码直接查询该表会被拒绝：

```
mysql >  select * from mysql.schemata;
ERROR 3554 (HY000): Access to data dictionary table 'mysql.schemata' is rejected.
```

但可以通过查询 information_schema 中的 schemata 视图从而得到同样的信息，代码如下：

```
mysql > select schema_name from information_schema.schemata;
+--------------------+
| SCHEMA_NAME        |
+--------------------+
| mysql              |
| information_schema |
| performance_schema |
| sys                |
| sbtest             |
| mydb               |
| tpcc1000           |
| sakila             |
| percona_schema     |
+--------------------+
9 rows in set (0.00 sec)
```

如上查询语句和 show databases 的效果一样，代码如下：

```
mysql > show databases;
+--------------------+
| Database           |
+--------------------+
| information_schema |
| mydb               |
| mysql              |
| percona_schema     |
| performance_schema |
| sakila             |
| sbtest             |
| sys                |
| tpcc1000           |
+--------------------+
9 rows in set (0.00 sec)
```

查看 information_schema. schemata 的定义如下：

```
mysql > show create table information_schema.schemata \G
*************************** 1. row ***************************
                View: SCHEMATA
         Create View: CREATE ALGORITHM = UNDEFINED
DEFINER = 'mysql.infoschema'@'localhost' SQL SECURITY DEFINER VIEW
'information_schema'.'SCHEMATA' AS
select 'cat'.'name' AS
'CATALOG_NAME','sch'.'name' AS 'SCHEMA_NAME','cs'.'name' AS 'DEFAULT_CHARACTER_SET_NAME','col'.
'name' AS 'DEFAULT_COLLATION_NAME',NULL
AS 'SQL_PATH','sch'.'default_encryption' AS 'DEFAULT_ENCRYPTION' from
((('mysql'.'schemata' 'sch' join 'mysql'.'catalogs' 'cat' on(('cat'.'id' =
'sch'.'catalog_id'))) join 'mysql'.'collations' 'col'
on(('sch'.'default_collation_id' = 'col'.'id'))) join
'mysql'.'character_sets' 'cs' on(('col'.'character_set_id' = 'cs'.'id')))
where (0 <> can_access_database('sch'.'name'))
character_set_client: utf8
collation_connection: utf8_general_ci
1 row in set (0.00 sec)
```

上面的 SQL 语句格式比较乱，把视图定义部分的 SQL 语句放到 MySQL 的图形工具 workbench 中进行格式整理，整理后的 SQL 语句如下：

```
SELECT
    'cat'.'name' AS 'CATALOG_NAME',
    'sch'.'name' AS 'SCHEMA_NAME',
    'cs'.'name' AS 'DEFAULT_CHARACTER_SET_NAME',
    'col'.'name' AS 'DEFAULT_COLLATION_NAME',
    NULL AS 'SQL_PATH',
    'sch'.'default_encryption' AS 'DEFAULT_ENCRYPTION'
FROM
    ((('mysql'.'schemata' 'sch'
    JOIN 'mysql'.'catalogs' 'cat' ON (('cat'.'id' = 'sch'.'catalog_id')))
    JOIN 'mysql'.'collations' 'col' ON (('sch'.'default_collation_id' = 'col'.'id')))
    JOIN 'mysql'.'character_sets' 'cs' ON (('col'.'character_set_id' = 'cs'.'id')))
WHERE
    (0 <> CAN_ACCESS_DATABASE('sch'.'name'))
```

可以看到表 information scchema. schemata 实际上是建立在 mysql. schemata 等表上的视图。

5.1.2 动态元数据

information_schema 数据库中除了静态元数据外，还有少量的动态元数据，这些动态元数据来源于数据库的存储引擎，如 innodb_metrics 表和统计信息表。与统计信息相关的字段如下：

```
STATISTICS.CARDINALITY
```

```
TABLES.AUTO_INCREMENT
TABLES.AVG_ROW_LENGTH
TABLES.CHECKSUM
TABLES.CHECK_TIME
TABLES.CREATE_TIME
TABLES.DATA_FREE
TABLES.DATA_LENGTH
TABLES.INDEX_LENGTH
TABLES.MAX_DATA_LENGTH
TABLES.TABLE_ROWS
TABLES.UPDATE_TIME
```

这些字段保存着与表和索引相关的动态统计信息，这些信息从存储引擎刷新到数据字典表 mysql.index_stats 和 mysql.table_stats 中(这两个表用户不能直接访问)，为了提高查询的效率，这两个表的信息都存放在缓存中，缓存刷新的时间由系统参数 information_schema_stats_expiry 决定，该参数的默认值是 86400 秒，也就是 24 小时。因此有时会出现查询到的统计信息较陈旧的情况，如果要查询实时的统计信息，可以把 information_schema_stats_expiry 设置为 0，这样每次查询的统计信息都来自存储引擎，这也是 MySQL 5.7 的做法。下面看一个例子。

首先查询表 sakila.actor 的统计信息 auto_increment、update_time 和 table_rows，相关代码及其运行结果如下：

```
mysql > select auto_increment, update_time, table_rows from information_schema.tables where
table_schema = 'sakila' and table_name = 'actor';
+----------------+---------------------+------------+
| AUTO_INCREMENT | UPDATE_TIME         | TABLE_ROWS |
+----------------+---------------------+------------+
|            201 | 2021-07-20 17:46:56 |        200 |
+----------------+---------------------+------------+
1 row in set (0.04 sec)
```

使用 show table status 命令查询的结果和从 information_schema.tables 表里查询的结果一样，代码如下：

```
mysql > show table status like 'actor'\G
*************************** 1. row ***************************
           Name: actor
         Engine: InnoDB
        Version: 10
     Row_format: Dynamic
           Rows: 200
 Avg_row_length: 81
    Data_length: 16384
Max_data_length: 0
   Index_length: 16384
      Data_free: 0
 Auto_increment: 201
    Create_time: 2021-07-20 17:46:43
    Update_time: 2021-07-20 17:46:56
```

```
      Check_time: NULL
       Collation: utf8mb4_0900_ai_ci
        Checksum: NULL
  Create_options:
         Comment:
1 row in set (0.01 sec)
```

然后向表 sakila.actor 中插入一条记录，命令如下：

```
mysql> insert into sakila.actor(first_name,last_name) values('si','Li');
Query OK, 1 row affected (0.00 sec)
```

第二次查询表 sakila.actor 的统计信息，相关代码及其运行结果如下：

```
mysql> select auto_increment,update_time,table_rows from information_schema.tables where
table_schema = 'sakila' and table_name = 'actor';
+----------------+---------------------+------------+
| AUTO_INCREMENT | UPDATE_TIME         | TABLE_ROWS |
+----------------+---------------------+------------+
|            201 | 2021-07-20 17:46:56 |        200 |
+----------------+---------------------+------------+
1 row in set (0.00 sec)
```

从以上运行结果可以看出没有变化，再使用 show table status 查询，代码如下：

```
mysql> show table status like 'actor'\G
*************************** 1. row ***************************
           Name: actor
         Engine: InnoDB
        Version: 10
     Row_format: Dynamic
           Rows: 200
 Avg_row_length: 81
    Data_length: 16384
Max_data_length: 0
   Index_length: 16384
      Data_free: 0
 Auto_increment: 201
    Create_time: 2021-07-20 17:46:43
    Update_time: 2021-07-20 17:46:56
     Check_time: NULL
      Collation: utf8mb4_0900_ai_ci
       Checksum: NULL
 Create_options:
        Comment:
1 row in set (0.01 sec)
```

结果还是没有变化。现在把 information_schema_stats_expiry 设置为 0 后，第三次查询表 sakila.actor 的统计信息，代码如下：

```
mysql> set session information_schema_stats_expiry = 0;
Query OK, 0 rows affected (0.00 sec)
```

```
mysql> select auto_increment,update_time,table_rows from information_schema.tables where
table_schema='sakila' and table_name='actor';
+----------------+---------------------+------------+
| AUTO_INCREMENT | UPDATE_TIME         | TABLE_ROWS |
+----------------+---------------------+------------+
|            202 | 2021-07-20 17:47:34 |        201 |
+----------------+---------------------+------------+
1 row in set (0.00 sec)

mysql> show table status like 'actor'\G
*************************** 1. row ***************************
           Name: actor
         Engine: InnoDB
        Version: 10
     Row_format: Dynamic
           Rows: 201
 Avg_row_length: 81
    Data_length: 16384
Max_data_length: 0
   Index_length: 16384
      Data_free: 0
 Auto_increment: 202
    Create_time: 2021-07-20 17:46:43
    Update_time: 2021-07-20 17:47:34
     Check_time: NULL
      Collation: utf8mb4_0900_ai_ci
       Checksum: NULL
 Create_options:
        Comment:
1 row in set (0.00 sec)
```

可以发现，actor表的统计信息已经更新到当前值了。该实验展示了系统参数information_schema_stats_expiry是如何影响查询动态统计信息结果的。另外，执行analyze table命令后，会立即更新mysql.index_stats和mysql.table_stats的缓存。

5.2 MySQL 8.0中的优化

information_schema从MySQL 5.0开始使用，长期以来用户一直抱怨该数据库的查询效率低。但这个问题在MySQL 8.0中彻底得到了解决。这得益于在MySQL 8.0中，数据字典中的数据被迁移到了事务性的表中。

5.2.1 数据字典的优化

在MySQL 8.0之前，数据字典的数据存放在文件系统中，包括如下这些类型文件。

- .frm文件：表定义文件。
- .par文件：分区定义文件。

- .trn 文件：触发器命名空间文件。
- .trg 文件：触发器参数文件。
- .isl 文件：指向不在 datadir 目录的表空间文件的链接。
- db.opt 文件：数据库配置文件。

这些文件在 MySQL 8.0 中都被取消了，文件中的数据全部放入系统数据库 mysql 中，这些表的存储引擎都是 InnoDB，表空间是 mysql.ibd，位于 datadir 目录下。

这样可以利用 InnoDB 的事务特性实现 DDL(data definition language，数据定义语言)的原子化，如下面的 DDL 语句：

```
mysql > drop table tab_exit,tab_not_exist;
ERROR 1051 (42S02): Unknown table 'sakila.tab_not_exist'
```

在 MySQL 8.0 中，第二个表删除失败，drop 语句作为一个事务失败，第一个表不会被删除；但在 MySQL 5.7 中，第二个表删除失败时，第一个表已经被删除了。

5.2.2 查询方式的变化

MySQL 8.0 数据字典的数据存放到事务性的表中带来了很多好处，如 DDL 语句的原子化、不受文件系统 Bug 的影响、崩溃恢复等，但最大的好处是解决了长期以来困扰用户的查询性能的问题。对于大型的数据库，可对数据字典的查询性能提高数十倍甚至数百倍。因为在 MySQL 8.0 中，数据字典的数据存放到表中，对它们进行查询时，可以充分利用 SQL 语句的特性，如索引、连接等。而在 MySQL 8.0 之前，MySQL 需要先从文件中读取数据，然后结合存储引擎里的信息组成临时表，再对临时表进行查询，这个过程效率很低。下面先看一个在 MySQL 5.7 中查询数据字典语句的执行计划：

```
mysql > explain select table_name from information_schema.tables where table_schema like
'sakila%' and table_name like 'ac%'\G
*************************** 1. row ***************************
           id: 1
  select_type: SIMPLE
        table: tables
   partitions: NULL
         type: ALL
possible_keys: NULL
          key: NULL
      key_len: NULL
          ref: NULL
         rows: NULL
     filtered: NULL
        Extra: Using where; Skip_open_table; Scanned all databases
1 row in set, 1 warning (0.00 sec)
```

从 type 字段中的 ALL 和 Extra 字段中的 Scanned all databases 可以看到，该执行计划需要对所有数据库和表定义文件进行扫描，这个成本非常高，再看看同样的查询在 MySQL 8.0 中的执行计划：

```
mysql >  explain select table_name from information_schema.tables where table_schema like
'sakila%' and table_name like 'ac%'\G
*************************** 1. row ***************************
           id: 1
  select_type: SIMPLE
        table: cat
   partitions: NULL
         type: index
possible_keys: PRIMARY
          key: name
      key_len: 194
          ref: NULL
         rows: 1
     filtered: 100.00
        Extra: Using index
*************************** 2. row ***************************
           id: 1
  select_type: SIMPLE
        table: sch
   partitions: NULL
         type: ref
possible_keys: PRIMARY,catalog_id
          key: catalog_id
      key_len: 8
          ref: mysql.cat.id
         rows: 10
     filtered: 11.11
        Extra: Using where; Using index
*************************** 3. row ***************************
           id: 1
  select_type: SIMPLE
        table: tbl
   partitions: NULL
         type: ref
possible_keys: schema_id
          key: schema_id
      key_len: 8
          ref: mysql.sch.id
         rows: 50
     filtered: 11.11
        Extra: Using where
*************************** 4. row ***************************
           id: 1
  select_type: SIMPLE
        table: col
   partitions: NULL
         type: eq_ref
possible_keys: PRIMARY
          key: PRIMARY
      key_len: 8
          ref: mysql.tbl.collation_id
```

```
         rows: 1
     filtered: 100.00
        Extra: Using index
*************************** 5. row ***************************
           id: 1
  select_type: SIMPLE
        table: ts
   partitions: NULL
         type: eq_ref
possible_keys: PRIMARY
          key: PRIMARY
      key_len: 8
          ref: mysql.tbl.tablespace_id
         rows: 1
     filtered: 100.00
        Extra: Using index
*************************** 6. row ***************************
           id: 1
  select_type: SIMPLE
        table: stat
   partitions: NULL
         type: eq_ref
possible_keys: PRIMARY
          key: PRIMARY
      key_len: 388
          ref: mysql.sch.name,mysql.tbl.name
         rows: 1
     filtered: 100.00
        Extra: Using index
6 rows in set, 1 warning (0.01 sec)
```

可以看到，在 MySQL 8.0 中，查询计划充分利用了索引和连接，对不需要数据的过滤率非常高。此时再使用 show warnings 命令看看被优化器重写了的 SQL 语句，如下所示：

```
mysql> show warnings\G
*************************** 1. row ***************************
  Level: Note
   Code: 1003
Message: /* select#1 */ select 'tbl'.'name' AS 'TABLE_NAME' from 'mysql'.'tables' 'tbl' join 'mysql'.'schemata' 'sch' join 'mysql'.'catalogs' 'cat' left join 'mysql'.'collations' 'col' on(('col'.'id' = 'tbl'.'collation_id')) left join 'mysql'.'tablespaces' 'ts' on(('ts'.'id' = 'tbl'.'tablespace_id')) left join 'mysql'.'table_stats' 'stat' on((('stat'.'schema_name' = 'sch'.'name') and ('stat'.'table_name' = 'tbl'.'name'))) where (('tbl'.'schema_id' = 'sch'.'id') and ('sch'.'catalog_id' = 'cat'.'id') and ('sch'.'name' like 'sakila%') and ('tbl'.'name' like 'ac%') and (0 <> can_access_table('sch'.'name','tbl'.'name')) and (0 <> is_visible_dd_object('tbl'.'hidden')))
1 row in set (0.00 sec)
```

可以看到，实际上是使用标准的 SQL 语句来查询 mysql 数据库中的表。通过这两个执行计划的对比可以看出，在 MySQL 8.0 中对数据字典的查询效率有了质的飞跃。

5.3 权限

用户对 information_schema 进行查询时，需要有相应的权限才能查询对应的对象。例如，对 information_schema 中的 tables 进行查询时，只能查询到用户有查询权限的表，如一个只有 usage 权限的用户查询 tables 表时，除了 information_schema 的表外，其他表都看不到。例如，如下代码查询当前用户的权限：

```
mysql > show grants;
+------------------------------------+
| Grants for lisi@ %                 |
+------------------------------------+
| GRANT USAGE ON * . * TO 'lisi'@'%' |
+------------------------------------+
1 row in set (0.00 sec)
```

可以发现，当前用户只有 usage 的权限，使用该用户查询 information_schema. tables 中的表：

```
mysql > select count( * ) from information_schema. tables where table_schema <>'information_
schema';
+----------+
| count( * ) |
+----------+
|        0 |
+----------+
1 row in set (0.00 sec)
```

可以发现，除了 information_schema 数据库中的表之外，一个表也查不到。现在把 sakila. actor 表的查询权限赋予只有 usage 权限的用户，命令如下：

```
mysql > grant select on sakila.actor to lisi;
```

再次查询 information_schema. tables 表，就可以看到 sakila. actor 表了，相关命令及其运行结果如下：

```
mysql > select table_schema,table_name from  information_schema. tables where table_schema <>
'information_schema';
+--------------+-------------+
| TABLE_SCHEMA | TABLE_NAME  |
+--------------+-------------+
| sakila       | actor       |
+--------------+-------------+
1 row in set (0.00 sec)
```

管理类的权限也一样，例如，对 InnoDB 表（以 innodb_开头的表）进行查询时，需要 PROCESS 权限，如果没有，会遇到下面的情况：

```
mysql >  select * from information_schema. innodb_tables;
ERROR 1227 (42000): Access denied; you need (at least one of) the PROCESS privilege(s) for this
operation
```

对 processlist 表进行查询时，如果没有 PROCESS 权限，只能看到当前用户的线程，其他用户的连接都看不到，相关命令及其运行结果如下：

```
mysql > select id,user,host from information_schema.processlist;
```

```
+-------+------+-----------+
| id    | user | host      |
+-------+------+-----------+
| 41264 | lisi | localhost |
+-------+------+-----------+
1 row in set (0.01 sec)
```

使用 show 语句进行查询时，对权限的要求和查询 information_schema 数据库的要求一样。

5.4　视图说明

按照 information_schema 中视图保存的信息，可以将这些视图分为四类，分别为数据库视图、实例视图、性能视图和权限视图。下面分别说明。

5.4.1　数据库视图

系统数据库 information_schema 中用于保存与数据库相关的视图占大部分，简单说明如下。

- CHECK_CONSTRAINTS：该视图在 MySQL 8.0.16 版本之后才有，它保存着表中定义的与 CHECK 限制相关的信息。
- COLUMNS：与表中字段相关的信息。
- COLUMN_STATISTICS：字段的直方图统计信息。
- EVENTS：定时任务(EVENT)的信息。
- FILES：表空间和对应文件的信息。
- KEY_COLUMN_USAGE：键字段相关的信息，包括主键、外键、唯一索引键相关的信息。
- INNODB_COLUMNS：InnoDB 表中字段的元信息。
- INNODB_DATAFILES：InnoDB 表空间 ID 和对应的文件。
- INNODB_FIELDS：InnoDB 索引对应的字段。
- INNODB_FOREIGN：InnoDB 表的外键信息。
- INNODB_FOREIGN_COLS：与 InnoDB 表外键相关的字段。
- INNODB_FT_CONFIG：InnoDB 表的全文索引的配置信息，查询该表时，先设置系统参数 innodb_ft_aux_table 指向要查询的包含全文索引的表，例如：

```
mysql > set global innodb_ft_aux_table = 'sakila/film_text';
Query OK, 0 rows affected (0.00 sec)

mysql > select *  from information_schema.innodb_ft_config;
```

```
+-----------------------------+--------+
| KEY                         | VALUE  |
+-----------------------------+--------+
| optimize_checkpoint_limit   | 180    |
| synced_doc_id               | 0      |
| stopword_table_name         |        |
| use_stopword                | 1      |
+-----------------------------+--------+
4 rows in set (0.18 sec)
```

- INNODB_FT_DELETED：保存InnoDB表的全文索引中被删除的行，为了提高性能，被删除的行并不是立刻从表中删除，而是保存到这个表中，当执行optimize table命令时才会被删除，查询该表之前也要先设置系统参数innodb_ft_aux_table指向要查的表。
- INNODB_FT_BEING_DELETED：当执行optimize table命令时，该表是INNODB_FT_DELETED表的一个快照。
- INNODB_FT_INDEX_CACHE：保存新插入的InnoDB表全文索引的行，为了提高性能，新插入的行会先保存到该表中，查询该表之前也要先设置系统参数innodb_ft_aux_table指向要查的表。
- INNODB_FT_INDEX_TABLE：InnoDB表的倒序全文索引，查询该表之前也要先设置系统参数innodb_ft_aux_table指向要查的表。
- INNODB_INDEXES：InnoDB表的索引信息。
- INNODB_TABLES：InnoDB表的信息。
- INNODB_TABLESPACES：InnoDB表空间的信息。
- INNODB_TABLESPACES_BRIEF：该视图的信息一部分来自INNODB_TABLESPACES，文件名和路径来自INNODB_DATAFILES。
- INNODB_TABLESTATS：InnoDB表的统计信息。
- INNODB_TEMP_TABLE_INFO：用户创建的InnoDB临时表的信息，不包括优化器创建的临时表。
- INNODB_VIRTUAL：InnoDB表的虚拟字段信息。
- PARAMETERS：存储过程和存储函数的参数，以及存储函数的返回值。
- PARTITIONS：表分区信息。
- REFERENTIAL_CONSTRAINTS：外键信息。
- ROUTINES：存储过程和存储函数的信息。
- SCHEMATA：数据库信息，SCHEMATA是SCHEMA复数的一种写法，另外一种更流行的写法是SCHEMAS。
- ST_GEOMETRY_COLUMNS：存储空间数据的字段信息。
- STATISTICS：索引定义和统计信息。
- TABLE_CONSTRAINTS：表约束的汇总信息，包括主键、唯一索引、外键、约束等。
- TABLES：表和视图的信息。如下SQL语句为统计实例中数据量最大的10个数据库的大小，这里是一个实验库，只展示两个应用的数据库：

```
mysql> select table_schema, count( * ) tables,
concat(round(sum(table_rows)/1000000,2),'m') table_rows, concat(round(sum(data_length)/
(1024 * 1024 * 1024),2),'g') data,
concat(round(sum(index_length)/(1024 * 1024 * 1024),2),'g') idx,
concat(round(sum(data_length + index_length)/(1024 * 1024 * 1024),2),'g') total_size
from information_schema.tables
where table_schema not in ("information_schema", "mysql", "performance_schema", "sys")
group by table_schema
order by sum(data_length + index_length) desc limit 10;
+--------------+--------+------------+-------+-------+------------+
| TABLE_SCHEMA | tables | table_rows | data  | idx   | total_size |
+--------------+--------+------------+-------+-------+------------+
| tpcc1000     |      9 | 5.32m      | 0.82g | 0.18g | 1.00g      |
| sbtest       |      3 | 0.02m      | 0.00g | 0.00g | 0.01g      |
+--------------+--------+------------+-------+-------+------------+
2 rows in set (0.00 sec)
```

如下 SQL 语句统计某库下最大的 10 个表的大小：

```
mysql>  select table_schema,table_name table_name,
concat(round(data_length / (1024 * 1024), 2),'m') data_length,
concat(round(index_length / (1024 * 1024), 2),'m') index_length,
concat(round(round(data_length + index_length) / (1024 * 1024),2),'m') total_size
from information_schema.tables
where table_schema = 'sakila'
order by (data_length + index_length) desc limit 10;
+--------------+---------------+-------------+--------------+------------+
| TABLE_SCHEMA | table_name    | data_length | index_length | total_size |
+--------------+---------------+-------------+--------------+------------+
| sakila       | rental        | 1.52M       | 1.14M        | 2.66M      |
| sakila       | payment       | 1.52M       | 0.61M        | 2.13M      |
| sakila       | inventory     | 0.17M       | 0.19M        | 0.36M      |
| sakila       | film          | 0.19M       | 0.08M        | 0.27M      |
| sakila       | film_actor    | 0.19M       | 0.08M        | 0.27M      |
| sakila       | film_text     | 0.17M       | 0.02M        | 0.19M      |
| sakila       | customer      | 0.08M       | 0.05M        | 0.13M      |
| sakila       | address       | 0.09M       | 0.02M        | 0.11M      |
| sakila       | staff         | 0.06M       | 0.03M        | 0.09M      |
| sakila       | film_category | 0.06M       | 0.02M        | 0.08M      |
+--------------+---------------+-------------+--------------+------------+
10 rows in set (0.02 sec)
```

如下 SQL 语句找出数据库中所有不是 InnoDB 引擎的表：

```
mysql>  select table_schema, table_name,engine
from information_schema.tables
where engine != 'innodb' and table_schema not in ('information_schema','sys','mysql','performance
_schema');
+--------------+------------+--------+
| TABLE_SCHEMA | TABLE_NAME | ENGINE |
+--------------+------------+--------+
| test         | test1      | MyISAM |
```

```
| test           | tba         | MyISAM  |
+----------------+-------------+---------+
2 rows in set (0.01 sec)
```

如下 SQL 语句找出没有主键或唯一索引的表：

```
mysql> select t1.table_schema,
t1.table_name
from information_schema.columns t1
join information_schema.tables t2
on t1.table_schema = t2.table_schema
and t1.table_name = t2.table_name
where t1.table_schema not in
('sys', 'mysql', 'information_schema', 'performance_schema')
and t2.table_type = 'BASE TABLE'
group by t1.table_schema, t1.table_name
having group_concat(column_key) not regexp 'PRI|UNI';
```

- TABLESPACES：该视图没有用，在将来的版本中将会被删除。
- TRIGGERS：触发器定义信息。
- VIEW_ROUTINE_USAGE：该视图从 MySQL 8.0.13 版本之后才有，它保存着视图定义里的存储函数的信息。
- VIEW_TABLE_USAGE：该视图从 MySQL 8.0.16 版本之后才有，它保存着视图定义里的表和视图的信息。
- VIEWS：视图的定义。

5.4.2 实例视图

实例视图保存与实例相关的信息，分别说明如下。

- CHARACTER_SETS：可用的字符集。
- COLLATIONS：每个字符集的排序规则。
- COLLATION_CHARACTER_SET_APPLICABILITY：排序规则和字符集的对应关系，该视图和 COLLATIONS 视图的前两个字段一样。
- ENGINES：存储引擎的相关说明和加载状态。如下 SQL 语句查询存储引擎的分布：

```
mysql> select table_schema,engine,count(*)
from information_schema.tables
where table_schema not in  ('sys','mysql','information_schema','performance_schema')
and table_type = 'BASE TABLE'
group by table_schema,engine;
+--------------+--------+----------+
| TABLE_SCHEMA | ENGINE | count(*) |
+--------------+--------+----------+
| test         | MyISAM |        2 |
| sakila       | InnoDB |       20 |
| sbtest       | InnoDB |        2 |
| mysqlslap    | InnoDB |        1 |
```

```
+--------------+--------+----------+
4 rows in set (0.01 sec)
```

- INNODB_FT_DEFAULT_STOPWORD：InnoDB表默认的全文索引的停止词。
- KEYWORDS：MySQL的关键词和是否保留。对于保留的关键词在引用时需要特别处理，如加上引号。
- PLUGINS：服务插件的信息。
- RESOURCE_GROUPS：资源组信息。
- ST_SPATIAL_REFERENCE_SYSTEMS：可用的空间参照系统清单。

5.4.3 性能视图

性能视图保存与实例相关的信息，分别说明如下。

- INNODB_BUFFER_PAGE：InnoDB缓存中的页清单和使用信息，查询该视图比较消耗系统资源，最好在测试机器上实现。如下SQL语句为查询缓存中系统数据占用的页数、总的页数、系统页的占比：

```
mysql > select
(
select count( * ) from information_schema.innodb_buffer_page
where table_name is null or (instr(table_name, '/') = 0
and instr(table_name, '.') = 0)
) as system_pages,
( select count( * ) from information_schema.innodb_buffer_page) as total_pages,
(select round((system_pages/total_pages) * 100)) as system_page_percentage;
+--------------+-------------+------------------------+
| system_pages | total_pages | system_page_percentage |
+--------------+-------------+------------------------+
|        15261 |       16384 |                     93 |
+--------------+-------------+------------------------+
1 row in set (0.37 sec)
```

如下SQL语句为查询缓存中的用户数据：

```
mysql > select distinct table_name
from information_schema.innodb_buffer_page
where table_name is not null and
(instr(table_name, '/') > 0 or instr(table_name, '.') > 0)
and table_name not like ''mysql'.'innodb_ % ';
+-----------------------------------------------------+
| table_name                                          |
+-----------------------------------------------------+
| 'mysql'.'columns'                                   |
| 'mysql'.'tables'                                    |
| 'mysql'.'index_column_usage'                        |
| 'mysql'.'schemata'                                  |
| 'mysql'.'tablespace_files'                          |
...
```

如下 SQL 语句为查询索引数据在缓存中的大小：

```
mysql> select index_name,count( * ) as pages,
round(sum(if(compressed_size = 0, @@global.innodb_page_size, compressed_size))/1024/
1024) as 'total data (mb)'
from information_schema.innodb_buffer_page
group by index_name order by pages desc;
+------------------------------+--------+-----------------+
| index_name                   | pages  | total data (mb) |
+------------------------------+--------+-----------------+
| NULL                         | 15256  |             238 |
| PRIMARY                      |   962  |              15 |
| table_id                     |    20  |               0 |
| name                         |    14  |               0 |
| idx_fk_staff_id              |    11  |               0 |
...
```

- INNODB_BUFFER_PAGE_LRU：InnoDB缓存中的页信息，特别是页在最近最少使用(least recently used,LRU)排序中的位置，这个位置决定了页被从缓存中驱逐的顺序，查询该视图也比较消耗系统资源，最好在测试机器上实现。
- INNODB_BUFFER_POOL_STATS：InnoDB缓存池(pool)的使用，该视图中的信息和SHOW ENGINE INNODB STATUS命令输出的BUFFER POOL AND MEMORY部分的内容一样。每个池在该视图里对应一条记录，如下为查询该视图的代码及其运行结果：

```
mysql> select * from information_schema.innodb_buffer_pool_stats \G
*************************** 1. row ***************************
                         POOL_ID: 0
                       POOL_SIZE: 8192
                    FREE_BUFFERS: 6216
                  DATABASE_PAGES: 1962
              OLD_DATABASE_PAGES: 704
         MODIFIED_DATABASE_PAGES: 0
              PENDING_DECOMPRESS: 0
                   PENDING_READS: 0
               PENDING_FLUSH_LRU: 0
              PENDING_FLUSH_LIST: 0
                PAGES_MADE_YOUNG: 332
            PAGES_NOT_MADE_YOUNG: 1160
           PAGES_MADE_YOUNG_RATE: 0
       PAGES_MADE_NOT_YOUNG_RATE: 0
               NUMBER_PAGES_READ: 1488
            NUMBER_PAGES_CREATED: 486
            NUMBER_PAGES_WRITTEN: 1241
                 PAGES_READ_RATE: 0
               PAGES_CREATE_RATE: 0
              PAGES_WRITTEN_RATE: 0
                NUMBER_PAGES_GET: 64392
                        HIT_RATE: 0
    YOUNG_MAKE_PER_THOUSAND_GETS: 0
NOT_YOUNG_MAKE_PER_THOUSAND_GETS: 0
         NUMBER_PAGES_READ_AHEAD: 251
```

```
        NUMBER_READ_AHEAD_EVICTED: 0
                  READ_AHEAD_RATE: 0
          READ_AHEAD_EVICTED_RATE: 0
                     LRU_IO_TOTAL: 0
                   LRU_IO_CURRENT: 0
                 UNCOMPRESS_TOTAL: 0
               UNCOMPRESS_CURRENT: 0
1 row in set (0.00 sec)
```

- INNODB_CACHED_INDEXES：InnoDB 缓存中的索引页信息。如下 SQL 语句为查询缓存中的索引名、表名和每个索引占用的页：

```
mysql> select tables.name as table_name, indexes.name as
index_name,cached.n_cached_pages as n_cached_pages from
information_schema.innodb_cached_indexes as cached,
information_schema.innodb_indexes as indexes,
information_schema.innodb_tables as tables
where cached.index_id = indexes.index_id
and indexes.table_id = tables.table_id
order by n_cached_pages desc;
+---------------------+---------------------+----------------+
| table_name          | index_name          | n_cached_pages |
+---------------------+---------------------+----------------+
| sbtest/sbtest1      | PRIMARY             |            139 |
| sbtest/sbtest2      | PRIMARY             |            139 |
| sakila/rental       | PRIMARY             |             47 |
| sakila/payment      | PRIMARY             |             44 |
...
```

- INNODB_CMP：对 InnoDB 压缩表进行操作的统计信息。
- INNODB_CMP_RESET：和视图 INNODB_CMP 记录的信息一样，但只记录上次查询后变化的信息。
- INNODB_CMP_PER_INDEX：保存的信息和视图 INNODB_CMP 一样，但按照索引进行分组。
- INNODB_CMP_PER_INDEX_RESET：和视图 INNODB_CMP_PER_INDEX 记录的信息一样，但只记录上次查询后变化的信息。
- INNODB_CMPMEM：在 InnoDB 缓存池中的压缩页信息。
- INNODB_CMPMEM_RESET：和视图 INNODB_CMPMEM 记录的信息一样，但只记录上次查询后变化的信息。
- INNODB_METRICS：与 InnoDB 相关的性能信息。记录的内容和全局状态变量类似，区别是只包括 InnoDB 的信息。每个记录对应一个性能状态值，每个记录都有一个 COMMENT 字段说明记录的内容。如下 SQL 语句为查询 InnoDB 中删除行的信息：

```
mysql> select * from information_schema.innodb_metrics
where name = 'dml_deletes'\G
*************************** 1. row ***************************
           NAME: dml_deletes
```

```
        SUBSYSTEM: dml
            COUNT: 11
        MAX_COUNT: 11
        MIN_COUNT: NULL
        AVG_COUNT: 0.000008523524151793117
      COUNT_RESET: 11
  MAX_COUNT_RESET: 11
  MIN_COUNT_RESET: NULL
  AVG_COUNT_RESET: NULL
     TIME_ENABLED: 2021-06-28 07:26:31
    TIME_DISABLED: NULL
     TIME_ELAPSED: 1290546
       TIME_RESET: NULL
           STATUS: enabled
             TYPE: status_counter
          COMMENT: Number of rows deleted
1 row in set (0.01 sec)
```

当记录的字段STATUS的值是disable时，该性能状态值没有收集。默认收集的信息很少，可以通过设置系统参数innodb_monitor_enable、innodb_monitor_disable、innodb_monitor_reset和innodb_monitor_reset_all来激活、停止和重置计数。

- INNODB_SESSION_TEMP_TABLESPACES：该视图是从MySQL 8.0.13版本后增加的，它保存着InnoDB临时表空间对应的文件信息。如下SQL语句为查询该视图中的内容：

```
mysql> select path, size, sstate from
information_schema.innodb_session_temp_tablespaces;
+----------------------------+----------+----------+
| path                       | size     | state    |
+----------------------------+----------+----------+
| ./#innodb_temp/temp_10.ibt | 15728640 | ACTIVE   |
| ./#innodb_temp/temp_1.ibt  |    81920 | INACTIVE |
| ./#innodb_temp/temp_2.ibt  |    81920 | INACTIVE |
| ./#innodb_temp/temp_3.ibt  |    81920 | INACTIVE |
| ./#innodb_temp/temp_4.ibt  |    81920 | INACTIVE |
| ./#innodb_temp/temp_5.ibt  |    81920 | INACTIVE |
| ./#innodb_temp/temp_6.ibt  |    81920 | INACTIVE |
| ./#innodb_temp/temp_7.ibt  |    81920 | INACTIVE |
| ./#innodb_temp/temp_8.ibt  |    81920 | INACTIVE |
| ./#innodb_temp/temp_9.ibt  |    81920 | INACTIVE |
+----------------------------+----------+----------+
10 rows in set (0.01 sec)
```

根据前面的查询内容，可以找到操作系统中的文件，代码如下：

```
mysql> system ls /var/lib/mysql/#innodb_temp/temp_10.ibt
/var/lib/mysql/#innodb_temp/temp_10.ibt
```

如果从操作系统层面看到某个临时文件太大，可以通过该视图找到文件对应的会话。

- INNODB_TRX：InnoDB的事务信息。如下SQL语句为查询超过10秒未提交的事务：

```
mysql> select trx_mysql_thread_id as thread_id ,
to_seconds(now()) - to_seconds(trx_started) as trx_last_time , user,host,db,trx_query
from information_schema.innodb_trx trx
join information_schema.processlist pcl
on trx.trx_mysql_thread_id = pcl.id
where trx_mysql_thread_id != connection_id()
and to_seconds(now()) - to_seconds(trx_started) >= 10;
+-----------+---------------+------+-----------+--------+-----------+
| thread_id | trx_last_time | user | host      | db     | trx_query |
+-----------+---------------+------+-----------+--------+-----------+
|     41266 |          1031 | root | localhost | sakila | NULL      |
+-----------+---------------+------+-----------+--------+-----------+
1 row in set (0.01 sec)
```

这样的事务如果一直不能提交，可以使用如下命令将会话"杀死"：

```
mysql> kill 41266;
```

- OPTIMIZER_TRACE：当激活了系统参数 OPTIMIZER_TRACE 后，该视图中保存着优化器的跟踪信息，这些信息可以帮助进行 SQL 语句的优化。
- PROCESSLIST：MySQL 中当前线程正在执行的操作。该视图中保存的信息和 SHOW FULL PROCESSLIST 命令输出的信息一样。查询该视图的代码及其运行效果如下：

```
mysql> select * from information_schema.processlist limit 1 \G
*************************** 1. row ***************************
     ID: 41265
   USER: root
   HOST: localhost
     DB: NULL
COMMAND: Query
   TIME: 0
  STATE: executing
   INFO: select * from information_schema.processlist limit 1
1 row in set (0.01 sec)
```

- PROFILING：当会话的参数 profiling 激活后，该视图中保存着 SQL 语句的概要信息，该视图在未来的版本中将被弃用。

5.4.4 权限视图

与权限相关的视图有四个，表 5.1 列出了每个权限视图的记录内容和对应的系统表。

表 5.1

权 限 视 图	记 录 内 容	对应系统表
user_privileges	用户拥有的权限	mysql.user 和 mysql.global_grants
schema_privileges	对数据库访问的权限	mysql.db
table_privileges	对表访问的权限	mysql.tables_priv
column_privileges	对字段访问的权限	mysql.columns_priv

从这些视图查询到的信息和show grants命令查询的信息一致，如下代码说明了用这两种方式查询账号scutech@localhost权限的结果：

方式一：

```
mysql> show grants for scutech@localhost;
+-----------------------------------------------------------+
| Grants for scutech@localhost                              |
+-----------------------------------------------------------+
| GRANT USAGE ON *.* TO 'scutech'@'localhost'               |
| GRANT AUDIT_ADMIN ON *.* TO 'scutech'@'localhost'         |
| GRANT INSERT ON 'sakila'.'actor' TO 'scutech'@'localhost' |
+-----------------------------------------------------------+
3 rows in set (0.00 sec)
```

方式二：

```
mysql> select privilege_type, is_grantable, null table_schema, null table_name from information
_schema.user_privileges where grantee = '''scutech''@''localhost'''
union
select privilege_type, is_grantable, table_schema, null table_name from information_schema.
schema_privileges where grantee = '''scutech''@''localhost'''
union
select privilege_type, is_grantable, table_schema, table_name from information_schema.table_
privileges where grantee = '''scutech''@''localhost'''
union
select privilege_type, is_grantable, table_schema, table_name from information_schema.column_
privileges where grantee = '''scutech''@''localhost''';
+----------------+--------------+--------------+------------+
| privilege_type | is_grantable | table_schema | table_name |
+----------------+--------------+--------------+------------+
| USAGE          | NO           | NULL         | NULL       |
| AUDIT_ADMIN    | NO           | NULL         | NULL       |
| INSERT         | NO           | sakila       | actor      |
+----------------+--------------+--------------+------------+
3 rows in set (0.00 sec)
```

从上面的查询可以看到，用show grants命令查询账号scutech@localhost权限是从这四个权限视图查询到的该账号的权限之和。

5.5 实验

MySQL 8.0中DDL语句原子化

视频演示

（1）对比同一个DDL语句在MySQL 8.0和MySQL 5.7中的执行结果，验证DDL语句在MySQL 8.0中实现了原子化。

（2）把DDL语句放在一个事务中，事务回滚后，观察DDL执行结果的变化。

第6章

performance_schema数据库

第 5 章介绍的 information_schema 数据库主要用于保存 MySQL 的静态元数据，本章介绍的 performance_schema 数据库主要用于保存 MySQL 的动态元数据，这些数据可以帮助用户了解 MySQL 的性能和内部正在执行的操作；本章的内容还包括该数据库的配置和典型用例。

6.1 作用和特点

performance_schema 数据库是 MySQL 性能监控信息的主要来源，它从底层监控 MySQL 的运行情况，通过分析该数据库中的信息可以回答如下问题：

- 哪些语句性能较差？
- 哪些操作用时较长？
- 哪类锁竞争经常发生？
- 会话内部正在执行什么操作？
- 内存是如何分配的？

performance_schema 数据库具有以下特点：

- performance_schema 数据库中的表使用 performance_schema 存储引擎，当这些表中的数据发生变化时不会被写入二进制日志中，也不会通过复制机制被复制到其他 MySQL 实例。
- performance_schema 数据库中的表在硬盘上只保存.sdi 文件，不保存数据，所有的数据都暂存在内存中，MySQL 实例关闭后所有的数据将丢失(包括配置表)。
- performance_schema 数据库通过监视 MySQL 服务的事件来实现监视 MySQL 服务内部运行情况，MySQL 服务所做的任何需要消耗时间并统计消耗时间的事情都是事件，包括函数调用、操作系统的等待、SQL 语句执行的阶段等。
- 可记录当前活跃事件、历史事件和汇总事件的相关信息。能提供某个事件的执行次数和花费的时间。进而可用于分析某个特定线程、特定对象(如 mutex 或文件)的活动。

- performance_schema 数据库的存储引擎使用 MySQL 源代码中的"检测点"来实现事件数据的收集。没有单独线程来收集这些数据,这与其他功能(如复制或事件计划程序)不同。
- 收集的事件数据存储在 performance_schema 数据库的表中,这些表可以使用 select 语句查询。可以使用 SQL 语句更改 performance_schema 数据库的配置表,该更改会立即生效,进而改变 MySQL 性能诊断信息的收集行为。
- 大部分 performance_schema 数据库的表都有索引,在访问这些表时通常不需要全表扫描。

要收集某个事件性能信息,必须同时启用与这个事件相关的性能计量(instrument)和消费者(consumer),但启用的计量和消费者会消耗内存和 CPU 的资源,通常收集的性能统计信息的粒度越细,对系统消耗的资源越多,在极端的情况下,可能会造成系统宕机。因此在启用性能计量和消费者时要谨慎,通常保持默认的设置。执行下面两条 SQL 语句可以启用所有的性能计量和消费者。

语句一:

```
mysql > update performance_schema.setup_instruments set enabled = 'yes', timed = 'yes';
Query OK, 928 rows affected (0.03 sec)
Rows matched: 1212  Changed: 928  Warnings: 0
```

语句二:

```
mysql > update performance_schema.setup_consumers set enabled = 'yes';
Query OK, 8 rows affected (0.00 sec)
Rows matched: 15  Changed: 8  Warnings: 0
```

再次说明,这样做对系统的性能影响非常大!

6.2 配置

6.2.1 启动配置

由于 performance_schema 数据库配置表里的记录在 MySQL 关闭后会丢失,因此 performance_schema 数据库的启动选项就显得更加重要了,通过如下代码可以查看 performance_schema 数据库的启动选项:

```
$ mysqld --verbose --help | grep performance-schema | grep -v '\-\-'
performance-schema                                           TRUE
performance-schema-accounts-size                             -1
performance-schema-consumer-events-stages-current            FALSE
performance-schema-consumer-events-stages-history            FALSE
performance-schema-consumer-events-stages-history-long       FALSE
performance-schema-consumer-events-statements-current        TRUE
performance-schema-consumer-events-statements-history        TRUE
```

```
performance-schema-consumer-events-statements-history-long     FALSE
performance-schema-consumer-events-transactions-current        TRUE
...
```

共有 61 个启动选项，下面对重要的启动选项进行说明。

选项 performance_schema 是控制 performance_schema 数据库功能的开关，默认值是 ON，如果 mysqld 在初始化 performance_schema 数据库时发现无法分配任何相关的内部缓冲区，则 performance_schema 数据库将自动禁用，并将 performance_schema 设置为 OFF。

以 performance-schema-consumer 开头的启动选项有 15 个，它们控制着 15 个消费者是否启动，这些选项在 mysqld 启动之后通过如下 show variables 命令无法查看，因为它们不属于系统变量：

```
mysql> show variables like 'performance_schema_consumer%';
Empty set (0.01 sec)
```

控制是否在 mysqld 启动时收集一个消费者的事件，使用下面的配置方法：

```
--performance-schema-consumer-consumer_name=value
```

这里的 *consumer_name* 是一个消费者的名字，消费者的名字可以通过查询 setup_consumers 表查到，例如，events_waits_history，value 可以有如下两个值。

- OFF、FALSE 或 0：不收集该消费者的事件。
- ON、TRUE 或 1：收集该消费者的事件。

例如，收集 events_waits_history 消费者的事件可使用如下配置：

```
--performance-schema-consumer-events-waits-history=ON
```

这里不允许使用通配符。这 15 个选项对应着配置表 setup_consumers 里的 15 条记录。当某个选项被设定为 TRUE 时，它对应在 setup_consumers 表里记录的 ENABLED 字段为 YES，表示为激活状态；当设置为 FALSE 时，对应的 ENABLED 字段为 NO，表示不激活，也就是不统计这些消费者的性能计量。

用于控制性能计量的启动选项是 performance_schema_instrument，它也不是系统变量，在 mysqld 启动之后也无法通过如下 show variables 命令查看：

```
mysql> show variables like 'performance_schema_instrument';
Empty set (0.01 sec)
```

该启动参数的配置方法如下：

```
--performance-schema-instrument='instrument_name=value'
```

这里的 *instrument_name* 是性能计量的名字，例如，wait/synch/mutex/sql/LOCK_open，*value* 可以有如下三个值。

- OFF、FALSE 或 0：不启动该性能计量。
- ON、TRUE 或 1：启动该性能计量并计时。
- COUNTED：启动该性能计量，但只计数，不计时。

一个--performance-schema-instrument 只能设定一个性能计量类，但可以使用通配符

“%”匹配多个性能计量类，也可以有多个--performance-schema-instrument，例如，如下的选项配置可关闭所有的性能计量：

```
-- performance - schema - instrument = '% = OFF'
```

该启动参数对应配置表 setup_instruments，控制着 1216 类（MySQL 8.0.22）事件的性能计量是否启动。

对于数字型的 performance_schema 系统变量，可以在 mysqld 启动时通过命令行或参数文件指定一个值，值为－1 时表示由 MySQL 自动调整，在 MySQL 运行时通过 show variables 可以查询系统变量的值。

6.2.2 运行配置

当 MySQL 实例运行时，对性能监控信息的收集行为由 performance_schema 的 5 个配置表来控制，如下所示：

```
$ mysqlshow performance_schema|grep setup
| setup_actors                                  |
| setup_consumers                               |
| setup_instruments                             |
| setup_objects                                 |
| setup_threads                                 |
```

这 5 个表可以通过 SQL 语句进行修改，修改后将立即影响性能监控信息的收集行为。分别说明如下。

（1）setup_actors：指定新前台线程的初始监视状态。

（2）setup_consumers：指定在消费阶段的过滤规则，同时也指定了记录性能诊断信息的表。

（3）setup_instruments：指定激活的性能计量类。

（4）setup_objects：指定收集性能信息的数据库对象，包括表和存储过程等。

（5）setup_threads：指定收集性能信息的服务线程。

在如上 5 个配置表中，setup_consumers 属于消费阶段的配置，其他 4 个表属于生产阶段的配置。但消费阶段的配置也会影响生产阶段，对于一个事件，如果在消费阶段没有存储信息，那么在生产阶段也不会收集这个事件的信息。

因为运行时的配置在 MySQL 重新启动后会被初始化，所以为了使 MySQL 在重新启动后再次使用之前的配置，可以将修改配置表的 SQL 语句存放在一个文件中，用系统参数 init_file 指向这个文件，MySQL 启动时会自动执行该文件，这样就恢复了之前的配置。也可以针对不同的场景准备不同的 SQL 语句文件，例如，生产系统的健康检查、故障排除、针对特定应用的监控等。

6.3 性能计量配置

性能计量名称由一系列用“/”字符分隔的元素组成，例如：

```
wait/io/file/myisam/log
wait/io/file/mysys/charset
wait/lock/table/sql/handler
wait/synch/cond/mysys/COND_alarm
wait/synch/cond/sql/BINLOG::update_cond
wait/synch/mutex/mysys/BITMAP_mutex
wait/synch/mutex/sql/LOCK_delete
wait/synch/rwlock/sql/Query_cache_query::lock
stage/sql/closing tables
stage/sql/Sorting result
statement/com/Execute
statement/com/Query
```

性能计量名中顶层的元素如下。

(1) idle：性能计量空闲事件，这类性能计量名没有子元素。

(2) error：性能计量错误事件，这类性能计量名也没有子元素。

(3) memory：性能计量内存事件。

(4) stage：性能计量阶段事件。

(5) statement：性能计量语句事件。

(6) transaction：性能计量事务事件，这类性能计量名也没有子元素。

(7) wait：性能计量等待事件。

性能计量配置表(setup_instruments)中列出了所有的性能计量名，如下所示：

```
mysql > select name, enabled, timed from performance_schema.setup_instruments;
+---------------------------------------------+----------+-------+
| NAME                                        | ENABLED  | TIMED |
+---------------------------------------------+----------+-------+
| wait/synch/mutex/pfs/LOCK_pfs_share_list    | YES      | YES   |
| wait/synch/mutex/sql/TC_LOG_MMAP::LOCK_tc   | YES      | YES   |
| wait/synch/mutex/sql/MYSQL_BIN_LOG::LOCK_commit | YES  | YES   |
...
| error                                       | YES      | NULL  |
+---------------------------------------------+----------+-------+
1212 rows in set (0.00 sec)
```

修改性能计量配置表可以控制性能计量的收集，把一个性能计量的 ENABLED 字段设置成 YES 会收集该性能计量的性能诊断信息；把 TIMED 字段设置为 YES 会收集该性能计量的运行时间信息。

运行如下命令不会启用所有性能计量的收集：

```
mysql > update performance_schema.setup_instruments set enabled = 'no';
```

运行如下命令不会启用所有文件性能计量的收集：

```
mysql > update performance_schema.setup_instruments set enabled = 'no' where name like 'wait/
io/file/%';
```

运行如下命令将启用除文件性能计量之外的所有性能计量的收集：

```
mysql > update performance_schema.setup_instruments set enabled = if(name like 'wait/io/
file/%', 'no', 'yes');
```

6.4 消费者配置

消费者处理性能计量生成的信息，并把这些信息记录到表中。消费者处理性能计量的结果是事件(event)，它说明了 MySQL 正在处理的事情。事件分为 4 类，这 4 类事件是分层的，其中事务事件处于最高层，包含最少的细节，等待事件处于最底层，包含最多的细节。

(1) 事务(transaction)：这类事件记录了事务的相关信息，包括事务的隔离级别和它的状态等，默认记录每个线程的当前事务和最近 10 个事务。

(2) 语句(statement)：这是最常用的事件类型，它记录了执行的 SQL 语句的相关信息，包括检查行数、返回行数、执行事件和索引使用情况等，默认记录每个线程的当前 SQL 语句和最近 10 个 SQL 语句。

(3) 阶段(stage)：这类事件记录了 SQL 语句的执行阶段，类似 show processlist 命令输出的 states 字段，这类事件默认没有被激活。

(4) 等待(wait)：这类事件是最底层的事件，它记录了 I/O 和线程互斥器(mutexes)的相关信息，这类事件默认没有被激活，收集这类事件的信息将对系统的性能影响最大。

每个事件按照它的生命周期对应如下 3 个消费者。

(1) current：线程当前正在执行的事件，对于空闲线程，是它最后执行的事件。

(2) history：每个线程最近执行的 10 个(默认值)事件，如果线程结束，事件将被删除。

(3) history_long：整个实例(不区分线程)最近执行的 10 000 个(默认值)事件，如果线程结束，事件仍然保留。

这 4 类事件和 3 类生命周期组成了 12 个消费者，它们在消费者配置表(setup_consumers)中的 name 字段中以 events_开头，该表列出了所有的消费者和它们是否被激活，该表的默认记录如下：

```
mysql > select * from setup_consumers;
+-----------------------------------+---------+
| NAME                              | ENABLED |
+-----------------------------------+---------+
| events_stages_current             | NO      |
| events_stages_history             | NO      |
| events_stages_history_long        | NO      |
| events_statements_current         | YES     |
| events_statements_history         | YES     |
| events_statements_history_long    | NO      |
| events_transactions_current       | YES     |
| events_transactions_history       | YES     |
| events_transactions_history_long  | NO      |
| events_waits_current              | NO      |
| events_waits_history              | NO      |
| events_waits_history_long         | NO      |
| global_instrumentation            | YES     |
```

```
| thread_instrumentation           | YES      |
| statements_digest                | YES      |
+----------------------------------+----------+
15 rows in set (0.00 sec)
```

修改这个表的相应记录会立即激活或不激活对应的消费者，但一个消费者的事件信息是否收集不仅仅由这个表的 enabled 字段决定，因为这 15 个消费者是有层次的，只有当该消费者对应的上层的消费者都处于收集状态和该消费者被激活时，才会收集该消费者的事件信息。这 15 个消费者的层次关系如表 6.1 所示。

表 6.1

global_instrumentation			
	thread_instrumentation	events_waits_current	
			events_waits_history events_waits_history_long
		events_stages_current	
			events_stages_history events_stages_history_long
		events_statements_current	
			events_statements_history events_statements_history_long
		events_transactions_current	
			events_transactions_history events_transactions_history_long
	statements_digest		

层次关系是从左到右的，而不是从上到下的。从上表中可以看出，如果要激活 events_statements_history，需要先把 global_instrumentation、thread_instrumentation 和 events_waits_current 都激活。其中 global_instrumentation 处于该层次的最顶层，如果该消费者没有被激活，其他消费者就都没有被激活，可以使用 sys 的函数 ps_is_consumer_enabled()来判断该消费者和它依赖的消费者是否被激活，代码如下：

```
mysql> select sys.ps_is_consumer_enabled('events_statements_history') as isenabled;
+-----------+
| isenabled |
+-----------+
| yes       |
+-----------+
1 row in set (0.00 sec)
```

初次查询 events_statements_history 消费者是启用的，然后把 global_instrumentation 设置成未被激活，代码如下：

```
mysql> update setup_consumers set enabled = 'no' where name = 'global_instrumentation';
query ok, 1 row affected (0.00 sec)
rows matched: 1  changed: 1  warnings: 0
```

再次查询，发现 events_statements_history 消费者变成了未启用，代码如下：

```
mysql> select sys.ps_is_consumer_enabled('events_statements_history') as isenabled;
+-----------+
| isenabled |
+-----------+
| no        |
+-----------+
1 row in set (0.00 sec)
```

从上面的例子可以看出，global_instrumentation 消费者可以控制包括 events_statements_history 在内的所有消费者的启用状态。

需要注意的是，statements_digest 这个消费者虽然处在第二层，但并没有任何消费者依赖于它，即使 statements_digest 没有被激活，events_statements_current、events_statements_history 和 events_statements_history_long 这三个消费者仍然可以被激活，只不过收集信息中的字段的 digest 和 digest_text 值是 null。

6.5 执行者配置

performance_schema 数据库还可以通过执行事件的账户来控制是否产生性能诊断信息，该控制是通过执行者配置表(setup_actors)来实现的，运行如下代码可得到该表的初始默认记录：

```
mysql> select * from setup_actors;
+------+------+------+---------+---------+
| host | user | role | enabled | history |
+------+------+------+---------+---------+
| %    | %    | %    | yes     | yes     |
+------+------+------+---------+---------+
1 row in set (0.00 sec)
```

配置表(setup_actors)中的 enabled 和 history 字段控制着匹配的账户是否激活性能计量和是否记录历史性能计量信息，从该表的默认记录可以看出，所有账户都激活性能计量和记录历史性能计量信息。这两个字段分别对应着 threads 表对应记录的 instrumented 和 history 字段。

例如，默认时，root 用户线程的 instrumented 和 history 字段都是 yes，如下所示：

```
mysql> select thread_id, instrumented, history from threads where processlist_user = 'root';
+-----------+--------------+---------+
| thread_id | instrumented | history |
+-----------+--------------+---------+
|        49 | yes          | yes     |
+-----------+--------------+---------+
1 row in set (0.00 sec)
```

向 setup_actors 表中插入一条记录，将 root 账户对应的 enabled 和 history 字段分别设置成 yes 和 no，代码如下：

```
mysql> insert into performance_schema.setup_actors (host,user,role,enabled,history) values
('localhost','root','%','yes','no');
query ok, 1 row affected (0.00 sec)
```

然后再登录一个 root 用户，可以看到新登录的 root 用户对应线程的 instrumented 和 history 字段已经被修改成对应的 setup_actors 表中的配置，旧的线程保持不变，代码如下：

```
mysql> select thread_id,instrumented,history from threads where processlist_user = 'root';
+-----------+--------------+---------+
| thread_id | instrumented | history |
+-----------+--------------+---------+
|        49 | yes          | yes     |
|        50 | yes          | no      |
+-----------+--------------+---------+
2 rows in set (0.00 sec)
```

运行如下代码可以不激活所有账户的性能计量和不记录历史性能计量信息：

```
mysql> update performance_schema.setup_actors set enabled = 'no', history = 'no' where host =
'%' and user = '%';
query ok, 1 row affected (0.00 sec)
rows matched: 1  changed: 1  warnings: 0
```

需要注意的是 setup_actors 表的配置只对前台线程起作用，后台线程在 threads 表对应记录的 instrumented 和 history 字段都是 yes。

6.6 对象配置

performance_schema 数据库还可以通过被执行的对象来控制性能诊断信息的产生，该控制是通过对象配置表（setup_objects）来实现的，运行如下代码可得到该表的初始默认记录：

```
mysql> select * from setup_objects;
+-------------+--------------------+-------------+---------+-------+
| OBJECT_TYPE | OBJECT_SCHEMA      | OBJECT_NAME | ENABLED | TIMED |
+-------------+--------------------+-------------+---------+-------+
| EVENT       | mysql              | %           | NO      | NO    |
| EVENT       | performance_schema | %           | NO      | NO    |
| EVENT       | information_schema | %           | NO      | NO    |
| EVENT       | %                  | %           | YES     | YES   |
| FUNCTION    | mysql              | %           | NO      | NO    |
| FUNCTION    | performance_schema | %           | NO      | NO    |
| FUNCTION    | information_schema | %           | NO      | NO    |
| FUNCTION    | %                  | %           | YES     | YES   |
| PROCEDURE   | mysql              | %           | NO      | NO    |
| PROCEDURE   | performance_schema | %           | NO      | NO    |
| PROCEDURE   | information_schema | %           | NO      | NO    |
| PROCEDURE   | %                  | %           | YES     | YES   |
| TABLE       | mysql              | %           | NO      | NO    |
| TABLE       | performance_schema | %           | NO      | NO    |
```

```
| TABLE       | information_schema | %           | NO       | NO     |
| TABLE       | %                  | %           | YES      | YES    |
| TRIGGER     | mysql              | %           | NO       | NO     |
| TRIGGER     | performance_schema | %           | NO       | NO     |
| TRIGGER     | information_schema | %           | NO       | NO     |
| TRIGGER     | %                  | %           | YES      | YES    |
+-------------+--------------------+-------------+----------+--------+
20 rows in set (0.01 sec)
```

其中，ENABLED和TIMED字段分别表示激活性能计量和记录执行的时间信息。从默认的记录看，除了默认的系统数据库外，所有的数据库对象都激活性能计量和记录执行的时间信息。

下面向对象配置表(setup_objects)中插入如下两条记录：

```
+-------------+---------------+--------------+----------+--------+
| OBJECT_TYPE | OBJECT_SCHEMA | OBJECT_NAME  | ENABLED  | TIMED  |
+-------------+---------------+--------------+----------+--------+
| TABLE       | db1           | t1           | YES      | YES    |
| TABLE       | db1           | %            | NO       | NO     |
+-------------+---------------+--------------+----------+--------+
```

如上两条记录的效果是，对于db1数据库中的表，除了t1外，所有的表都不激活性能计量和不记录执行的时间信息。

6.7 典型用例

6.7.1 监控SQL语句的执行性能

可以通过查询如下3个表来监控SQL语句的执行性能。

(1) events_statements_current：默认激活，为每个线程保存一条当前正在执行的SQL语句的相关信息。

(2) events_statements_history：默认激活，为每个线程保存N条(N默认为10)最近执行的SQL语句的相关信息，当线程退出时，与它相关的记录将会被删除。

(3) events_statements_history_long：默认未激活，为整个实例保存N条(N默认为10 000)最近执行的SQL语句的相关信息。

如下代码用于找出等待时间最长的3个SQL语句：

```
mysql> select thread_id,event_name,source,
sys.format_time(timer_wait),
sys.format_time(lock_time),sql_text,
current_schema,message_text,rows_affected,rows_sent,rows_examined
from events_statements_history
where current_schema!= 'performance_schema'
order by timer_wait desc limit 3\G
***************************** 1. row *****************************
                THREAD_ID: 309
               EVENT_NAME: statement/sql/select
```

```
                    SOURCE: init_net_server_extension.cc: 95
sys.format_time(TIMER_WAIT): 572.57 ms
sys.format_time(LOCK_TIME): 314 us
                  SQL_TEXT: select actor.last_name,rental.staff_id from rental,actor
            CURRENT_SCHEMA: sakila
              MESSAGE_TEXT: NULL
             ROWS_AFFECTED: 0
                 ROWS_SENT: 3208800
             ROWS_EXAMINED: 16244
*************************** 2. row ***************************
                 THREAD_ID: 309
                EVENT_NAME: statement/com/Field List
                    SOURCE: init_net_server_extension.cc: 95
sys.format_time(TIMER_WAIT): 1.07 ms
sys.format_time(LOCK_TIME): 0 ps
                  SQL_TEXT: NULL
            CURRENT_SCHEMA: sakila
              MESSAGE_TEXT: SELECT command denied to user ''@'%' for column 'manager_staff_
id' in table 'store'
             ROWS_AFFECTED: 0
                 ROWS_SENT: 0
             ROWS_EXAMINED: 0
*************************** 3. row ***************************
                 THREAD_ID: 309
                EVENT_NAME: statement/com/Field List
                    SOURCE: init_net_server_extension.cc: 95
sys.format_time(TIMER_WAIT): 767.56 us
sys.format_time(LOCK_TIME): 0 ps
                  SQL_TEXT: NULL
            CURRENT_SCHEMA: sakila
              MESSAGE_TEXT: NULL
             ROWS_AFFECTED: 0
                 ROWS_SENT: 0
             ROWS_EXAMINED: 0
3 rows in set (0.01 sec)
```

如下代码用于查询没有使用索引的 SQL 语句：

```
mysql > select thread_id as tid,
substr(sql_text, 1, 50)  as sql_text,
rows_sent , rows_examined ,
created_tmp_tables, no_index_used,
no_good_index_used
from performance_schema.events_statements_history
where no_index_used = 1  or no_good_index_used = 1\G
*************************** 1. row ***************************
               TID: 304
          SQL_TEXT: select amount from payment
         ROWS_SENT: 16049
     ROWS_EXAMINED: 16049
CREATED_TMP_TABLES: 0
     NO_INDEX_USED: 1
NO_GOOD_INDEX_USED: 0
1 row in set (0.00 sec)
```

除了是否使用索引外，对这 3 个表中的如下字段进行检查也有助于发现 SQL 语句执行的瓶颈：

- CREATED_TMP_DISK_TABLES
- CREATED_TMP_TABLES
- SELECT_FULL_JOIN
- SELECT_RANGE_CHECK
- SELECT_SCAN
- SORT_MERGE_PASSES
- SORT_SCAN

对于一个执行速度慢的 SQL 语句，可能需要知道它到底在哪个阶段执行的速度慢，可以用 event_id 字段把 events_stages_* 表和 events_statements_* 连接起来进行查询，但 events_stages_* 视图默认没有被激活，在查询这些视图前先要运行如下代码以激活这些视图：

```
mysql> call sys.ps_setup_enable_consumer('events_stages');
+---------------------+
| summary             |
+---------------------+
| Enabled 3 consumers |
+---------------------+
1 row in set (0.00 sec)

Query OK, 0 rows affected (0.00 sec)
```

运行如下代码以找出执行慢的阶段：

```
mysql> select eshl.event_name, sql_text,
eshl.timer_wait/1000000000000 w_s
from performance_schema.events_stages_history_long eshl
join performance_schema.events_statements_history_long esthl
on (eshl.nesting_event_id = esthl.event_id)
where eshl.timer_wait > 1 * 100000000000\G
*************************** 1. row ***************************
event_name: stage/sql/executing
  sql_text: select actor.last_name,rental.staff_id from rental,actor
       w_s: 0.5994
1 row in set (0.01 sec)
```

在 events_stages_* 视图中还有如下其他需要关注的事件：

- 与临时表相关的事件，事件类型是"stage/sql/%tmp%"。
- 与锁相关的事件，事件类型是"stage/sql/%lock%"。
- 正在等待的事件，事件类型是"stage/%/Waiting for%"。

在 SQL 语句中其他容易成为瓶颈的事件还包括：

- EVENT NAME='stage/sql/end'
- EVENT NAME='stage/sql/freeing items'
- EVENT NAME='stage/sql/Sending data'

- EVENT NAME='stage/sql/cleaning up'
- EVENT NAME='stage/sql/closing tables'

6.7.2 监控锁

记录锁信息的视图共有 4 个，分别为 data_locks、data_lock_waits、metadata_locks、data_handles。

1. data_locks

data_locks 视图记录着持有和请求数据锁的信息。对于数据锁，没有性能计量控制这类诊断信息是否被激活，这类信息已经在 MySQL 服务中，收集这类信息也不消耗 CPU 和内存，这类信息保存在 data_locks 表中，下面是查询数据锁的一个例子：

```
mysql > begin;
Query OK, 0 rows affected (0.00 sec)

mysql > update sakila.actor set first_name = 'a' where actor_id = 1;
Query OK, 1 row affected (0.01 sec)
Rows matched: 1  Changed: 1  Warnings: 0

mysql > select * from performance_schema.data_locks\G
*************************** 1. row ***************************
               ENGINE: INNODB
       ENGINE_LOCK_ID: 139961059929496:4024:139960971770400
ENGINE_TRANSACTION_ID: 1150803
            THREAD_ID: 74
             EVENT_ID: 5
        OBJECT_SCHEMA: sakila
          OBJECT_NAME: actor
       PARTITION_NAME: NULL
    SUBPARTITION_NAME: NULL
           INDEX_NAME: NULL
OBJECT_INSTANCE_BEGIN: 139960971770400
            LOCK_TYPE: TABLE    -- 表级锁
            LOCK_MODE: IX    -- 意向独占锁
          LOCK_STATUS: GRANTED
            LOCK_DATA: NULL
*************************** 2. row ***************************
               ENGINE: INNODB
       ENGINE_LOCK_ID: 139961059929496:2700:4:2:139960971767488
ENGINE_TRANSACTION_ID: 1150803
            THREAD_ID: 74
             EVENT_ID: 5
        OBJECT_SCHEMA: sakila
          OBJECT_NAME: actor
       PARTITION_NAME: NULL
    SUBPARTITION_NAME: NULL
           INDEX_NAME: PRIMARY
OBJECT_INSTANCE_BEGIN: 139960971767488
            LOCK_TYPE: RECORD    -- 行级锁
            LOCK_MODE: X,REC_NOT_GAP    -- 独占锁
```

```
        LOCK_STATUS: GRANTED
          LOCK_DATA: 1
2 rows in set (0.00 sec)
```

可以看出，74 号线程在 sakila.actor 上有两个锁，一个是表级的意向独占锁(IX)，另一个是行级的独占锁(X)。

2. data_lock_waits

data_lock_waits 视图记录着数据锁持有者和被这些持有者阻止的数据锁请求者之间的关系。接着前面的实验，这个时候如果再有一个线程要修改同一个记录就会被 74 号线程阻止，在 data_lock_waits 表中可以查到下面的记录：

```
mysql> select * from performance_schema.data_lock_waits\G
*************************** 1. row ***************************
                          ENGINE: INNODB
       REQUESTING_ENGINE_LOCK_ID: 139961059930352:2700:4:2:139960971773632
REQUESTING_ENGINE_TRANSACTION_ID: 1150804
            REQUESTING_THREAD_ID: 75
             REQUESTING_EVENT_ID: 4
REQUESTING_OBJECT_INSTANCE_BEGIN: 139960971773632
         BLOCKING_ENGINE_LOCK_ID: 139961059929496:2700:4:2:139960971767488
  BLOCKING_ENGINE_TRANSACTION_ID: 1150803
              BLOCKING_THREAD_ID: 74
               BLOCKING_EVENT_ID: 5
  BLOCKING_OBJECT_INSTANCE_BEGIN: 139960971767488
1 row in set (0.00 sec)
```

从 data_lock_waits 表中可以看到 75 号线程被 74 号线程阻止了，在 sys.innodb_lock_waits 中可以看到更加详细的记录，如下所示：

```
mysql> select * from sys.innodb_lock_waits \G
*************************** 1. row ***************************
                wait_started: 2021-04-22 05:39:58
                    wait_age: 10:30:09
               wait_age_secs: 37809
                locked_table: 'sakila'.'actor'
         locked_table_schema: sakila
           locked_table_name: actor
      locked_table_partition: NULL
   locked_table_subpartition: NULL
                locked_index: PRIMARY
                 locked_type: RECORD
              waiting_trx_id: 1150805
         waiting_trx_started: 2021-04-22 05:39:58
             waiting_trx_age: 10:30:09
     waiting_trx_rows_locked: 1
   waiting_trx_rows_modified: 0
                 waiting_pid: 32
               waiting_query: update sakila.actor set first_name = 'a' where actor_id = 1
             waiting_lock_id: 139961059930352:2700:4:2:139960971773632
           waiting_lock_mode: X,REC_NOT_GAP
```

```
           blocking_trx_id: 1150803
               blocking_pid: 31
            blocking_query: select * from sys.innodb_lock_waits
           blocking_lock_id: 139961059929496: 2700: 4: 2: 139960971767488
         blocking_lock_mode: X,REC_NOT_GAP
      blocking_trx_started: 2021 - 04 - 22 05: 33: 19
           blocking_trx_age: 10: 36: 48
   blocking_trx_rows_locked: 1
 blocking_trx_rows_modified: 1
     sql_kill_blocking_query: KILL QUERY 31
sql_kill_blocking_connection: KILL 31
1 row in set (0.00 sec)
```

可以看出,最后两个字段给出为了解决锁冲突而“杀死”查询或连接的语句。

3. metadata_locks

metadata_locks 视图记录着持有和请求元数据锁的信息。元数据锁记录对应的性能计量是 wait/lock/metadata/sql/mdl,在 MySQL 8.0 里默认是激活的,之前的版本没有默认激活这个性能计量。元数据锁用于管理对同一个数据库对象的访问,保护对象的元数据。这种锁很容易被忽视,如查询一个表时,会获得一个共享的元数据锁,并且该元数据锁会一直保持到事务结束,这时不能对该表执行任何 DDL 语句,包括 OPTIMIZE TABLE,而且如果一个 DDL 语句被元数据锁阻止,那么它又会阻塞所有新的查询使用该表。

图 6.1 显示了元数据锁阻止 SQL 运行的例子。

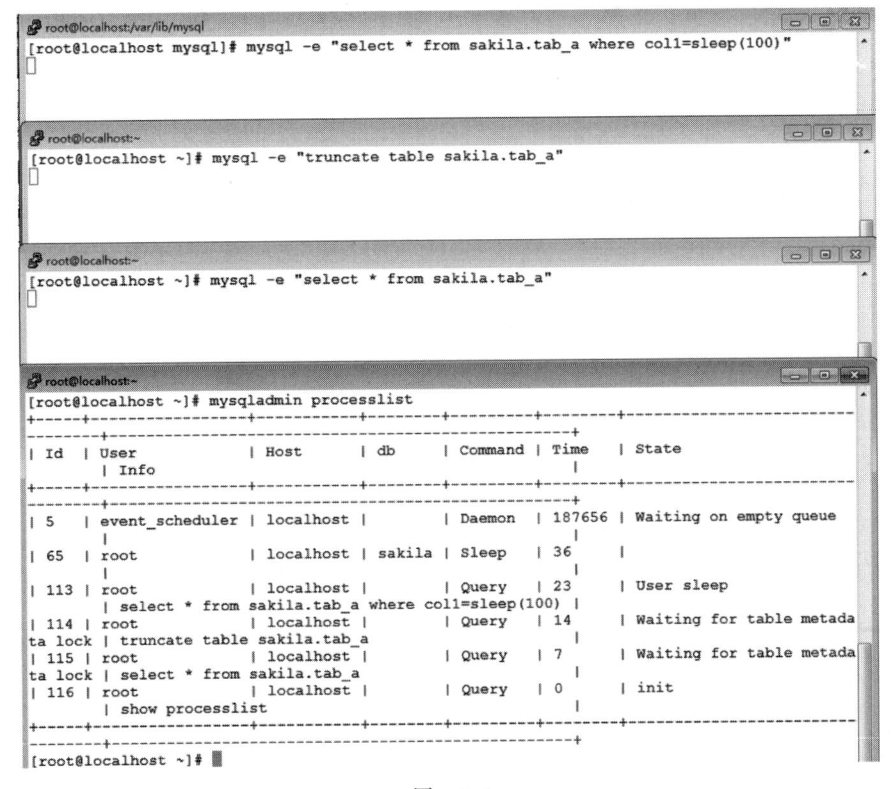

图 6.1

从图 6.1 中可以看出:

(1) 会话 113 里执行一个运行时间长的 select 语句。

(2) 会话 114 里执行对同一个对象的 DDL 语句,该 DDL 语句被元数据锁阻止。注意,不仅修改字段类型、增加减少字段、创建或删除索引的语句是 DDL 语句,truncate 语句也是 DDL 语句,它和 delete 语句不同。

(3) 会话 115 里执行对同一个对象的 select 语句也被阻止。

运行如下代码可以在 metadata_locks 表中查询到相应的元数据锁:

```
mysql > select object_schema,object_name,lock_type,
lock_status,owner_thread_id id
from performance_schema.metadata_locks
where object_name = 'tab_a';
+---------------+-------------+-------------+-------------+------+
| object_schema | object_name | lock_type   | lock_status | id   |
+---------------+-------------+-------------+-------------+------+
| sakila        | tab_a       | EXCLUSIVE   | PENDING     | 193  |
| sakila        | tab_a       | SHARED_READ | PENDING     | 194  |
| sakila        | tab_a       | SHARED_READ | GRANTED     | 192  |
+---------------+-------------+-------------+-------------+------+
3 rows in set (0.00 sec)
```

从查询结果可以看出,192 号线程拿到了一个共享读的元数据锁,阻止了 193 号线程申请独占锁,又阻止了 194 号语句申请共享读锁的元数据锁。

元数据锁的目的是保护数据库对象的元数据,因此事务里的 SQL 语句在执行完成后仍然会保留元数据锁直到事务结束,例如,如下的 SQL 语句在事务结束前会一直持有 tab_a 表的元数据锁:

```
mysql > begin;
Query OK, 0 rows affected (0.00 sec)

mysql > select * from tab_a;
+------+
| col1 |
+------+
|    1 |
|    1 |
+------+
2 rows in set (0.00 sec)
```

这时如果有 DDL 语句被阻止,在 sys. session 里将找不到阻止它的 SQL 语句,因为该 SQL 语句已经被执行完成了。这时可以根据线程号在 events_transactions_current 视图里找到当前活跃的事务的 event_id,然后根据该 event_id 在 events_statements_history 查找它嵌套的事件,从而找到执行过的 SQL 语句,代码如下:

```
mysql > select event_id, current_schema, sql_text from
performance_schema.events_statements_history a
where thread_id = 57 and
nesting_event_type = 'transaction'
```

```
and nesting_event_id in
( select event_id from
performance_schema.events_transactions_current b
where a.thread_id = b.thread_id);
+----------+----------------+-------------------------------------+
| event_id | current_schema | sql_text                            |
+----------+----------------+-------------------------------------+
|       37 | sakila         | select * from tab_a                 |
|       38 | sakila         | select * from tab_a                 |
|       39 | sakila         | select * from actor where actor_id = 1 |
+----------+----------------+-------------------------------------+
3 rows in set (0.00 sec)
```

从如上代码中可以看到该事务已经执行过的 3 个 SQL 语句。

除了在 performance_schema.metadata_locks 可以查询元数据锁之外，在 sys.schema_table_lock_waits 里也可以查到正在等待元数据锁的会话，例如：

```
mysql> select * from sys.schema_table_lock_waits\G
*************************** 1. row ***************************
               object_schema: sakila
                 object_name: tab_a
           waiting_thread_id: 59
                 waiting_pid: 20
             waiting_account: root@localhost
           waiting_lock_type: EXCLUSIVE
       waiting_lock_duration: TRANSACTION
               waiting_query: truncate tab_a
          waiting_query_secs: 4857
 waiting_query_rows_affected: 0
 waiting_query_rows_examined: 0
          blocking_thread_id: 57
                blocking_pid: 18
            blocking_account: root@localhost
          blocking_lock_type: SHARED_READ
      blocking_lock_duration: TRANSACTION
     sql_kill_blocking_query: KILL QUERY 18
sql_kill_blocking_connection: KILL 18
*************************** 2. row ***************************
               object_schema: sakila
                 object_name: tab_a
           waiting_thread_id: 60
                 waiting_pid: 21
             waiting_account: root@localhost
           waiting_lock_type: SHARED_READ
       waiting_lock_duration: TRANSACTION
               waiting_query: select * from sakila.tab_a
          waiting_query_secs: 4727
 waiting_query_rows_affected: 0
 waiting_query_rows_examined: 0
          blocking_thread_id: 57
                blocking_pid: 18
```

```
          blocking_account: root@localhost
        blocking_lock_type: SHARED_READ
    blocking_lock_duration: TRANSACTION
   sql_kill_blocking_query: KILL QUERY 18
sql_kill_blocking_connection: KILL 18
2 rows in set (0.00 sec)
```

元数据锁的超时时间由系统参数 lock_wait_timeout 控制，默认是 31 536 000 秒，也就是一年。因此如果一个活跃的事务长期不提交，它持有的元数据锁就一直不会释放，实际生产中的一个常见现象是在与客户端交互的过程中发起了一个事务，因为客户操作的时间很长，元数据锁在这个过程中一直被持有，可能会引起连锁反应从而造成灾难性的后果，因此在开发过程中要尽量把长事务拆分成短事务。另外一个可能会遇到的问题是有些客户端工具默认没有激活自动提交(auto commit)，这样发起一个简单的 select 语句也会开启一个事务，而且不会自动提交，也可能会引起麻烦。

通过 kill connection 或 kill query 命令可以终止正在执行的 SQL 语句并释放元数据锁，需要注意的是，kill 命令默认是 kill connection，kill connection 会“杀死”指定的连接，因此会终止该线程正在执行的事务和 SQL 语句，kill query 只会“杀死”指定的连接正在执行的 SQL 语句，不会“杀死”连接和终止事务。使用 kill 命令时要特别小心，因为被“杀死”的 SQL 语句会回滚，通常一个 SQL 语句回滚用的时间比进行变更的时间还长，查询视图 information_schema.innodb_trx 中的字段 trx_rows_modified 可得到该事务的修改行数，如果该值较大，不要轻易“杀死”这个 SQL 语句。

4. table_handles

table_handles 视图记录着持有和请求表锁的信息。表级锁对应的性能计量是 wait/lock/table/sql/handler，默认是激活的。如下是一个查询表级锁的例子：

```
mysql > lock table sakila.actor  read;
Query OK, 0 rows affected (0.00 sec)

mysql > select object_type, object_schema, object_name, owner_thread_id from performance_schema.table_handles where owner_thread_id is not null;
+-------------+---------------+-------------+-----------------+
| OBJECT_TYPE | OBJECT_SCHEMA | OBJECT_NAME | OWNER_THREAD_ID |
+-------------+---------------+-------------+-----------------+
| TABLE       | sakila        | actor       |            3240 |
+-------------+---------------+-------------+-----------------+
1 row in set (0.00 sec)
```

6.7.3 查询当前等待事件

查询当前等待事件可以找出系统当前的瓶颈，当前等待事件由消费者 events_waits_current 记录，该消费者默认没有被激活，可使用如下 SQL 语句激活该消费者 events_waits_current：

```
mysql > update setup_consumers set enabled = 'yes'
```

```
where name = 'events_waits_current';
```

运行如下代码可找出当前等待事件中等待时间最长的事件：

```
mysql > select thread_id,event_name,
format_pico_time(timer_wait) wait_time,operation
from events_waits_current
where event_name!= 'idle' and
event_name!= 'wait/synch/cond/mysqlx/scheduler_dynamic_worker_pending'
order by timer_wait desc limit 5;
+-----------+-----------------------------------+---------+--------+
| thread_id | event_name                        | wait_time | operation |
+-----------+-----------------------------------+---------+--------+
|      1812 | wait/io/file/sql/binlog           | 10.45 ms | sync    |
|        12 | wait/io/file/innodb/innodb_data_file | 10.41 ms | sync  |
|        16 | wait/io/file/innodb/innodb_log_file | 9.55 ms | sync    |
|        37 | wait/synch/mutex/innodb/srv_sys_mutex | 44.41 us | lock  |
|       307 | wait/synch/mutex/sql/THD::LOCK_query_plan | 620.5 ns | lock |
+-----------+-----------------------------------+---------+--------+
5 rows in set (0.00 sec)
```

可以看出，当前的等待事件最长的是写二进制日志、数据文件和日志文件，显然当前的瓶颈是对硬盘的写操作。

性能视图里的时间通常是以皮秒(picosecond)为单位，一皮秒等于一万亿分之一秒(即1皮秒$=1\times10^{-12}$秒)，显示时需要使用函数 sys. format_time()(MySQL 8.0.15 之前)或 format_pico_time()(MySQL 8.0.15 之后)对时间进行格式化。

6.7.4 查询错误语句

MySQL 中用于记录错误的视图有如下 5 个：

- events_errors_summary_by_account_by_error
- events_errors_summary_by_host_by_error
- events_errors_summary_by_thread_by_error
- events_errors_summary_by_user_by_error
- events_errors_summary_global_by_error

如上视图记录的内容相同，只是分组的方法不同，从视图的名字可以推断出它们的分组方法。具体错误代码的含义可以登录如下网站查询：https://dev.mysql.com/doc/mysql-errors/8.0/en/server-error-reference.html。

从这些视图中可以查询出错信息，例如，先执行如下一个错误的 SQL 语句：

```
mysql > select * from no_exit_table;
ERROR 1146 (42S02): Table 'performance_schema.no_exit_table' doesn't exist
```

然后根据错误号 1146 可以在记录错误信息的视图中查询出错的信息，相关命令及其运行结果如下：

```
mysql > select user,host,error_name,sql_state,last_seen from
```

```
performance_schema.events_errors_summary_by_account_by_error where
error_number = '1146' order by last_seen desc limit 1;
+------+-----------+-------------------+----------+--------------------+
| USER | HOST      | ERROR_NAME        | SQL_STATE | LAST_SEEN          |
+------+-----------+-------------------+----------+--------------------+
| root | localhost | ER_NO_SUCH_TABLE  | 42S02    | 2021-07-21 17:20:11 |
+------+-----------+-------------------+----------+--------------------+
1 row in set (0.05 sec)
```

也可以根据错误号从 events_statements_history 或 events_statements_history_long 表中查到出错的 SQL 语句，相关命令及其运行结果如下：

```
mysql> select thread_id, sql_text, message_text from events_statements_history where mysql_
errno = 1146 \G
*************************** 1. row ***************************
   thread_id: 48
    sql_text: select * from no_exit_table
message_text: Table 'performance_schema.no_exit_table' doesn't exist
1 row in set (0.01 sec)
```

在日常工作中，很多时候前端反映 SQL 语句出错，但并不知道错误号，遇到这种情况时通过查询字段 errors 大于零也可以找到出错的语句，例如：

```
mysql> select * from events_statements_history_long where errors > 0 \G
```

如下 SQL 语句可以查询按账号分组的死锁信息：

```
mysql> select user, host, sql_state from
performance_schema.events_errors_summary_by_account_by_error
where error_name = 'er_lock_deadlock';
```

6.7.5 监控表 I/O

与表 I/O 相关的性能视图有如下两个。

- table_io_waits_summary_by_table：以表进行分组记录表 I/O 信息，这些信息由计量 wait/io/table/sql/handler 生成。
- table_io_waits_summary_by_index_usage：以索引进行分组记录表 I/O 信息，这些信息也是由计量 wait/io/table/sql/handler 生成。

这两个视图里记录的 I/O 不是磁盘 I/O，只是表示对表的读写，可能是从缓存中，也可能是从磁盘。如下 SQL 语句可查询按表分组的一个表的 I/O：

```
mysql> select * from
performance_schema.table_io_waits_summary_by_table
where object_schema = 'sakila' and object_name = 'actor'\G
*************************** 1. row ***************************
    OBJECT_TYPE: TABLE
  OBJECT_SCHEMA: sakila
    OBJECT_NAME: actor
     COUNT_STAR: 752
```

```
   SUM_TIMER_WAIT: 38694446229
   MIN_TIMER_WAIT: 2198424
   AVG_TIMER_WAIT: 51455298
   MAX_TIMER_WAIT: 22532026212
       COUNT_READ: 748
   SUM_TIMER_READ: 3029876964
   MIN_TIMER_READ: 2198424
   AVG_TIMER_READ: 4050321
   MAX_TIMER_READ: 453108864
      COUNT_WRITE: 4
  SUM_TIMER_WRITE: 35664569265
  MIN_TIMER_WRITE: 112371075
  AVG_TIMER_WRITE: 8916142212
  MAX_TIMER_WRITE: 22532026212
      COUNT_FETCH: 748
  SUM_TIMER_FETCH: 3029876964
  MIN_TIMER_FETCH: 2198424
  AVG_TIMER_FETCH: 4050321
  MAX_TIMER_FETCH: 453108864
     COUNT_INSERT: 0
 SUM_TIMER_INSERT: 0
 MIN_TIMER_INSERT: 0
 AVG_TIMER_INSERT: 0
 MAX_TIMER_INSERT: 0
     COUNT_UPDATE: 3
 SUM_TIMER_UPDATE: 22792343883
 MIN_TIMER_UPDATE: 112371075
 AVG_TIMER_UPDATE: 7597447683
 MAX_TIMER_UPDATE: 22532026212
     COUNT_DELETE: 1
 SUM_TIMER_DELETE: 12872225382
 MIN_TIMER_DELETE: 12872225382
 AVG_TIMER_DELETE: 12872225382
 MAX_TIMER_DELETE: 12872225382
1 row in set (0.01 sec)
```

如下 SQL 语句可查询按索引分组的一个表的 I/O：

```
mysql> select  index_name,count_read,count_update,count_delete from
performance_schema.table_io_waits_summary_by_index_usage
where object_schema = 'sakila' and object_name = 'actor';
+---------------------+------------+--------------+--------------+
| index_name          | count_read | count_update | count_delete |
+---------------------+------------+--------------+--------------+
| PRIMARY             |        101 |            1 |            1 |
| idx_actor_last_name |         41 |            2 |            0 |
| NULL                |        606 |            0 |            0 |
+---------------------+------------+--------------+--------------+
3 rows in set (0.02 sec)
```

可以看到，该表有两个索引，一个是主键，另一个是二级索引，分成三类进行统计，其中 index_

name 为 NULL 的记录是不使用索引进行访问的信息，经过计算可以发现，这三类字段的汇总值等于视图 table_io_waits_summary_by_table 中对应的字段值。该视图里还有一些其他字段，例如，与 I/O 用时相关的字段，这里为了节约篇幅只列出了一小部分字段。

如上两个视图可以帮助用户监测表的繁忙程度，例如，可以把访问多的表迁移到较快的磁盘上。

6.7.6 监控文件 I/O

文件 I/O 记录的是访问磁盘的 I/O 信息，不包括对缓存的访问，这和前面介绍的表 I/O 不同，在 performance_schema 数据库中有如下三个视图可以用于获得文件 I/O 的信息。

- events_waits_summary_global_by_event_name：记录着按事件名进行汇总的事件等待信息，查询事件名以 wait/io/file/开头的事件可以查询到文件 I/O 的信息，这样的事件名有 51 个，例如，wait/io/file/sql/relaylog 记录着中继日志的 I/O 信息，这些事件的计量默认都是激活的。
- file_summary_by_event_name：记录内容和 events_waits_summary_global_by_event_name 类似，也记录了 51 类文件 I/O 的事件信息，只是更加详细。
- file_summary_by_instance：记录着按硬盘上实际文件进行分组的文件 I/O 的信息。

如上三个视图都很有用，如果要查询某个事件的 I/O 信息，如二进制日志的 I/O 信息，可以查询前面两个视图中的 wait/io/file/sql/binlog 事件；如果要查询某个具体文件的 I/O 信息，可以查询第三个视图，如二进制日志文件，该视图里包括所有二进制日志文件的 I/O 信息。

events_waits_summary_global_by_event_name 中记录了 I/O 的汇总信息，而 file_summary_by_event_name 和 file_summary_by_instance 把 I/O 分成了三类进行统计：读、写和杂项，前两类好理解，杂项是指创建、打开、关闭、刷新和访问元数据等。

例如，如下 SQL 语句可找出系统中 I/O 最繁忙的事件：

```
mysql> select * from
performance_schema.events_waits_summary_global_by_event_name
where event_name like 'wait/io/file/%'
order by sum_timer_wait desc limit 1\G
*************************** 1. row ***************************
    EVENT_NAME: wait/io/file/innodb/innodb_log_file
    COUNT_STAR: 6657
SUM_TIMER_WAIT: 4757510679504
MIN_TIMER_WAIT: 949509
AVG_TIMER_WAIT: 714662523
MAX_TIMER_WAIT: 137717550972
1 row in set (0.04 sec)
```

可以看到，最繁忙的 I/O 事件是对 InnoDB 日志的访问，使用如下 SQL 语句可查看对具体 InnoDB 日志文件 I/O 信息：

```
mysql> select file_name,count_star,sum_timer_wait from
file_summary_by_instance where
```

```
event_name = 'wait/io/file/innodb/innodb_log_file';
+------------------------------+------------+-----------------+
| FILE_NAME                    | COUNT_STAR | SUM_TIMER_WAIT  |
+------------------------------+------------+-----------------+
| /var/lib/mysql/ib_logfile0   |         51 |     49819222269 |
| /var/lib/mysql/ib_logfile1   |       6606 |   4707691457235 |
+------------------------------+------------+-----------------+
2 rows in set (0.00 sec)
```

可以看到,有两个 InnoDB 日志文件,其中第二个文件使用较多,这两个文件的 I/O 汇总正好等于 events_waits_summary_global_by_event_name 中记录的 wait/io/file/innodb/innodb_log_file 事件的 I/O 信息。

通过查询这些 I/O 视图,可以找出系统的性能瓶颈,可能是二进制日志、InnoDB 日志、中继日志、慢查询日志或者某个具体的表空间文件。

访问 InnoDB 表空间的事件名是 wait/io/file/innodb/innodb_data_file,使用该事件名可找出 MySQL 中最繁忙的 10 个表空间,相关命令及其运行结果如下:

```
mysql> select file_name from file_summary_by_instance
where event_name = 'wait/io/file/innodb/innodb_data_file'
order by sum_timer_wait desc limit 10;
+----------------------------------------+
| file_name                              |
+----------------------------------------+
| /var/lib/mysql/undo_001                |
| /var/lib/mysql/mysql.ibd               |
| /var/lib/mysql/undo_002                |
| /var/lib/mysql/ibtmp1                  |
| /var/lib/mysql/ibdata1                 |
| /var/lib/mysql/sakila/actor.ibd        |
| /var/lib/mysql/sakila/film_actor.ibd   |
| /var/lib/mysql/sakila/store.ibd        |
| /var/lib/mysql/mysql/backup_history.ibd|
| /var/lib/mysql/sbtest/sbtest1.ibd      |
+----------------------------------------+
10 rows in set (0.00 sec)
```

通过查询这些 I/O 视图有时还可以帮助用户发现设置上的错误,例如,通用查询日志通常是不打开的,这时事件名 wait/io/file/sql/query_log 对应的 I/O 计量应该是零,如果打开后忘了关闭,检查时会发现该事件对应的 I/O 计量不是零,相关命令及其运行结果如下:

```
mysql> select count_star,format_pico_time(sum_timer_wait) time_sum
from performance_schema.file_summary_by_event_name
where event_name = 'wait/io/file/sql/query_log';
+------------+-----------+
| count_star | time_sum  |
+------------+-----------+
|       7275 | 515.33 ms |
+------------+-----------+
1 row in set (0.00 sec)
```

如上检查结果和如下代码按文件名进行检查的结果一致：

```
mysql > select event_name,count_star,
format_pico_time(sum_timer_wait) time_sum
from performance_schema.file_summary_by_instance
where file_name like '/var/lib/mysql/scutech.log';
+----------------------------+------------+------------+
| event_name                 | count_star | time_sum   |
+----------------------------+------------+------------+
| wait/io/file/sql/query_log |       7276 | 515.49 ms  |
+----------------------------+------------+------------+
1 row in set (0.00 sec)
```

这里的 count_star 比前面的多 1，time_sum 也略多，这是由于多执行一个 SQL 语句造成的。

6.7.7 查询连接情况

如下 SQL 语句为按账号查询连接的情况：

```
mysql > select * from accounts;
+-----------------+-----------+---------------------+-------------------+
| USER            | HOST      | CURRENT_CONNECTIONS | TOTAL_CONNECTIONS |
+-----------------+-----------+---------------------+-------------------+
| NULL            | NULL      |                  40 |                51 |
| event_scheduler | localhost |                   1 |                 1 |
| root            | localhost |                   1 |                19 |
+-----------------+-----------+---------------------+-------------------+
3 rows in set (0.00 sec)
```

如下 SQL 语句为查询连接的属性：

```
mysql > select * from session_account_connect_attrs;
+----------------+-----------------+------------+------------------+
| PROCESSLIST_ID | ATTR_NAME       | ATTR_VALUE | ORDINAL_POSITION |
+----------------+-----------------+------------+------------------+
|             37 | _pid            | 3372       |                0 |
|             37 | _platform       | x86_64     |                1 |
|             37 | _os             | Linux      |                2 |
|             37 | _client_name    | libmysql   |                3 |
|             37 | os_user         | root       |                4 |
|             37 | _client_version | 8.0.26     |                5 |
|             37 | program_name    | mysql      |                6 |
+----------------+-----------------+------------+------------------+
7 rows in set (0.00 sec)
```

6.7.8 其他监控信息

关于事务的信息可以在如下表中查询到：

- events_transactions_current
- events_transactions_history
- events_transactions_history_long
- events_transactions_summary_by_account_by_event_name
- events_transactions_summary_by_host_by_event_name
- events_transactions_summary_by_thread_by_event_name
- events_transactions_summary_by_user_by_event_name
- events_transactions_summary_global_by_event_name

例如，如下命令可以查询当前的活动事务：

```
mysql > select thread_id, event_name, state, access_mode, isolation_level from events_
transactions_current where state = 'active';
+-----------+-------------+---------+-------------+-----------------+
| thread_id | event_name  | state   | access_mode | isolation_level |
+-----------+-------------+---------+-------------+-----------------+
|      3241 | transaction | ACTIVE  | READ WRITE  | REPEATABLE READ |
+-----------+-------------+---------+-------------+-----------------+
1 row in set (0.00 sec)
```

关于内存的使用情况可以在如下表中查询到：

- memory_summary_by_account_by_event_name
- memory_summary_by_host_by_event_name
- memory_summary_by_thread_by_event_name
- memory_summary_by_user_by_event_name
- memory_summary_global_by_event_name

performance_schema 数据库可以监控的信息包括 MySQL 运行的各个方面，它有一千多种性能计量，保存在一百多个表中，这里只是列出了常用的性能诊断信息，用户可以根据自己的需要在日常工作中逐步熟悉这些性能诊断信息。

6.8 实验

使用 performance_schema 视图监控元数据锁

试用元数据锁阻止 DDL 语句，DDL 语句又阻止对同一个对象的查询，观察这个过程中 performance_schema 视图记录的相关信息。

视频演示

第7章

sys数据库

第 5 章和第 6 章分别介绍了 information_schema 数据库和 performance_schema 数据库，本章介绍 sys 数据库，它是建立在前面两个数据库基础上的数据库，它包括一些视图、存储过程和函数，用于方便 DBA(database administrator，数据库管理员)、开发人员和操作人员的日常工作。

7.1 简介

sys 数据库的前身是 ps_helper 项目，最开始由 Mark Leith 开发。在 MySQL 8.0.18 版本之前，sys 数据库的源程序和 MySQL 的源程序是分开维护的，它位于 https://github.com/mysql/mysql-sys。sys 数据库主要由一组视图、存储过程和函数组成，运行如下代码可查询它的版本和 MySQL 的版本：

```
mysql > select * from sys.version;
+-------------+---------------+
| sys_version | mysql_version |
+-------------+---------------+
| 2.1.1       | 8.0.22        |
+-------------+---------------+
1 row in set (0.00 sec)
```

如下代码为查询 sys 数据库中的对象：

```
mysql > select * from sys.schema_object_overview  where db = 'sys';
+-----+---------------+-------+
| db  | object_type   | count |
+-----+---------------+-------+
| sys | BASE TABLE    |     1 |
| sys | FUNCTION      |    22 |
```

```
| sys    | INDEX (BTREE)     |    1  |
| sys    | PROCEDURE         |   26  |
| sys    | TRIGGER           |    2  |
| sys    | VIEW              |  100  |
+-----+---------------+-------+
6 rows in set (0.06 sec)
```

sys 数据库存在的目的是方便 DBA、开发人员和操作人员的日常工作。sys 数据库的视图主要建立在 performance_schema 数据库的表上，也用到了少量的 information_schema 数据库中的表。

为了使 sys 数据库收集的信息更完整，可以根据需要使用如下存储过程激活 performance_schema 数据库中非默认的计量和消费者：

```
mysql > call sys.ps_setup_enable_instrument('wait');
mysql > call sys.ps_setup_enable_instrument('stage');
mysql > call sys.ps_setup_enable_instrument('statement');
mysql > call sys.ps_setup_enable_consumer('current');
mysql > call sys.ps_setup_enable_consumer('history');
mysql > call sys.ps_setup_enable_consumer('history_long');
```

收集更多的信息将会消耗更多的系统资源，大多数情况下，默认的配置是恰当的，调用如下存储过程可以恢复默认的配置：

```
mysql > call sys.ps_setup_reset_to_default(true);
```

sys 数据库中的表除了 sys_config 外，都是视图，这些视图大部分是成对出现的，一个格式化了的视图是用于用户查询，对这样的视图大部分时候使用 select * 查询即可；另一个视图的名字是在前面的视图名前加上 x$，里面保存着原始数据，这类视图可用于程序访问，如 workbench 工具就是查询这类视图。这两类视图的底层数据是一样的，只是展现方式不同。

7.2 配置参数

sys 数据库的配置参数保存在 sys_config 表中，运行如下代码可得到该表中的参数和默认值：

```
mysql > select variable,value from sys.sys_config;
+-----------------------------------------+-------+
| variable                                | value |
+-----------------------------------------+-------+
| diagnostics.allow_i_s_tables            | OFF   |
| diagnostics.include_raw                 | OFF   |
| ps_thread_trx_info.max_length           | 65535 |
| statement_performance_analyzer.limit    | 100   |
| statement_performance_analyzer.view     | NULL  |
| statement_truncate_len                  | 64    |
+-----------------------------------------+-------+
6 rows in set (0.00 sec)
```

这些配置参数中，最有可能被修改的参数是 statement_truncate_len，它控制着 format_statement()函数返回的 SQL 语句的最大长度，sys 数据库中的很多视图都是调用该函数对输出的 SQL 语句进行格式化，为了让格式化后的 SQL 语句能适应屏幕的宽度，该参数的默认值是 64。用户可以直接修改 sys_config 表来改变参数的设置，由于 sys_config 表是一个普通的 InnoDB 表，修改的结果在 MySQL 下次启动后仍然有效。

每个 sys_config 表中的参数有一个对应的用户变量，用户变量名是在参数前面加上 sys. 的前缀。如参数 statement_truncate_len 对应的用户变量是 sys. statement_truncate_len。如果当前会话中设置了用户变量，用户变量的优先级比 sys_config 表中参数的优先级高。因此如果只想修改当前会话的参数就设置用户变量，如果要修改全局的参数就修改 sys_config 表。如下是 format_statement()函数格式化 SQL 语句的一个例子：

```
mysql > set @sql = 'select actor_id, first_name, last_name from sakila. actor';
Query OK, 0 rows affected (0.00 sec)

mysql > select sys. format_statement(@sql);
+-----------------------------------------------------------+
| sys. format_statement(@sql)                               |
+-----------------------------------------------------------+
| select actor_id, first_name, last_name from sakila. actor |
+-----------------------------------------------------------+
1 row in set (0.00 sec)
```

可以看到，在默认的设置下，format_statement 函数完整地输出该 SQL 语句，下面把用户变量 sys. statement_truncate_len 设置成 32：

```
mysql > set @sys. statement_truncate_len = 32;
query ok, 0 rows affected (0.00 sec)
```

再次调用 format_statement()函数输出该 SQL 语句如下：

```
mysql > select sys. format_statement(@sql);
+-----------------------------------+
| sys. format_statement(@sql)       |
+-----------------------------------+
| select actor_i ... m sakila. actor |
+-----------------------------------+
1 row in set (0.00 sec)
```

这时 SQL 语句中的部分内容被用省略号代替，总长度被限制在 32 位。

7.3 存储过程

sys 数据库中包含了一些实用的存储过程和函数，这些存储过程和函数的使用方法在 information_schema. routines 表的 routine_comment 字段中可以查到，下面介绍常用的存储过程。

1. ps_setup_reset_to_default() 存储过程

把 performance_schema 的收集性能诊断设置恢复到默认设置。收集性能诊断信息对

MySQL 的性能影响较大，通常默认设置是合理的，如果进行了修改，可以使用该存储过程在诊断完成后恢复默认值。该存储过程只有一个布尔型的输入参数，指定是否显示设置过程中的信息。

2. diagnostics()存储过程

diagnostics()存储过程会生成一个关于当前 MySQL 实例整体性能的诊断报告，该存储过程有如下 3 个参数。

(1) in_max_runtime：以秒为单位设置最长运行时间，设置为 null 时，默认是 60 秒。

(2) in_interval：以秒为单位设置两次收集之间的休眠时间，设置为 null 时，默认是 30 秒。

(3) in_auto_config：设置启用计量和消费者。该参数是枚举类型，包括：

- current，保持当前的设置不变。
- medium，启用部分计量和消费者。
- full，启用全部计量和消费者。

影响 diagnostics()存储过程执行行为的有 sys_config 表里的配置选项或其相应的用户定义变量。

- debug 或@sys. debug：如果此选项为 ON，则产生调试输出。默认值为 OFF。
- diagnostics. allow_i_s_tables 或@sys. diagnostics. allow_i_s_tables：如果此选项为 ON，diagnostics()将被允许扫描 information_schema. tables 里的表，如果有很多表，该操作的成本可能很高。默认值为 OFF。
- diagnostics. include_raw 或@sys. diagnostics. include_raw：如果此选项为 ON，则 diagnostics()过程输出包括查询 metrics 视图的原始输出，默认值为 OFF。
- statement_truncate_len 或@sys. statement_truncate_len：该配置选项前面已经介绍，此处不再赘述。

报告的内容非常丰富，包括 sys 数据库中的很多视图(如 metrics 视图)，还包括这些视图在诊断报告收集期间的记录值的变化量。如下是调用 diagnostics()存储过程并将报告输出到文件的一个例子：

```
mysql > tee diagnostics.log
Logging to file 'diagnostics.log'
mysql > call sys.diagnostics(null,null,'current');
+--------------------+
| summary            |
+--------------------+
| Disabled 1 thread  |
+--------------------+
1 row in set (0.00 sec)

+-------------------------+---------------------------------------+
| Name                    | Value                                 |
+-------------------------+---------------------------------------+
| Hostname                | redhat7.novalocal                     |
| Port                    | 3306                                  |
```

```
| Socket                   | /var/lib/mysql/mysql.sock                |
| Datadir                  | /disk1/data/                             |
| Server UUID              | 38b9af85-a31b-11eb-a0cb-fa163e368ff4     |
| ------------------------ | ---------------------------------------- |
| MySQL Version            | 8.0.23                                   |
| Sys Schema Version       | 2.1.1                                    |
| Version Comment          | MySQL Community Server - GPL             |
| Version Compile OS       | Linux                                    |
| Version Compile Machine  | x86_64                                   |
| ------------------------ | ---------------------------------------- |
| UTC Time                 | 2021-07-25 01:22:56                      |
| Local Time               | 2021-07-25 09:22:56                      |
| Time Zone                | SYSTEM                                   |
| System Time Zone         | CST                                      |
| Time Zone Offset         | 08:00:00                                 |
+--------------------------+------------------------------------------+
...

Query OK, 0 rows affected (31.41 sec)
mysql> notee;
Outfile disabled.
```

3. execute_prepared_stmt()存储过程

execute_prepared_stmt()存储过程会把一个给定的字符串当成SQL语句来执行，如下是一个例子：

```
mysql> call sys.execute_prepared_stmt('select count(*) from sakila.actor');
+----------+
| COUNT(*) |
+----------+
|      200 |
+----------+
1 row in set (0.02 sec)

Query OK, 0 rows affected (0.02 sec)
```

4. 控制和显示收集诊断信息的存储过程

如下存储过程用于控制和显示收集诊断信息。

- ps_setup_disable_background_threads()：不启用所有后台线程的诊断信息收集。
- ps_setup_disable_consumer()：使用like匹配输入参数不启用指定的消费者。
- ps_setup_disable_instrument()：使用like匹配输入参数不启用指定的计量。
- ps_setup_disable_thread()：不启用指定线程的诊断信息收集。
- ps_setup_enable_background_threads()：启用所有后台线程的诊断信息收集。
- ps_setup_enable_consumer()：使用like匹配输入参数启动指定的消费者。
- ps_setup_enable_instrument()：使用like匹配输入参数启用指定的计量。

- ps_setup_enable_thread()：启用指定线程的诊断信息收集。
- ps_setup_show_disabled()：显示当前处于未启用状态的 performance_schema 数据库的设置。
- ps_setup_show_disabled_consumers()：显示当前未启用的消费者。
- ps_setup_show_disabled_instruments()：显示当前未启用的计量。
- ps_setup_show_enabled()：显示当前处于启用状态的 performance_schema 数据库的设置。
- ps_setup_show_enabled_consumers()：显示当前启用的消费者。
- ps_setup_show_enabled_instruments()：显示当前启用的计量。

5. 保存和恢复 performance_schema 数据库的设置

存储过程 sys.ps_setup_save()和 sys.ps_setup_reload_saved()分别用于保存和恢复 performance_schema 数据库的设置，这两个功能在需要临时对 performance_schema 数据库的设置进行调整时很有用，例如，在调整 performance_schema 数据库的设置之前，使用如下命令先保存设置：

```
mysql > call sys.ps_setup_save(10);
```

其中的输入参数 10 表示等待锁的时间为 10 秒。保存完设置后就可以进行 performance_schema 数据库设置的调整，调整完成后再使用如下命令恢复设置：

```
mysql > call sys.ps_setup_reload_saved();
```

6. ps_trace_statement_digest()存储过程

ps_trace_statement_digest()存储过程可以根据提供的 SQL 语句摘要哈希值跟踪收集这些 SQL 语句执行过程中的性能诊断信息，该存储过程有如下 5 个输入参数。

(1) in_digest：需要跟踪的 SQL 语句摘要的哈希值。

(2) in_runtime：跟踪这些 SQL 语句的时间跨度(以秒为单位)。

(3) in_interval：两次收集信息之间的间隔(以秒为单位)，可以是小数。

(4) in_start_fresh：布尔值，为 true 时，在开始时会截断 performance_schema 数据库的 events_statements_history_long 和 events_stages_history_long 表；为 false 时不截断。

(5) in_auto_enable：布尔值，为 true 时会自动激活需要的消费者；为 false 时不激活。

例如，可以使用 ps_trace_statement_digest()存储过程跟踪收集如下 SQL 语句的性能诊断信息：

```
select * from actor where last_name = 'DEAN';
select * from actor where last_name = 'ALLEN';
select * from actor where last_name = 'CLOSE';
```

如上 SQL 语句的摘要值是一样的，可以使用 statement_digest()函数获得 SQL 语句的摘要值，也可以在一些性能视图，如 events_statements_summary_by_digest 中查到 SQL 语句的摘要值，把 SQL 语句的摘要值赋给一个变量的代码如下：

```
mysql > set @digest = statement_digest('select * from actor where last_name = ''DEAN''');
```

下面的例子是在60秒内跟踪指定的SQL语句,每0.1秒收集一次性能诊断信息:

```
mysql> call sys.ps_trace_statement_digest(@digest, 60, 0.1, true, true);
+--------------------+
| summary            |
+--------------------+
| Disabled 1 thread  |
+--------------------+
1 row in set (0.00 sec)

+--------------------+
| SUMMARY STATISTICS |
+--------------------+
| SUMMARY STATISTICS |
+--------------------+
1 row in set (59.43 sec)

+------------+-----------+-----------+-----------+---------------+---------------+----
-----------+-----------+-----------+
| executions | exec_time | lock_time | rows_sent | rows_affected | rows_examined | tmp_
tables | full_scans |
+------------+-----------+-----------+-----------+---------------+---------------+----
-----------+-----------+-----------+
|          6 | 4.33 ms   | 1.70 ms   |          7 |             0 |
7 |          0 |          0 |
+------------+-----------+-----------+-----------+---------------+---------------+----
-----------+-----------+-----------+
1 row in set (59.43 sec)

+----------------------------------------------------+-------+-----------+
| event_name                                         | count | latency   |
+----------------------------------------------------+-------+-----------+
| stage/sql/statistics                               |     6 | 1.16 ms   |
| stage/sql/starting                                 |    12 | 1.04 ms   |
| stage/sql/executing                                |     6 | 690.42 us |
| stage/sql/Opening tables                           |     6 | 580.33 us |
| stage/sql/preparing                                |     6 | 175.80 us |
| stage/sql/freeing items                            |     6 | 141.26 us |
| stage/sql/optimizing                               |     6 | 121.50 us |
| stage/sql/System lock                              |     6 | 105.67 us |
| stage/sql/closing tables                           |     6 | 80.32 us  |
| stage/sql/waiting for handler commit               |     6 | 58.87 us  |
| stage/sql/checking permissions                     |     6 | 51.54 us  |
| stage/sql/init                                     |     6 | 45.50 us  |
| stage/sql/Executing hook on transaction begin.     |     6 | 14.67 us  |
| stage/sql/end                                      |     6 | 10.97 us  |
| stage/sql/query end                                |     6 | 10.54 us  |
```

```
| stage/sql/cleaning up                                  |      6  | 5.35 us     |
+--------------------------------------------------------+---------+-------------+
16 rows in set (59.44 sec)

+----------------------------+
| LONGEST RUNNING STATEMENT  |
+----------------------------+
| LONGEST RUNNING STATEMENT  |
+----------------------------+
1 row in set (59.44 sec)

+-----------+-----------+-----------+-----------+---------------+-----
----------+------------+-----------+
| thread_id | exec_time | lock_time | rows_sent | rows_affected | rows_examined | tmp_tables |
full_scan |
+-----------+-----------+-----------+-----------+---------------+-----
----------+------------+-----------+
|        59 | 1.05 ms   | 373.00 us |         2 |             0 |            2 |
  0 |         0 |
+-----------+-----------+-----------+-----------+---------------+-----
----------+------------+-----------+
1 row in set (59.44 sec)

+------------------------------------------+
| sql_text                                 |
+------------------------------------------+
| select * from actor where last_name = 'DEAN'  |
+------------------------------------------+
1 row in set (59.44 sec)

+------------------------------------------------------+-----------+
| event_name                                           | latency   |
+------------------------------------------------------+-----------+
| stage/sql/starting                                   | 208.58 us |
| stage/sql/Executing hook on transaction begin.       | 5.81 us   |
| stage/sql/starting                                   | 13.62 us  |
| stage/sql/checking permissions                       | 7.74 us   |
| stage/sql/Opening tables                             | 135.56 us |
| stage/sql/init                                       | 14.20 us  |
| stage/sql/System lock                                | 24.63 us  |
| stage/sql/optimizing                                 | 13.40 us  |
| stage/sql/statistics                                 | 314.94 us |
| stage/sql/preparing                                  | 31.54 us  |
| stage/sql/executing                                  | 175.76 us |
| stage/sql/end                                        | 1.66 us   |
| stage/sql/query end                                  | 1.72 us   |
| stage/sql/waiting for handler commit                 | 8.95 us   |
```

```
| stage/sql/closing tables                                | 24.29 us    |
| stage/sql/freeing items                                 | 45.79 us    |
| stage/sql/cleaning up                                   | 750.00 ns   |
+---------------------------------------------------------+-------------+
17 rows in set (59.44 sec)

+-------------------+
| summary           |
+-------------------+
| Enabled 1 thread  |
+-------------------+
1 row in set (59.45 sec)

Query OK, 0 rows affected (59.45 sec)
```

7. statement_performance_analyzer()存储过程

statement_performance_analyzer()存储过程可以生成当前 MySQL 实例中正在运行的 SQL 语句的两个快照，并对比这两个快照，生成增量报告。

statement_performance_analyzer 存储过程使用如下 3 个输入参数。

(1) in_action 参数是枚举类型，包括如下成员。

- snapshot：创建快照，默认基于 events_statements_summary_by_digest 视图，可通过 in_table 参数修改，快照信息保存在 sys. tmp_digests 临时表中。
- overall：基于 in_table 参数指定的表生成分析报告。
- delta：生成增量分析报告，增量是基于 in_table 参数与已有的快照进行对比生成。
- create_tmp：创建临时表，可用于后续计算增量。
- create_table：创建普通表，可用于后续计算增量。
- save：保存快照到 in_table 参数指定的表中。
- cleanup：移除用于快照和增量的临时表。

(2) in_table varchar(129)：指定需要表名的操作。

(3) in_views：指定操作包含的性能视图，该参数是集合类型，可以指定下面的一个或多个成员，默认是包括除了 custom 之外的所有成员。

- with_runtimes_in_95th_percentile：使用 statements_with_runtimes_in_95th_percentile 视图。
- analysis：使用 statement_analysis 视图。
- with_errors_or_warnings：使用 statements_with_errors_or_warnings 视图。
- with_full_table_scans：使用 statements_with_full_table_scans 视图。
- with_sorting：使用 statements_with_sorting 视图。
- with_temp_tables：使用 statements_with_temp_tables 视图。
- custom：使用配置项 statement_performance_analyzer. view 指定视图或一个查询语句。

sys 库的配置选项 statement_performance_analyzer. limit 或用户变量@sys. statement_performance_analyzer. limit 指定生成报告里包括的 SQL 语句最多数量，默认是 100。

statement_performance_analyzer()存储过程的典型应用是创建两个快照，并生成这个两个快照之间的变化报告。下面是整个执行过程的一个例子。

(1) 不收集当前线程的性能计量，以免干扰生成报告的准确性，命令如下：

```
mysql > call sys.ps_setup_disable_thread(connection_id());
```

(2) 创建存放快照的 schema，命令如下：

```
mysql > create database if not exists analysisdb;
```

(3) 创建保存快照的临时表，命令如下：

```
mysql > call sys.statement_performance_analyzer('create_tmp', 'analysisdb.tmp_ana', null);
```

(4) 生成初始化快照，代码如下：

```
mysql > call sys.statement_performance_analyzer('snapshot', null, null);
```

(5) 把初始化快照保存到临时表中，代码如下：

```
mysql > call sys.statement_performance_analyzer('save', 'analysisdb.tmp_ana', null);
```

(6) 在当前会话中等待一分钟，代码如下：

```
mysql > do sleep(60);
```

(7) 在这段时间，在其他会话中向数据施加一些负载，例如：

```
$ mysqlslap --concurrency=5 --iterations=200 --number-int-cols=2 --number-char
-cols=3 --auto-generate-sql
```

(8) 创建新快照，代码如下：

```
mysql > call sys.statement_performance_analyzer('snapshot', null, null);
```

(9) 基于新快照与初始化快照进行增量性能分析，代码如下：

```
mysql > call sys.statement_performance_analyzer('delta', 'analysisdb.tmp_ana', 'analysis')\G
*************************** 1. row ***************************
Next Output: Top 100 Queries Ordered by Total Latency
1 row in set (0.01 sec)

*************************** 1. row ***************************
         query: INSERT INTO 't1' VALUES (...)
            db: mysqlslap
     full_scan:
    exec_count: 25800
     err_count: 0
    warn_count: 0
 total_latency: 1.05 min
   max_latency: 75.22 ms
   avg_latency: 2.45 ms
  lock_latency: 1.84 s
     rows_sent: 0
...
```

这一部分是这个存储过程的主体部分，输出了这个时间的TOP SQL，不超过100个，可以看到在这段时间执行最多的SQL是insert语句，执行了25800次，用时1.05分钟，还有其他一些信息。通过这部分的报告，可以了解这段时间实例执行的TOP SQL的情况，方便找到系统的瓶颈进行针对性优化。

（10）清理分析过程中产生的临时表，代码如下：

```
mysql > call sys.statement_performance_analyzer('cleanup', null, null);
```

（11）删除创建的临时表，代码如下：

```
mysql > drop temporary table analysisdb.tmp_ana;
```

（12）激活当前线程的性能计量收集，代码如下：

```
mysql > call sys.ps_setup_enable_thread(connection_id());
```

8. ps_trace_thread()存储过程

ps_trace_thread()存储过程可以跟踪某个线程的执行过程，把该线程执行的所有SQL语句的性能信息都记录下来，并输出报告。该存储过程适合用于执行存储过程或多个SQL语句的线程。输入参数如下。

- in_thread_id INT：被跟踪的线程号。
- in_outfile VARCHAR(255)：输出的包含性能信息的文件，该文件是DOT格式的文件，可以通过graphviz工具集转换成图片或pdf文档。
- in_max_runtime DECIMAL(20,2)：跟踪和收集性能信息的时间长度，默认为60秒。
- in_interval DECIMAL(20,2)：两次收集性能信息之间的时间间隔，默认为1秒。
- in_start_fresh BOOLEAN：布尔值，在收集性能信息之前是否重置性能数据，建议设置为true，生成报告不受历史数据的干扰。
- in_auto_setup BOOLEAN：布尔值，是否激活所有用到的计量和消费者，这些修改将在该存储过程执行完成后恢复。设置为true时对系统性能的影响很大，通常设置为false。
- in_debug BOOLEAN：布尔值，是否激活debug模式。通常设置为false，debug模式的报告通常是给内部开发人员看的。

为了将DOT格式的文件转换成图片，需要先安装graphviz工具集，在RedHat上的安装命令如下：

```
$ sudo yum install graphviz
```

下面是一个例子。

首先手工激活消费者events_statements_history_long，也可以把输入参数in_auto_setup设置为true，自动激活所有用到的计量和消费者，但这样跟踪的信息很多，对系统的性能影响较大，生成的图片也很大，代码如下：

```
mysql > update performance_schema.setup_consumers set enabled = 'yes' where name = 'events_
statements_history_long';
```

获取需要跟踪的线程号，代码如下：

```
mysql> select ps_current_thread_id();
+-------------------------+
| ps_current_thread_id()  |
+-------------------------+
|                     51  |
+-------------------------+
1 row in set (0.00 sec)
```

在另外一个会话里采用当前的配置启动对51号线程的跟踪，跟踪60秒，每秒收集一次性能信息，生成性能报告文件，代码如下：

```
mysql> call sys.ps_trace_thread(51,'/tmp/td_51.dot',null, null, true,false, false);

+-------------------+
| summary           |
+-------------------+
| Disabled 1 thread |
+-------------------+
1 row in set (0.01 sec)

+--------------------------------------------+
| Info                                       |
+--------------------------------------------+
| Data collection starting for THREAD_ID = 51 |
+--------------------------------------------+
1 row in set (0.06 sec)

+-----------------------------------+
| Info                              |
+-----------------------------------+
| Stack trace written to /tmp/td_51.dot |
+-----------------------------------+
1 row in set (1 min 0.16 sec)

+----------------------------------------------+
| Convert to PDF                               |
+----------------------------------------------+
| dot -Tpdf -o /tmp/stack_51.pdf /tmp/td_51.dot |
+----------------------------------------------+
1 row in set (1 min 0.16 sec)

+----------------------------------------------+
| Convert to PNG                               |
+----------------------------------------------+
| dot -Tpng -o /tmp/stack_51.png /tmp/td_51.dot |
+----------------------------------------------+
```

```
1 row in set (1 min 0.16 sec)

+-------------------+
| summary           |
+-------------------+
| Enabled 1 thread  |
+-------------------+
1 row in set (1 min 0.18 sec)

Query OK, 0 rows affected (1 min 0.18 sec)
```

在收集过程中，在被跟踪锁线程上执行如下 SQL 语句：

```
mysql> select * from sakila.actor where last_name = 'BALL';
```

根据给出的操作方法生成.png 和.pdf 文件，代码如下：

```
$ dot -Tpdf -o /tmp/stack_51.pdf /tmp/td_51.dot
$ dot -Tpng -o /tmp/stack_51.png /tmp/td_51.dot
```

.png 和.pdf 文件的内容一样，.png 文件如图 7.1 所示。

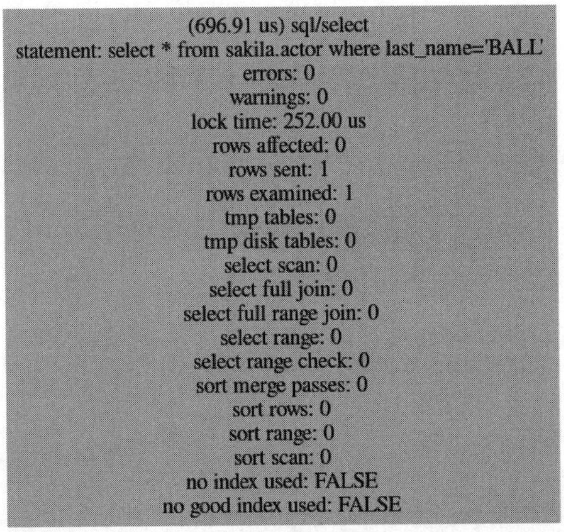

(696.91 us) sql/select
statement: select * from sakila.actor where last_name='BALL'
errors: 0
warnings: 0
lock time: 252.00 us
rows affected: 0
rows sent: 1
rows examined: 1
tmp tables: 0
tmp disk tables: 0
select scan: 0
select full join: 0
select full range join: 0
select range: 0
select range check: 0
sort merge passes: 0
sort rows: 0
sort range: 0
sort scan: 0
no index used: FALSE
no good index used: FALSE

图 7.1

根据当前的设置，该报告里只包括了 SQL 语句事件。如果要激活所有用到的计量和消费者对线程的跟踪使用如下命令：

```
mysql> call sys.ps_trace_thread(54,concat('/var/lib/mysql-files/thread-', replace(now(),
' ', '-'), '.dot'), null, null, true, true, false);
```

该命令与前面命令的不同之处是倒数第二个参数是 true，这样生成的.png 和.pdf 文件内容要丰富得多，不仅包括了 SQL 语句事件，还包括阶段事件和等待事件，因为篇幅所限只截取了生成图形的一部分，如图 7.2 所示。

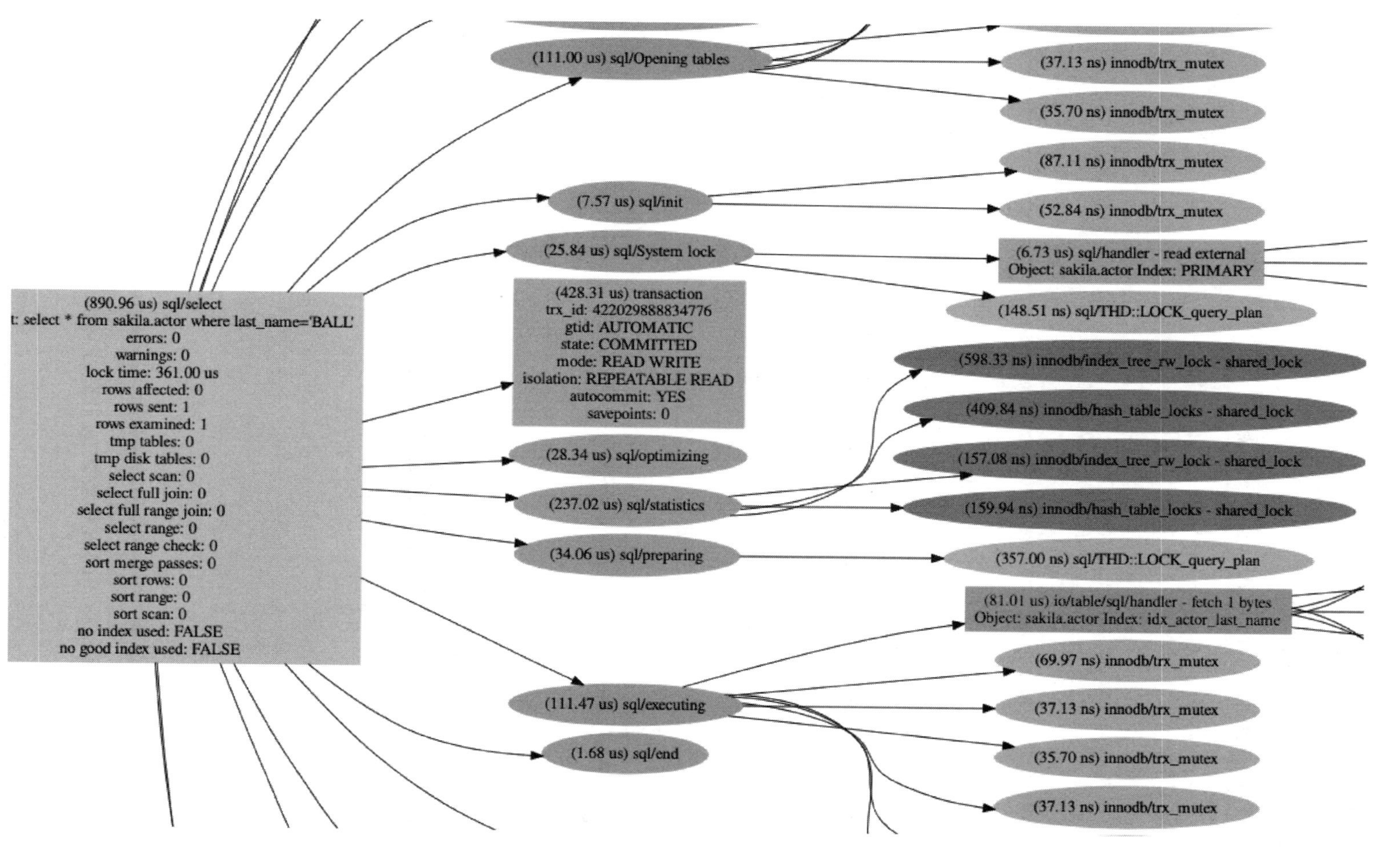
(890.96 us) sql/select
t: select * from sakila.actor where last_name='BALL'
errors: 0
warnings: 0
lock time: 361.00 us
rows affected: 0
rows sent: 1
rows examined: 1
tmp tables: 0
tmp disk tables: 0
select scan: 0
select full join: 0
select full range join: 0
select range: 0
select range check: 0
sort merge passes: 0
sort rows: 0
sort range: 0
sort scan: 0
no index used: FALSE
no good index used: FALSE
(111.00 us) sql/Opening tables
(7.57 us) sql/init
(25.84 us) sql/System lock
(428.31 us) transaction
trx_id: 422029888834776
gtid: AUTOMATIC
state: COMMITTED
mode: READ WRITE
isolation: REPEATABLE READ
autocommit: YES
savepoints: 0
(28.34 us) sql/optimizing
(237.02 us) sql/statistics
(34.06 us) sql/preparing
(111.47 us) sql/executing
(1.68 us) sql/end
(37.13 ns) innodb/trx_mutex
(35.70 ns) innodb/trx_mutex
(87.11 ns) innodb/trx_mutex
(52.84 ns) innodb/trx_mutex
(6.73 us) sql/handler - read external
Object: sakila.actor Index: PRIMARY
(148.51 ns) sql/THD::LOCK_query_plan
(598.33 ns) innodb/index_tree_rw_lock - shared_lock
(409.84 ns) innodb/hash_table_locks - shared_lock
(157.08 ns) innodb/index_tree_rw_lock - shared_lock
(159.94 ns) innodb/hash_table_locks - shared_lock
(357.00 ns) sql/THD::LOCK_query_plan
(81.01 us) io/table/sql/handler - fetch 1 bytes
Object: sakila.actor Index: idx_actor_last_name
(69.97 ns) innodb/trx_mutex
(37.13 ns) innodb/trx_mutex
(35.70 ns) innodb/trx_mutex
(37.13 ns) innodb/trx_mutex

图 7.2

7.4 函数

sys数据库中常用的函数如下。

1. extract_schema_from_file_name()函数

给定一个带路径的文件名，extract_schema_from_file_name()函数将返回该文件对应的schema名，例如：

```
mysql> select
sys.extract_schema_from_file_name('/var/lib/mysql/sakila/actor.ibd');
+--------------------------------------------------------------------+
| sys.extract_schema_from_file_name('/var/lib/mysql/sakila/actor.ibd') |
+--------------------------------------------------------------------+
| sakila                                                             |
+--------------------------------------------------------------------+
1 row in set (0.01 sec)
```

2. extract_table_from_file_name()函数

给定一个带路径的文件名，extract_table_from_file_name()函数将返回该文件对应的表名，例如：

```
mysql> select
extract_table_from_file_name('/var/lib/mysql/sakila/actor.ibd');
+--------------------------------------------------------------------+
| extract_table_from_file_name('/var/lib/mysql/sakila/actor.ibd')     |
+--------------------------------------------------------------------+
| actor                                                              |
+--------------------------------------------------------------------+
1 row in set (0.00 sec)
```

3. format_bytes()函数

format_bytes()函数把以字节计数的数字转换成用户可阅读的格式输出，例如：

```
mysql> select sys.format_bytes(20000), sys.format_bytes(999999999);
+-------------------------+-----------------------------+
| sys.format_bytes(20000) | sys.format_bytes(999999999) |
+-------------------------+-----------------------------+
| 19.53 KiB               | 953.67 MiB                  |
+-------------------------+-----------------------------+
1 row in set (0.00 sec)
```

4. format_time() 函数

format_time() 函数把输入的皮秒转换成用户可阅读的时间，转换结果是数值加上时间单位，例如：

```
mysql> select sys.format_time(2399), sys.format_time(90327840099830);
+-----------------------+---------------------------------+
```

```
| sys.format_time(2399)     | sys.format_time(90327840099830)     |
+---------------------------+-------------------------------------+
| 2.4 ns                    | 1.51 m                              |
+---------------------------+-------------------------------------+
1 row in set (0.00 sec)
```

format_time()函数在 MySQL 8.0.16 以后被废弃，建议使用 format_pico_time()函数。

5. format_pico_time()函数

format_pico_time()函数和 format_time() 函数功能类似，把输入的皮秒转换成用户可阅读的时间，例如：

```
mysql> select format_pico_time(2399), format_pico_time(90327840099830);
+---------------------------+-------------------------------------+
| format_pico_time(2399)    | format_pico_time(90327840099830)    |
+---------------------------+-------------------------------------+
| 2.40 ns                   | 1.51 min                            |
+---------------------------+-------------------------------------+
1 row in set (0.00 sec)
```

format_pico_time()函数很重要，因为性能视图里的时间很多都是皮秒，在输出时可以使用该函数转换成符合阅读习惯的格式。

6. list_add()函数和 list_drop()函数

list_add()函数可以向一个以逗号分隔的清单中增加条目，该功能经常用于对 sql_mode 的设置，例如，检查当前 sql_mode 的设置如下：

```
mysql> select @@sql_mode\G
*************************** 1. row ***************************
@@sql_mode:
ONLY_FULL_GROUP_BY,STRICT_TRANS_TABLES,NO_ZERO_IN_DATE,NO_ZERO_DATE,ERROR_FOR_DIVISION_BY
_ZERO
1 row in set (0.00 sec)
```

向 sql_mode 中增加一个条目的代码如下：

```
mysql> set @@sql_mode = sys.list_add(@@sql_mode, 'NO_ENGINE_SUBSTITUTION');
Query OK, 0 rows affected (0.00 sec)
```

再次检查当前 sql_mode 的设置如下：

```
mysql> select @@sql_mode\G
*************************** 1. row ***************************
@@sql_mode:
ONLY_FULL_GROUP_BY,STRICT_TRANS_TABLES,NO_ZERO_IN_DATE,NO_ZERO_DATE,ERROR_FOR_DIVISION_BY_
ZERO,NO_ENGINE_SUBSTITUTION
1 row in set (0.01 sec)
```

list_drop()函数的功能和 list_add()函数正好相反，用于从一个以逗号分隔的清单中减少条目。

7. 查询 performance_schema 数据库配置的函数

- ps_is_account_enabled()：查询对某个账户是否启用性能诊断信息收集。
- ps_is_consumer_enabled()：查询对某个消费者是否启用性能诊断信息收集。
- ps_is_instrument_default_enabled()：查询是否默认启用某个计量。
- ps_is_instrument_default_timed()：查询是否默认启用某个计量的计时。
- ps_is_thread_instrumented()：查询对某个连接是否启用性能诊断信息收集。
- ps_thread_account()：查询对某个账户是否启用性能诊断信息收集。
- ps_thread_stack()：以 JSON 格式返回指定线程的 SQL 语句、事件和阶段信息。
- ps_thread_trx_info()：以 JSON 格式返回指定线程的事务和 SQL 语句信息。

8. 查询 MySQL 版本号的函数

查询 MySQL 版本号的函数有如下 3 个。

- version_major()：返回当前 MySQL 的大版本号。
- version_minor()：返回当前 MySQL 的小版本号。
- version_patch()：返回当前 MySQL 的补丁号。

下面是一个使用的例子：

```
mysql > select version(), version_major(),version_minor(),version_patch() \G
*************************** 1. row ***************************
      VERSION(): 8.0.22
version_major(): 8
version_minor(): 0
version_patch(): 22
1 row in set (0.01 sec)
```

7.5 视图

7.5.1 视图清单

sys 数据库中的视图清单如下。

- host_summary、x$host_summary：按主机分组的语句、文件 I/O 和连接信息。
- host_summary_by_file_io、x$host_summary_by_file_io：按主机分组的文件 I/O 信息。
- host_summary_by_file_io_type、x$host_summary_by_file_io_type：按主机和事件类型分组的文件 I/O 信息。
- host_summary_by_stages、x$host_summary_by_stages：按主机分组的语句阶段。
- host_summary_by_statement_latency、x$host_summary_by_statement_latency：按主机分组的语句统计信息。
- host_summary_by_statement_type、x$host_summary_by_statement_type：按主机和语句类型统计的执行语句的汇总信息。
- innodb_buffer_stats_by_schema、x$innodb_buffer_stats_by_schema：按 schema

分组的 InnoDB 缓存信息。

- innodb_buffer_stats_by_table、x$innodb_buffer_stats_by_table：按 schema 和表分组的 InnoDB 缓存信息。
- innodb_lock_waits、x$innodb_lock_waits：InnoDB 锁信息。
- io_by_thread_by_latency、x$io_by_thread_by_latency：按线程分组的 I/O 信息。
- io_global_by_file_by_bytes、x$io_global_by_file_by_bytes：按文件和字节数分组的全局 I/O 消费者。
- io_global_by_file_by_latency、x$io_global_by_file_by_latency：按文件和延迟分组的全局 I/O 消费者。
- io_global_by_wait_by_bytes、x$io_global_by_wait_by_bytes：按字节数分组的全局 I/O 消费者。
- io_global_by_wait_by_latency、x$io_global_by_wait_by_latency：按延迟分组的全局 I/O 消费者。
- latest_file_io、x$latest_file_io：按文件和线程分组的最近 I/O 信息。
- memory_by_host_by_current_bytes、x$memory_by_host_by_current_bytes：按主机分组的内存使用情况。
- memory_by_thread_by_current_bytes、x$memory_by_thread_by_current_bytes：按线程分组的内存使用情况。
- memory_by_user_by_current_bytes、x$memory_by_user_by_current_bytes：按用户分组的内存使用情况。
- memory_global_by_current_bytes、x$memory_global_by_current_bytes：按类型分组的内存使用情况。
- memory_global_total、x$memory_global_total：总内存使用情况。
- metrics：服务计量信息。
- processlist、x$processlist：线程清单信息。
- ps_check_lost_instrumentation：有关丢失的计量信息，以指示计量是否无法监视的数据。
- schema_auto_increment_columns：自增字段信息。
- schema_index_statistics、x$schema_index_statistics：索引统计信息。
- schema_object_overview：每个 schema 的对象汇总。
- schema_redundant_indexes：重复索引。
- schema_table_lock_waits、x$schema_table_lock_waits：等待元数据锁的会话。
- schema_table_statistics、x$schema_table_statistics：表统计信息。
- schema_table_statistics_with_buffer、x$schema_table_statistics_with_buffer：包括 InnoDB 缓存池的表统计信息。
- schema_tables_with_full_table_scans、x$schema_tables_with_full_table_scans：被以全表扫描方式访问的表。
- schema_unused_indexes：没有用到的索引。
- session、x$session：用户会话信息。

- session_ssl_status：SSL 连接信息。
- statement_analysis、x $ statement_analysis：语句汇总统计信息。
- statements_with_errors_or_warnings、x $ statements_with_errors_or_warnings：出错或报警语句。
- statements_with_full_table_scans、x $ statements_with_full_table_scans：执行全表扫描的语句。
- statements_with_runtimes_in_95th_percentile、x $ statements_with_runtimes_in_95th_percentile：运行时间最长的 5%的语句。
- statements_with_sorting、x $ statements_with_sorting：执行排序的语句。
- statements_with_temp_tables、x $ statements_with_temp_tables：使用临时表的语句。
- user_summary、x $ user_summary：用户执行的语句和连接。
- user_summary_by_file_io、x $ user_summary_by_file_io：按用户分组的文件 I/O。
- user_summary_by_file_io_type、x $ user_summary_by_file_io_type：按用户和事件分组的文件 I/O。
- user_summary_by_stages、x $ user_summary_by_stages：按用户分组的事件阶段。
- user_summary_by_statement_latency、x $ user_summary_by_statement_latency：按用户分组的语句汇总信息。
- user_summary_by_statement_type、x $ user_summary_by_statement_type：按用户和语句类型分组的执行语句统计。
- version：当前 MySQL 和 sys 版本号。
- wait_classes_global_by_avg_latency、x $ wait_classes_global_by_avg_latency：按事件类型分组的等待类型平均延迟统计信息。
- wait_classes_global_by_latency、x $ wait_classes_global_by_latency：按事件类型分组的等待类型总延迟统计信息。
- waits_by_host_by_latency、x $ waits_by_host_by_latency：按主机和事件分组的等待事件延迟统计信息。
- waits_by_user_by_latency、x $ waits_by_user_by_latency：按用户和事件分组的等待事件延迟统计信息。
- waits_global_by_latency、x $ waits_global_by_latency：按事件分组的等待事件延迟统计信息。
- x $ ps_digest_95th_percentile_by_avg_us：95th-percentile 视图的帮助视图。
- x $ ps_digest_avg_latency_distribution：95th-percentile 视图的帮助视图。
- x $ ps_schema_table_statistics_io：table-statistics 视图的帮助视图。
- x $ schema_flattened_keys：schema_redundant_indexes 视图的帮助视图。

7.5.2 视图用例

sys 数据库视图的典型用例如下。

1. 用户或主机分组的汇总视图

可以方便地按用户或主机查询性能和负载情况查询根据用户或主机分组的汇总视图，例如，查询按主机分组的汇总信息的SQL语句及其运行结果如下：

```
mysql> select * from host_summary \G
*************************** 1. row ***************************
                  host: 192.168.17.103
            statements: 22207
     statement_latency: 2.43 min
 statement_avg_latency: 6.57 ms
           table_scans: 9999
              file_ios: 46722
       file_io_latency: 14.22 s
   current_connections: 2
     total_connections: 4
          unique_users: 1
        current_memory: 239.98 KiB
total_memory_allocated: 111.47 MiB
*************************** 2. row ***************************
                  host: localhost
            statements: 653402
     statement_latency: 2.54 h
 statement_avg_latency: 14.00 ms
           table_scans: 205572
              file_ios: 1132650
       file_io_latency: 7.40 min
   current_connections: 2
     total_connections: 845
          unique_users: 2
        current_memory: 62.24 MiB
total_memory_allocated: 6.86 GiB
2 rows in set (0.02 sec)
```

查询按主机分组的SQL语句的延时汇总信息的SQL语句和运行结果如下：

```
mysql> select * from host_summary_by_statement_latency\G
*************************** 1. row ***************************
         host: localhost
        total: 326723
total_latency: 1.27 h
  max_latency: 37.14 s
 lock_latency: 33.33 s
    rows_sent: 2008463327
rows_examined: 2006964853
rows_affected: 203622
   full_scans: 102791
*************************** 2. row ***************************
         host: 192.168.17.103
        total: 50457
total_latency: 13.19 min
```

```
   max_latency: 239.59 ms
  lock_latency: 6.70 s
     rows_sent: 215479590
 rows_examined: 215479590
 rows_affected: 27609
    full_scans: 22840
```

2. 查询进度

对于执行时间长的事务，很多时候需要知道它们的执行进度，sys 数据库的下列 4 个表可以提供进度查询：

- processlist
- session
- x$processlist
- x$session

能够查询进度的阶段包括：

- stage/sql/Copying to tmp table
- stage/innodb/alter table (end)
- stage/innodb/alter table (flush)
- stage/innodb/alter table (insert)
- stage/innodb/alter table (log apply index)
- stage/innodb/alter table (log apply table)
- stage/innodb/alter table (merge sort)
- stage/innodb/alter table (read PK and internal sort)
- stage/innodb/buffer pool load

例如，查询当前增加索引的进度的 SQL 语句及其运行结果如下：

```
mysql>  select * from sys.session where conn_id!= connection_id()\G;
*************************** 1. row ***************************
                thd_id: 45
               conn_id: 4
...
                 state: alter table (merge sort)
                  time: 30
     current_statement: alter table sbtest1 add index i_1(c)
     statement_latency: 21.12 s
              progress: 46.40    -- 进度百分比
          lock_latency: 2.34 ms
         rows_examined: 0
             rows_sent: 0
         rows_affected: 0
            tmp_tables: 0
       tmp_disk_tables: 0
             full_scan: NO
...
```

3. I/O 汇总视图

对于 MySQL 数据库，很多时候瓶颈在 I/O，查询 I/O 汇总视图可以了解 I/O 的性能，例如，查询哪个线程 I/O 最忙的 SQL 语句及其运行结果如下：

```
mysql> select * from io_by_thread_by_latency\G
*************************** 1. row ***************************
          user: log_flusher_thread
         total: 474786
 total_latency: 10.48 min
   min_latency: 28.99 us
   avg_latency: 1.32 ms
   max_latency: 483.98 ms
     thread_id: 16
processlist_id: NULL
*************************** 2. row ***************************
          user: root@192.168.17.103
         total: 105302
 total_latency: 35.78 s
   min_latency: 7.01 us
   avg_latency: 730.19 us
   max_latency: 95.62 ms
     thread_id: 941
processlist_id: 858
...
```

查询 I/O 最忙的两个文件的 SQL 语句及其运行结果如下：

```
mysql> select * from io_global_by_file_by_latency limit 2\G
*************************** 1. row ***************************
         file: @@datadir/ib_logfile1
        total: 617116
total_latency: 5.80 min
   count_read: 2
 read_latency: 166.55 us
  count_write: 360089
write_latency: 6.54 s
   count_misc: 257025
 misc_latency: 5.69 min
*************************** 2. row ***************************
         file: @@datadir/ib_logfile0
        total: 541091
total_latency: 5.15 min
   count_read: 6
 read_latency: 33.92 us
  count_write: 315141
write_latency: 5.91 s
   count_misc: 225944
 misc_latency: 5.06 min
2 rows in set (0.01 sec)
```

4. 全表扫描

找出全表扫描语句的 SQL 语句及其运行结果如下：

```
mysql> select * from statements_with_full_table_scans where db<>'sys'\G
*************************** 1. row ***************************
                   query: SELECT 'cat' . 'name' AS 'CATA ... database' ( 'sch' . 'name' ) )
                      db: sakila
              exec_count: 2
           total_latency: 22.30 ms
     no_index_used_count: 2
no_good_index_used_count: 0
       no_index_used_pct: 100
               rows_sent: 13
           rows_examined: 54
           rows_sent_avg: 7
       rows_examined_avg: 27
              first_seen: 2021-03-23 11:38:18.965826
               last_seen: 2021-03-24 17:11:08.544267
                  digest: db22765f86fdb091d2faced7479895f8497b40140b15fad577fcd224bfc4586f
*************************** 2. row ***************************
                   query: SELECT 'intcol1', 'intcol2', ... arcol2', 'charcol3' FROM 't1'
                      db: mysqlslap
              exec_count: 142562
           total_latency: 1.49 h
     no_index_used_count: 142562
no_good_index_used_count: 0
       no_index_used_pct: 100
               rows_sent: 2492504704
           rows_examined: 2492504704
           rows_sent_avg: 17484
       rows_examined_avg: 17484
              first_seen: 2021-03-23 10:37:01.942116
               last_seen: 2021-03-26 15:28:47.206534
                  digest: 4a81b5a003501db5d438cd8cc20e888fd278413cc821cd223ecb42c68e1a0912
2 rows in set (0.01 sec)
```

找出全表扫描的表的SQL语句及其运行结果如下：

```
mysql> select * from schema_tables_with_full_table_scans limit 1;
+---------------+-------------+------------------+-----------+
| object_schema | object_name | rows_full_scanned | latency  |
+---------------+-------------+------------------+-----------+
| mysqlslap     | t1          |          68324464 | 3.58 min |
+---------------+-------------+------------------+-----------+
1 row in set (0.02 sec)
```

5. 内存

查询MySQL分配的总内存的SQL语句及其运行结果如下：

```
mysql> select * from memory_global_total;
+-----------------+
| total_allocated |
+-----------------+
| 495.06 MiB      |
+-----------------+
1 row in set (0.00 sec)
```

查询按线程分配的内存的SQL语句及其运行结果如下：

```
mysql> select * from  memory_by_thread_by_current_bytes limit 3 \G
*************************** 1. row ***************************
         thread_id: 873
              user: root@localhost
current_count_used: 267
current_allocated: 1.44 MiB
current_avg_alloc: 5.52 KiB
current_max_alloc: 862.42 KiB
   total_allocated: 65.18 MiB
*************************** 2. row ***************************
         thread_id: 33
              user: innodb/clone_gtid_thread
current_count_used: 88
current_allocated: 264.76 KiB
current_avg_alloc: 3.01 KiB
current_max_alloc: 118.20 KiB
   total_allocated: 234.49 MiB
*************************** 3. row ***************************
         thread_id: 943
              user: root@192.168.17.103
current_count_used: 45
current_allocated: 86.71 KiB
current_avg_alloc: 1.93 KiB
current_max_alloc: 43.79 KiB
   total_allocated: 1.37 GiB
3 rows in set (0.08 sec)
```

查询按主机分配的内存的SQL语句及其运行结果如下：

```
mysql> select * from  memory_by_host_by_current_bytes limit 3 \G
*************************** 1. row ***************************
              host: background
current_count_used: 7357
current_allocated: 80.28 MiB
current_avg_alloc: 11.17 KiB
current_max_alloc: 24.00 MiB
   total_allocated: 2.08 GiB
*************************** 2. row ***************************
              host: localhost
current_count_used: 7522
current_allocated: 30.78 MiB
current_avg_alloc: 4.19 KiB
current_max_alloc: 10.58 MiB
   total_allocated: 3.47 GiB
*************************** 3. row ***************************
              host: 192.168.17.103
current_count_used: 153
current_allocated: 309.57 KiB
```

```
current_avg_alloc: 2.02 KiB
current_max_alloc: 96.14 KiB
   total_allocated: 2.32 GiB
3 rows in set (0.01 sec)
```

7.6 实验

视频演示

1. diagnostics()存储过程

使用 diagnostics()存储过程生成当前 MySQL 实例整体性能的诊断报告(参考 7.3 节)。

视频演示

2. ps_trace_statement_digest()存储过程

使用 ps_trace_statement_digest()存储过程跟踪收集 SQL 语句执行过程中的性能诊断信息(参考 7.3 节)。

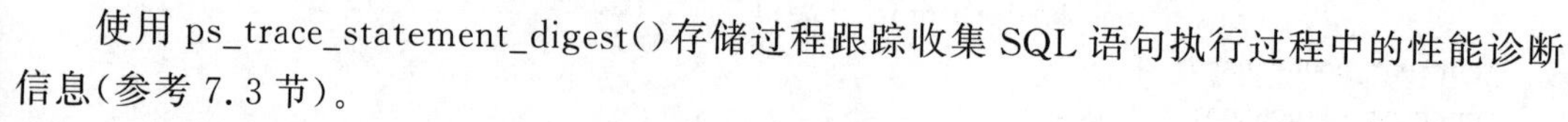

3. statement_performance_analyzer()存储过程

视频演示

使用 statement_performance_analyzer()存储过程生成当前 MySQL 实例中正在运行的 SQL 语句的性能报告(参考 7.3 节)。

4. ps_trace_thread()存储过程

视频演示

使用 ps_trace_thread()存储过程跟踪线程执行的 SQL 语句,输出性能报告(参考 7.3 节)。

第三部分

备份恢复

第8章

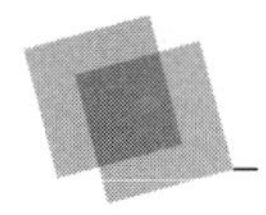

逻 辑 备 份

MySQL 数据库的备份分为逻辑备份和物理备份，本章介绍 MySQL 逻辑备份的相关知识和 4 种常用的 MySQL 逻辑备份工具：mysqldump、mysqlpump、mydumper 和 MySQL Shell。

8.1 逻辑备份和物理备份的区别

MySQL 数据库的备份可以分为两类：逻辑备份和物理备份。逻辑备份是把数据备份成 MySQL 能够解析的格式：SQL 语句或文本文件；物理备份是采用直接复制操作系统层的数据文件的方式进行备份。

逻辑备份的优点如下：

- 用户可以查看甚至修改备份出来的文件，更加容易观察备份的过程和把握备份的正确性。
- 更加灵活，可以很容易地在备份过程中对数据进行过滤甚至修改。
- 可以发现数据文件的损坏，如果内存中仍然有正确的数据，逻辑备份可以生成正确的备份集，但物理备份很难发现和规避数据文件的损坏。
- 可以用于在云或容器上的不能访问底层数据文件的数据库的备份。
- 很容易实现跨平台或存储引擎的数据迁移。

逻辑备份的缺点如下：

- 备份时间长，不适用于大型数据库。
- 备份和恢复过程都需要消耗较多的系统资源。
- 数据库必须启动后才能进行备份。
- 备份集可能比原始数据还要大。

物理备份最大的优点是备份和恢复速度都很快，接近从操作系统层复制文件的速度；其缺点主要是较难发现数据文件的损坏，而且当不能访问底层数据文件时无法进行备份。

逻辑备份和物理备份互有优缺点，在设计备份策略时应扬长避短，将两种备份方法结合起来使用。

8.2 mysqldump

mysqldump 是 MySQL 自带的经典逻辑备份工具，其备份原理是通过客户端连接到 MySQL 实例，查询需要备份的数据并将这些数据保存成 create 和 insert 语句，还原时执行这些语句即可将对应的数据还原。

8.2.1 调用语法

mysqldump 主要有如下三种调用方式。

(1) 导出一组表，代码如下：

```
$ mysqldump [options] db_name [tbl_name ...]
```

(2) 导出一组数据库，代码如下：

```
$ mysqldump [options] --databases db_name ...
```

(3) 导出整个 MySQL 数据库，代码如下：

```
$ mysqldump [options] --all-databases
```

8.2.2 选项

mysqldump 的选项众多，分类说明如下。

1. 连接选项

连接选项决定了 mysqldump 如何连接到 MySQL 实例，mysqldump 有如下连接选项：

- --bind-address=ip_address
- --compress，-C
- --compression-algorithms=value
- --default-auth=plugin
- --enable-cleartext-plugin
- --get-server-public-key
- --host=host_name，-h host_name
- --login-path=name
- --password[=password]，-p[password]
- --pipe，-W
- --plugin-dir=dir_name
- --port=port_num，-P port_num
- --protocol={TCP|SOCKET|PIPE|MEMORY}

- --server-public-key-path=file_name
- --socket=path,-S path
- --ssl *
- --tls-ciphersuites=ciphersuite_list
- --tls-version=protocol_list
- --user=user_name,-u user_name
- --zstd-compression-level=level

如上选项和控制 mysql 客户端连接到 MySQL 的选项类似,此处不再一一说明。

2. 选项文件选项

选项文件选项决定了 mysqldump 如何读取选项文件,mysqldump 有如下选项文件选项:

- --defaults-extra-file=file_name
- --defaults-file=file_name
- --defaults-group-suffix=string
- --no-defaults
- --print-defaults

如上选项也和控制 mysql 客户端读取选项文件的选项类似,此处不再一一说明。

3. DDL 选项

DDL 选项决定了导出文件中的 DDL 语句,这些语句将在导入时发挥作用,mysqldump 有如下 DDL 选项。

- --add-drop-database:每个数据库创建之前添加 drop 数据库语句。
- --add-drop-table:每个数据表创建之前添加 drop 数据表语句,该选项默认是打开的,使用--skip-add-drop-table 取消选项。
- --add-drop-trigger:每个触发器创建之前添加 drop 触发器语句。
- --all-tablespaces,-Y:该选项只用于 NDB 集群表,创建导出 NDB 集群表需要的表空间。
- --no-create-db,-n:不输出创建数据库的语句。
- --no-create-info,-t:不输出创建表的语句。
- --no-tablespaces,-y:不输出 create logfile group 和 create tablespace 的语句。
- --replace:使用 replace 语句代替 insert 语句。

4. debug 选项

这类选项和 debug 相关,mysqldump 有如下 debug 选项。

- --allow-keywords:通过在字段名前加上表名的方式允许字段名使用关键字。
- --comments,-i:输出注释信息,如程序版本、服务版本和主机名等。该选项默认是打开的,使用--skip-comment 可取消该选项。
- --debug[=debug_options],-# [debug_options]:输出 debug 日志,该选项只能用于采用 WITH_DEBUG 方式编译的 MySQL。
- --debug-check:程序退出时输出 debug 信息,该选项只能用于采用 WITH_DEBUG

方式编译的 MySQL。

- --debug-info：程序退出时输出 CPU 和内存使用信息，该选项只能用于采用 WITH_DEBUG 方式编译的 MySQL。
- --dump-date：当使用--comment 选项时，该选项输出导出完成时的日期。该选项默认是打开的，使用--skip-dump-date 可取消该选项。
- --force,-f：出错时仍然继续。
- --log-error=file_name：向指定的文件输出警告和错误信息，该选项默认不打开。
- --verbose,-v：输出更多信息。

5. 帮助选项

- --help,-?：输出帮助信息并退出。
- --version,-V：输出版本信息并退出。

6. 国际化选项

- --character-sets-dir=dir_name：指定安装的字符集的目录。
- --default-character-set=charset_name：指定默认的字符集。
- --set-charset：输出 SET NAMES default_character_set 语句，该选项默认是打开的，使用--skip-set-charset 或--no-set-names 可取消该选项。

7. 复制选项

- --apply-slave-statements：在使用--dump-slave 导出的文件中，在 CHANGE REPLICATION SOURCE TO 或 CHANGE MASTER 语句前添加 STOP SLAVE，并且在导出的文件最后添加 START SLAVE。
- --delete-master-logs：主库备份后删除日志，该参数将自动激活--master-data。
- --dump-slave[=value]：该选项和--master-data 类似，只是用于在备库创建备份集。
- --include-master-host-port：在使用--dump-slave 导出的文件中，在 CHANGE REPLICATION SOURCE TO 或 CHANGE MASTER 语句中增加主机名和端口号。
- --master-data[=value]：为 1 时，在导出文件中增加 CHANGE REPLICATION SOURCE TO 或 CHANGE MASTER 语句说明导出时的二进制文件位置；为 2 时，采用注释的形式把这样的语句添加到导出的文件中。该选项将屏蔽--lock-tables 选项，除非指定了--single-transaction 选项，它将会激活--lock-all-tables 选项。
- --set-gtid-purged=value：该参数决定是否在导出的文件里输出 set @@global. gtid_purged 为源库的 gtid_executed 和 set @@session. sql_log_bin=0，有如下 4 种类型。
- Auto：两个值都输出。
- Off：两个值都不输出。
- On：当源库的 GTID 没有被激活时，报错；当源库的 GTID 被激活时，和 auto 一样。
- Commented：两个值都输出，但 set @@global. gtid_purged 放在注释中。

8. 格式化选项

- --compact：输出更少的信息，去掉了注释和头尾等。该选项激活了--skip-add-drop-

table、--skip-add-locks、--skip-comments、--skip-disable-keys 和--skip-set-charset 等选项。

- --compatible=name：导出的数据将和其他数据库或旧版本的 MySQL 相兼容。该值只能设置为 ansi，意义和 SQL mode 的选项一样。
- --complete-insert，-c：使用完整的 insert 语句（包含列名称）。
- --create-options：在 CREATE TABLE 语句中包括所有 MySQL 特性选项，默认为打开状态。
- --fields-terminated-by=…，--fields-enclosed-by=…，--fields-optionally-enclosed-by=…，--fields-escaped-by=…，--lines-terminated-by=…：这些选项和--tab 一起使用，它们的意义和 LOAD DATA 语句中 FIELDS 子句中的选项一样。
- --hex-blob：使用十六进制格式导出二进制字符串字段。
- --quote-names，-Q：使用'（反引号）或"（双引号）引起表和列名。默认为打开状态，使用--skip-quote-names 取消该选项。
- --result-file=file_name，-r file_name：直接输出到指定文件中。在 Windows 系统上使用该选项防止\n 被转换成\r\n。
- --show-create-skip-secondary-engine = value：在 create table 语句中不输出 secondary engine 子句。
- --tab=dir_name，-T dir_name：在指定路径为每个表创建用制表符分割的文本文件。注意：仅仅用于 mysqldump 和 mysqld 服务器运行在同一台机器上时。
- --tz-utc：在导出文件顶部设置时区 TIME_ZONE='+00:00'，以保证在不同时区导出的 TIMESTAMP 数据或者数据被移动到其他时区时的正确性。该选项默认是被激活的，使用--skip-tz-utc 可屏蔽该选项。
- --xml，-X：输出内容采用 xml 格式。

9. 过滤选项

- --all-databases，-A：导出全部数据库。
- --databases，-B：导出多个数据库，参数后面的所有名字参数都被看作数据库名。
- --events，-E：导出事件。
- --ignore-error=error[，error]…：忽略指定的错误。
- --ignore-table=db_name.tbl_name：不导出指定表。
- --no-data，-d：不导出任何数据，只导出数据库表结构。
- --routines，-R：导出存储过程和自定义函数。
- --tables：屏蔽--databases (-B)参数，指定需要导出的表名。
- --triggers：导出触发器。该选项默认是被激活的，用--skip-triggers 可屏蔽该选项。
- --where='where_condition'，-w 'where_condition'：只导出符合给定 WHERE 条件的记录。

10. 性能选项

- --column-statistics：输出 ANALYZE TABLE 语句，为导出的表在导入时生成直方图统计信息，该选项默认没有被激活，因为对大表生产直方图统计信息耗时较长。

- --disable-keys,-K：对于每个表,用"/*! 40000 ALTER TABLE tbl_name DISABLE KEYS */;"和"/*! 40000 ALTER TABLE tbl_name ENABLE KEYS */;"语句引用 INSERT 语句。这样导入时速度更快,因为在插入所有行后才创建索引。该选项只适合 MyISAM 表,默认为打开状态。
- --extended-insert,-e：使用具有多个 VALUES 列的 INSERT 语法。这样使导出文件更小,并在导入时更快。默认为打开状态,使用--skip-extended-insert 可取消该选项。
- --insert-ignore：在插入行时使用 INSERT IGNORE 语句而不是 INSERT 语句。
- --max-allowed-packet=value：设置服务器和客户端之间通信的缓冲区的最大值,默认是 24MB。
- --net-buffer-length=value：设置服务器和客户端之间通信的缓冲区的初始值。
- --network-timeout,-M：在大表导出时设置--max-allowed-packet 和网络超时时间到更大的值,默认为打开状态,使用--skip-network-timeout 可取消该选项。
- --opt：该选项相当于同时设置了--add-drop-table、--add-locks、--create-options、--disable-keys、--extended-insert、--lock-tables、--quick、--set-charset,该选项默认为开启状态,可以用--skip-opt 取消该选项。
- --quick,-q：不缓冲查询结果,直接导出到标准输出。默认为打开状态,使用--skip-quick 可以取消该选项,这时在导出时将使用缓存,这对大表是不合适的。

11. 事务选项

- --add-locks：在每个表导出之前增加 LOCK TABLES 语句并在之后增加 UNLOCK TABLE 语句,这样可以加快导入速度。默认为打开状态,使用--skip-add-locks 可取消该选项。
- --flush-logs,-F：开始导出之前刷新日志。如果一次导出多个数据库(使用选项--databases 或者--all-databases),将会逐个刷新数据库日志,除非使用了--lock-all-tables、--master-data 或--single-transaction。使用这些选项时,日志将只会刷新一次。
- --flush-privileges：在导出 mysql 系统数据库之后,输出一条 FLUSH PRIVILEGES 语句。为了正确恢复,该选项应该用于导出 mysql 系统数据库和依赖 mysql 系统数据库数据的任何时候。
- --lock-all-tables,-x：锁定所有数据库中的所有表。该选项将屏蔽--single-transaction 和--lock-tables 选项。
- --lock-tables,-l：当导出某个数据库时,锁定该数据库中的所有表。该选项是分库进行锁表,因此不能保证不同数据库之间的数据是一致的。
- --no-autocommit：对于每个导出的表,使用 SET autocommit = 0 和 COMMIT 语句将 INSERT 语句包裹住。
- --order-by-primary：在导出时使用表的主键或唯一键对记录进行排序。
- --shared-memory-base-name=name：在 Windows 平台上,采用共享内存连接时,指定共享内存名,默认是 MYSQL。
- --single-transaction：该选项设定事务的隔离级别为可重复读,在导出数据之前提交一个 START TRANSACTION 语句开启一个事务，事务在数据导出完成后结束，InnoDB 的表可以保证导出的数据是一致的，但这可能会导致出现一个非常长的事务，将消耗很多的锁和 undo 表空间，在某些负载下这样的大开销是不可接受的。

8.2.3 典型用例

1. 备份成 SQL 格式的备份文件

备份 sakila 数据库的命令如下：

```
$ mysqldump sakila > sakila.sql
```

备份 sakila 数据库中 actor 表的命令如下：

```
$ mysqldump sakila actor > sakila_actor.sql
```

创建一个全量备份，包括所有数据、存储过程和事件的命令如下：

```
$ mysqldump --all-databases --routines --events > all_databases.sql
```

对于 InnoDB 的表，创建一个一致备份集的命令如下：

```
$ mysqldump --all-databases --master-data --single-transaction > all_databases.sql
```

--single-transaction 在开始时会把事务的隔离级别设置为可重复读(repeatable read)。在导出数据之前提交一个 start transaction 语句开启事务，利用 InnoDB 的 MVCC 特性创建一个一致的备份集。

--master-data 在开始备份时获取 flush tables with read lock 锁，一旦获得了该锁，就会读取二进制日志坐标，然后释放锁。这样后面的备份过程中就没有锁阻塞其他对数据库的访问。

如果在备库上进行这样的备份，把--master-data 换成--dump-slave。

如果生产环境和测试环境的数据库结构一致，并且测试环境已经有部分的数据，使用如下命令可以将生产环境的数据导出到测试环境：

```
$ mysqldump --database sakila --skip-add-drop-tables --no-create-info --replace >
to_test.sql
```

--replace 会替换重复的记录，--skip-add-drop-tables 不会添加 drop 表的语句，--no-create-info 不会创建表。

2. 备份带分隔符的文本文件格式的备份文件

以分隔符的文本文件格式备份数据库 sakila 的命令如下：

```
$ mysqldump --tab=/var/lib/mysql-files/ sakila
```

参数--secure-file-priv 用于控制导入导出数据的访问路径，查询该参数值的 SQL 语句及其运行结果如下：

```
mysql> show variables like 'secure_file_priv';
```

```
+-------------------+------------------------+
| Variable_name     | Value                  |
+-------------------+------------------------+
| secure_file_priv  | /var/lib/mysql-files/  |
+-------------------+------------------------+
1 row in set (0.00 sec)
```

当导出的目录不是--secure-file-priv 指定的目录时会遇到下面的错误信息：

```
mysqldump: Got error: 1290: The MySQL server is running with the
--secure-file-priv option so it cannot execute this statement when executing
'SELECT INTO OUTFILE'
```

这个备份命令会为每个表生成两个文件，一个是用制表符分隔的数据文件，以.txt 结尾；另一个是由 SQL 语句组成的建表文件，以.sql 结尾，对于视图，只有一个创建视图的 SQL 语句文件，如下所示：

```
/var/lib/mysql-files$ ls
actor_info.sql   category.txt        customer.sql        film_list.sql
inventory.txt                   rental.sql                staff.txt
actor.sql        city.sql            customer.txt        film.sql
language.sql                    rental.txt                store.sql
actor.txt        city.txt            film_actor.sql      film_text.sql
language.txt                    sales_by_film_category.sql  store.txt
address.sql      country.sql         film_actor.txt      film_text.txt
nicer_but_slower_film_list.sql  sales_by_store.sql
address.txt      country.txt         film_category.sql   film.txt
payment.sql                     staff_list.sql
category.sql     customer_list.sql   film_category.txt   inventory.sql
payment.txt                     staff.sql
```

查询数据文件的内容如下：

```
/var/lib/mysql-files$ cat -T actor.txt |head
1^IPENELOPE^IGUINESS^I2006-02-15 08:04:33
2^INICK^IWAHLBERG^I2006-02-15 08:04:33
3^IED^ICHASE^I2006-02-15 08:04:33
4^IJENNIFER^IDAVIS^I2006-02-15 08:04:33
5^IJOHNNY^ILOLLOBRIGIDA^I2006-02-15 08:04:33
6^IBETTE^INICHOLSON^I2006-02-15 08:04:33
7^IGRACE^IMOSTEL^I2006-02-15 08:04:33
8^IMATTHEW^IJOHANSSON^I2006-02-15 08:04:33
9^IJOE^ISWANK^I2006-02-15 08:04:33
10^ICHRISTIAN^IGABLE^I2006-02-15 08:04:33
```

其中^I 为制表符。

去掉注释后的 SQL 语句文件内容如下：

```
/var/lib/mysql-files$ cat actor.sql|grep -v '^[-\/$]'
DROP TABLE IF EXISTS 'actor';
CREATE TABLE 'actor' (
  'actor_id' smallint unsigned NOT NULL AUTO_INCREMENT,
```

```
  'first_name' varchar(45) NOT NULL,
  'last_name' varchar(45) NOT NULL,
  'last_update' timestamp NOT NULL DEFAULT CURRENT_TIMESTAMP ON UPDATE CURRENT_TIMESTAMP,
  PRIMARY KEY ('actor_id'),
  KEY 'idx_actor_last_name' ('last_name')
) ENGINE = InnoDB AUTO_INCREMENT = 201 DEFAULT CHARSET = utf8mb4 COLLATE = utf8mb4_0900_ai_ci;
```

3. 修改备份文件里的对象名

mysqldump在导出的过程中不能修改对象名,如果要把一个数据库或表在导入时换成另外一个数据库或表,需要手动修改数据库名或表名,常用的工具是sed。例如,对于使用如下命令备份的sakila数据库:

```
$ mysqldump sakila > sakila.sql
```

可以使用sed命令将其中的actor表名修改成newactor表名:

```
$ sed -i 's/'actor'/'newactor'/g' sakila.sql
```

8.2.4 恢复

恢复过程分为两类:一类是恢复SQL格式的备份文件;另一类是恢复带分隔符的文本文件格式的备份文件。

1. 恢复SQL格式的备份文件

通过mysqldump备份的文件,如果用了--all-databases或--databases选项,则在备份文件中包含CREATE DATABASE和USE语句,故并不需要指定一个默认的数据库去恢复备份文件。

在操作系统命令行中,使用如下命令进行恢复:

```
$ mysql  <  sakila.sql
```

或者在mysql客户端里,使用如下source命令执行备份文件进行恢复:

```
mysql > source sakila.sql;
```

如果备份文件中没有建库的语句,需要指定一个默认的数据库,例如:

```
$ mysql sakila < sakila.sql
```

或者:

```
$ mysql -e "source sakila.sql" sakila
```

2. 恢复带分隔符的文本文件格式的备份文件

首先用mysql命令执行.sql文件创建表结构,命令如下:

```
$ mysql sakila < actor.sql
```

然后使用mysqlimport命令加载.txt文件,命令如下:

```
$ mysqlimport sakila /var/lib/mysql-files/actor.txt
sakila.actor: Records: 200  Deleted: 0  Skipped: 0  Warnings: 0
```

也可以用 mysql 中的 load data infile 代替 mysqlimport 加载数据，命令如下：

```
mysql> use sakila;
mysql> load data infile '/var/lib/mysql-files/actor.txt' into table actor;
Query OK, 200 rows affected (0.02 sec)
Records: 200  Deleted: 0  Skipped: 0  Warnings: 0
```

mysqlimport 实际上是 load data infile 的一个包装。

8.3 mysqlpump

mysqlpump 是在 MySQL 5.7 中推出的一个新的逻辑备份工具，它和 mysqldump 的语法基本兼容，功能更强大，下面说明它的功能增强部分。

8.3.1 并行备份

mysqlpump 支持并行备份，它可以在数据库之间进行并行备份，也可以在同一个数据库内部的不同对象之间进行并行备份。对并行度的控制由如下两个参数完成。

- --default-parallelism＝N：为每个队列指定线程的个数，默认为 2。
- --parallel-schemas＝[N：]db_list：创建新的备份队列，N 是该队列的线程数，db_list 是该队列中包括的数据库名，可以使用%和_对数据库名进行匹配。

如下命令创建了 3 个备份队列，第 1 个队列备份 db1 和 db2，第 2 个队列备份 db3，第 3 个队列备份其他的数据库，这 3 个队列的并行度都是 4：

```
$ mysqlpump --parallel-schemas=db1,db2 --parallel-schemas=db3
--default-parallelism=4
```

如下命令和上面的命令类似，只是并行度不同，第 1 个队列的并行度是 4，第 2 个队列的并行度是 3，第 3 个队列的并行度是 2：

```
$ mysqlpump --parallel-schemas=4:db1,db2 --parallel-schemas=3:db3
```

第 3 个队列的并行度的设置在如上命令中没有看到，这是因为默认的并行度就是 2。

8.3.2 进度显示

mysqlpump 在备份的过程中会显示备份的进度，例如：

```
$ mysqlpump -B sbtest --result-file=/tmp/sbtest
Dump progress: 0/1 tables, 250/986422 rows
Dump progress: 0/2 tables, 560500/1973754 rows
Dump progress: 0/2 tables, 1091000/1973754 rows
Dump progress: 0/2 tables, 1382000/1973754 rows
```

```
Dump progress: 0/2 tables, 1460500/1973754 rows
Dump progress: 0/2 tables, 1547500/1973754 rows
Dump progress: 0/2 tables, 1659750/1973754 rows
Dump progress: 1/2 tables, 1843500/1973754 rows
Dump completed in 8530
```

8.3.3 使用正则表达式指定数据库对象

mysqlpump 中可以使用如下选项指定或排除数据库对象：

- --include-databases
- --exclude-databases
- --include-tables
- --exclude-tables
- --include-triggers
- --exclude-triggers
- --include-routines
- --exclude-routine
- --include-events
- --exclude-events
- --include-users
- --exclude-users

指定数据库对象名时支持如下通配符：

- %：匹配 0 个或多个字符。
- _：匹配任意单个字符。

8.3.4 备份用户

在 mysqldump 中通过备份系统数据库 mysql 中的相关表完成用户和权限的备份，恢复时加上--flush-privileges 将用户和权限刷新到 MySQL 实例中。

在 mysqlpump 中可以使用--users 选项将账户备份为账户管理语句：CREATE USER 和 GRANT。例如：

```
$ mysqlpump --exclude-databases=% --users
...
CREATE USER ''@'%' IDENTIFIED WITH 'caching_sha2_password' AS
'$A$005$&ryQ%E)\'L?aphDlL5osUiakQyt5K0l6LBvoiZkxoJLVDt4Op9kHytVp7' REQUIRE
NONE PASSWORD EXPIRE DEFAULT ACCOUNT UNLOCK PASSWORD HISTORY DEFAULT PASSWORD
REUSE INTERVAL DEFAULT PASSWORD REQUIRE CURRENT DEFAULT;
GRANT USAGE ON *.* TO ''@'%';
GRANT PROXY ON 'manager'@'%' TO 'client'@'%';
CREATE USER 'manager'@'%' IDENTIFIED WITH 'mysql_no_login' REQUIRE NONE
PASSWORD EXPIRE DEFAULT ACCOUNT LOCK PASSWORD HISTORY DEFAULT PASSWORD REUSE
```

```
INTERVAL DEFAULT PASSWORD REQUIRE CURRENT DEFAULT;
GRANT USAGE ON *.* TO 'manager'@'%';
...
```

还可以通过--exclude-users 排除部分用户,代码如下:

```
$ mysqlpump --exclude-databases=% --exclude-users=root --users
```

8.3.5 加载后创建索引

选项--defer-table-indexes 默认是激活的,它表示在 mysqlpump 的输出中,创建索引是在表的数据加载完成后才进行的,由于 InnoDB 表是按主键排序保存数据的,推迟创建索引是指二级索引。因为导入的数据通常不是按照二级索引进行排序的,如果一开始就创建二级索引,在整个导入过程中索引会不断地发生页分裂,因此在导入完成后再创建索引可以加快数据的导入。mysqldump 没有该选项,但可以手工修改导出文件实现这个功能。

可以使用--skip-defer-table-indexes 屏蔽该选项。

8.4 mydumper

mydumper 是一款和 mysqldump、mysqlpump 类似的 MySQL 逻辑备份工具,它有以下特点:

- 并行和一致性。mydumper 在并行进行备份的同时能保证不同线程之间数据的一致性。
- 输出更易于管理。mydumper 输出的元数据和实际数据分开到不同文件中进行保存。方便查看和解析数据,mysqlpump 将所有数据都输出到一个文件中。
- 使用正则表达式包含和排除数据库对象。

8.4.1 安装

mydumper 是一款开源软件(https://github.com/maxbube/mydumper),需要单独安装。

在 RedHat 和 Centos 平台的安装方法如下:

```
$ sudo yum install
https://github.com/maxbube/mydumper/releases/download/v0.9.5/mydumper-0.9.5-1.el7.x86_64.rpm
```

在 Ubuntu 和 Debian 平台的安装方法如下:

```
$ wget
https://github.com/maxbube/mydumper/releases/download/v0.9.5/mydumper_0.9.5-1.xenial_amd64.deb
$ sudo dpkg -i mydumper_0.9.5-1.xenial_amd64.deb
Preparing to unpack mydumper_0.10.1-2.bionic_amd64.deb ...
```

```
Unpacking mydumper (0.10.1-2) ...
Setting up mydumper (0.10.1-2) ...
```

安装后会有两个可执行程序，mydumper 用于将数据导出，myloader 用于将数据导入。

8.4.2 使用

--help 选项可以查看 mydumper 的帮助信息，如下所示：

```
$ mydumper --help
Usage:
  mydumper [OPTION?] multi-threaded MySQL dumping

Help Options:
  -?, --help                  Show help options

Application Options:
  -B, --database              Database to dump
  -T, --tables-list            Comma delimited table list to dump (does not exclude regex
option)
...
```

导出数据库 sakila 数据的代码如下：

```
$ mydumper -u root -p dingjia -P 3306 -h localhost -B sakila -o ~/output
```

备份完成后生成的文件如下：

```
$ ls ~/output
metadata  sakila.actor-schema.sql  sakila.actor.sql
sakila-schema-create.sql ...
```

其中，metadata 文件中记录了备份时间点的二进制日志坐标，如下所示：

```
$ cat metadata
Started dump at: 2021-02-21 03:53:10
SHOW MASTER STATUS:
    Log: slave-binlog.000057
    Pos: 16043
    GTID: 20193e38-53a7-11ea-b94a-fa163e585706:1-79315,
93bf2ecc-4e4b-11ea-a17a-fa163e0ec694:1-66

SHOW SLAVE STATUS:
    Host: 192.168.17.40
    Log: master-binlog.000003
    Pos: 195
    GTID: 20193e38-53a7-11ea-b94a-fa163e585706:1-79315,
93bf2ecc-4e4b-11ea-a17a-fa163e0ec694:1-66

Finished dump at: 2021-02-21 03:53:10
```

保存创建数据库语句的文件，如 sakila schema-create. sql：

```
$ cat sakila-schema-create.sql
CREATE DATABASE 'sakila' /*!40100 DEFAULT CHARACTER SET utf8mb4 COLLATE utf8mb4_0900_ai_ci
*/ /*!80016 DEFAULT ENCRYPTION='N' */;
```

每个表包括如下两个备份文件。

(1) 表结构文件，保存 create table 语句，如 sakila.actor-schema.sql：

```
$ cat sakila.actor-schema.sql
/*!40101 SET NAMES binary*/;
/*!40014 SET FOREIGN_KEY_CHECKS=0*/;

/*!40103 SET TIME_ZONE='+00:00' */;
CREATE TABLE 'actor' (
  'actor_id' smallint unsigned NOT NULL AUTO_INCREMENT,
  'first_name' varchar(45) NOT NULL,
  'last_name' varchar(45) NOT NULL,
  'last_update' timestamp NOT NULL DEFAULT CURRENT_TIMESTAMP ON UPDATE CURRENT_TIMESTAMP,
  PRIMARY KEY ('actor_id'),
  KEY 'idx_actor_last_name' ('last_name')
) ENGINE=InnoDB AUTO_INCREMENT=201 DEFAULT CHARSET=utf8mb4 COLLATE=utf8mb4_0900_ai_ci;
```

(2) 表数据文件，以 insert 语句的形式保存数据，如 sakila.actor.sql：

```
$ cat sakila.actor.sql|head
/*!40101 SET NAMES binary*/;
/*!40014 SET FOREIGN_KEY_CHECKS=0*/;
/*!40103 SET TIME_ZONE='+00:00' */;
INSERT INTO 'actor' ('actor_id','first_name','last_name') VALUES
(1,"PENELOPE","GUINESS"),
(2,"NICK","WAHLBERG"),
(3,"ED","CHASE"),
(4,"JENNIFER","DAVIS"),
(5,"JOHNNY","LOLLOBRIGIDA"),
(6,"BETTE","NICHOLSON"),
```

恢复 sakila 数据库的命令如下：

```
$ myloader  -u root -p dingjia -P 3306 -h localhost -B sakila -d ./output
```

8.4.3 并行和一致性

mydumper 可以对单个大表进行并行导出，并行的单位块(chunk)可以用行数或文件大小来定义，并行导出表数据的命令如下，每 100 000 行为一个单位：

```
$  mydumper -u root -p dingjia -P 3306 -h localhost -B sbtest -T sbtest1 -t 8
--rows=100000 --trx-consistency-only -o ./output
```

其中，-t 指定线程的数量为 8 个；--trx-consistency-only 指明这里只包括 InnoDB 的表，可以减少锁的时间；--row 指明并行的单位，这里每个线程一次处理 100 000 行。

每个块会创建一个名为 database_name.table_name.number.sql 的文件，其中 number

是用0补全的5位顺序数字。如上的导出命令会生成下面的文件：

```
$ ls -lh
total 192M
-rw-r--r-- 1 root root 404 Feb 21 04:24 metadata
-rw-r--r-- 1 root root 22M Feb 21 04:24 sbtest.sbtest1.00000.sql
-rw-r--r-- 1 root root 22M Feb 21 04:24 sbtest.sbtest1.00001.sql
-rw-r--r-- 1 root root 22M Feb 21 04:24 sbtest.sbtest1.00002.sql
-rw-r--r-- 1 root root 22M Feb 21 04:24 sbtest.sbtest1.00003.sql
-rw-r--r-- 1 root root 22M Feb 21 04:24 sbtest.sbtest1.00004.sql
-rw-r--r-- 1 root root 22M Feb 21 04:24 sbtest.sbtest1.00005.sql
-rw-r--r-- 1 root root 22M Feb 21 04:24 sbtest.sbtest1.00006.sql
-rw-r--r-- 1 root root 22M Feb 21 04:24 sbtest.sbtest1.00007.sql
-rw-r--r-- 1 root root 22M Feb 21 04:24 sbtest.sbtest1.00008.sql
-rw-r--r-- 1 root root 402 Feb 21 04:24 sbtest.sbtest1-schema.sql
-rw-r--r-- 1 root root 130 Feb 21 04:24 sbtest-schema-create.sql
```

还可以指定文件大小来决定并行的单位，例如，如下命令指定并行的单位为20MB：

```
$ mydumper -u root -p dingjia -P 3306 -h localhost -B sbtest -T sbtest1
--chunk-filesize 20 --trx-consistency-only -o ./output
```

如上命令生成的文件如下：

```
$ ll -lh output/
total 192M
drwx------ 2 root  root  4.0K Feb 21 04:31 ./
drwxr-x--- 3 mysql mysql 4.0K Feb 21 04:31 ../
-rw-r--r-- 1 root  root   404 Feb 21 04:31 metadata
-rw-r--r-- 1 root  root   21M Feb 21 04:31 sbtest.sbtest1.00001.sql
-rw-r--r-- 1 root  root   21M Feb 21 04:31 sbtest.sbtest1.00002.sql
-rw-r--r-- 1 root  root   21M Feb 21 04:31 sbtest.sbtest1.00003.sql
-rw-r--r-- 1 root  root   21M Feb 21 04:31 sbtest.sbtest1.00004.sql
-rw-r--r-- 1 root  root   21M Feb 21 04:31 sbtest.sbtest1.00005.sql
-rw-r--r-- 1 root  root   21M Feb 21 04:31 sbtest.sbtest1.00006.sql
-rw-r--r-- 1 root  root   21M Feb 21 04:31 sbtest.sbtest1.00007.sql
-rw-r--r-- 1 root  root   21M Feb 21 04:31 sbtest.sbtest1.00008.sql
-rw-r--r-- 1 root  root   21M Feb 21 04:31 sbtest.sbtest1.00009.sql
-rw-r--r-- 1 root  root   12M Feb 21 04:31 sbtest.sbtest1.00010.sql
-rw-r--r-- 1 root  root   402 Feb 21 04:31 sbtest.sbtest1-schema.sql
-rw-r--r-- 1 root  root   130 Feb 21 04:31 sbtest-schema-create.sql
```

mydumper对于InnoDB表在进行并行备份的同时还能保证数据的一致性，其实现原理和步骤如下。

(1) 主线程获得全局读锁：

```
2021-02-21T08:04:24.827026Z     145 Query   FLUSH TABLES WITH READ LOCK
2021-02-21T08:04:24.837600Z     145 Query   START TRANSACTION /*!40108 WITH CONSISTENT
SNAPSHOT */
2021-02-21T08:04:24.837864Z   145 Query   /*!40101 SET NAMES binary*/
```

(2) 主线程获得二进制日志文件的坐标：

```
2021-02-21T08:04:24.838083Z     145 Query   SHOW MASTER STATUS
2021-02-21T08:04:24.838315Z     145 Query   SHOW SLAVE STATUS
```

(3) 多个从线程连接到 MySQL，并创建快照：

```
2021-02-21T08:04:24.843456Z    146 Query  SET SESSION wait_timeout = 2147483
2021-02-21T08:04:24.843786Z    146 Query   SET SESSION TRANSACTION ISOLATION LEVEL
REPEATABLE READ
2021-02-21T08:04:24.843926Z  146 Query   START TRANSACTION /*!40108 WITH CONSISTENT
SNAPSHOT */
2021-02-21T08:04:24.845271Z    147 Connect  root@localhost on  using Socket
2021-02-21T08:04:24.845471Z    147 Query  SET SESSION wait_timeout = 2147483
2021-02-21T08:04:24.845718Z    147 Query   SET SESSION TRANSACTION ISOLATION LEVEL
REPEATABLE READ
2021-02-21T08:04:24.845803Z  147 Query   START TRANSACTION /*!40108 WITH CONSISTENT
SNAPSHOT */
...
```

(4) 当所有从线程的快照创建完成后，主线程释放锁：

```
2021-02-21T08:04:24.850687Z  145 Query  UNLOCK TABLES /* trx-only */
```

(5) 各个线程开始进行并行数据导出：

```
2021-02-21T08:04:24.865331Z    149 Query  SELECT /*!40001 SQL_NO_CACHE */ *
FROM 'sbtest'.'sbtest1'
```

(6) 各线程导出完成后退出：

```
2021-02-21T08:04:24.864442Z    147 Quit
2021-02-21T08:04:27.658937Z    149 Quit
```

8.4.4 使用正则表达式指定数据库对象

可以使用--regex 选项包含或排除指定的数据库，例如，如下命令不备份 mysql 和 sbtest 数据库：

```
$ mydumper -u root -p dingjia -P 3306 -h localhost --regex '^(?!(mysql\.|sbtest\.))' -o
./output
```

如下命令只备份 mysql 和 sbtest 数据库：

```
$ mydumper -u root -p dingjia -P 3306 -h localhost --regex '^(mysql\.|sbtest\.)' -o
./output
```

如下命令不备份 sbtest 开头的数据库：

```
$ mydumper -u root -p dingjia -P 3306 -h localhost --regex '^(?!(sbtest))' -o ./output
```

8.5 MySQL Shell 中的备份恢复工具

8.5.1 简介

MySQL Shell 8.0 中增加了如下新的逻辑备份恢复工具。

- util.dumpInstance()：备份整个实例。
- util.dumpSchemas()：备份指定的数据库。
- util.dumpTables()：备份指定的表或视图。
- util.loadDump()：恢复备份。

如果需要导出实例中的大部分数据库，可以使用 util.dumpInstance()指定导出整个实例中的数据，然后用 excludeSchemas 选项指定排除的数据库。同样的，如果需要导出一个数据库中的大部分表，可以使用 util.dumpSchemas()指定整个数据库，然后用 excludeTables 选项指定排除的表。在导出整个实例时，不会导出 information_schema、mysql、ndbinfo、performance_schema 和 sys 数据库。

这些工具有以下特点：

- 默认使用 zstd 实时压缩。
- 支持分块并行导出，每块默认为 64MB。
- 支持并行导入。
- 支持备份到云的对象存储。

8.5.2 导出

导出工具的使用语法如下：

```
util.dumpInstance(outputUrl[, options])
util.dumpSchemas(schemas, outputUrl[, options])
util.dumpTables(schema, tables, outputUrl[, options])
```

当导出到本地时，outputUrl 是存放导出数据的目录，该目录在导出时必须为空。

查询 util.dump_schemas 的帮助信息如下：

```
MySQL  localhost:33060+ ssl  Py > \h util.dump_schemas
NAME
      dump_schemas - Dumps the specified schemas to the files in the output
                     directory.

SYNTAX
      util.dump_schemas(schemas, outputUrl[, options])

WHERE
      schemas: List of schemas to be dumped.
      outputUrl: Target directory to store the dump files.
      options: Dictionary with the dump options.

DESCRIPTION
      The schemas parameter cannot be an empty list.

      The outputUrl specifies where the dump is going to be stored.
...
```

导出 sakila 数据库到 datadump 目录的代码如下：

```
MySQL  localhost: 33060 + ssl  Py > util.dump_schemas(["sakila"],"datadump")
Acquiring global read lock
Global read lock acquired
Gathering information - done
All transactions have been started
Locking instance for backup
Global read lock has been released
Writing global DDL files
Preparing data dump for table 'sakila'.'address'
Data dump for table 'sakila'.'address' will be chunked using column 'address_id'
Preparing data dump for table 'sakila'.'category'
Data dump for table 'sakila'.'category' will be chunked using column 'category_id'
Writing DDL for schema 'sakila'
Preparing data dump for table 'sakila'.'city'
...
```

8.5.3 导入

使用 util.loadDump 工具把导出的数据导入，查询 util.load_dump 的帮助信息如下：

```
MySQL  localhost: 33060 + ssl  Py > \? util.load_dump
NAME
      load_dump - Loads database dumps created by MySQL Shell.

SYNTAX
      util.load_dump(url[, options])9

WHERE
      url: URL or path to the dump directory
      options: Dictionary with load options

DESCRIPTION
      url can be one of:

       - /path/to/file - Path to a locally or remotely (e.g. in OCI Object
         Storage) accessible file or directory
       - file:///path/to/file - Path to a locally accessible file or directory
       - http[s]://host.domain[: port]/path/to/file - Location of a remote file
         accessible through HTTP(s) (import_table() only)
...
```

导入 datadump 目录中数据的命令如下：

```
MySQL  localhost: 33060 + ssl  Py > util.load_dump("datadump")
Loading DDL and Data from 'datadump' using 4 threads.
Opening dump...
Target is MySQL 8.0.22. Dump was produced from MySQL 8.0.22
Checking for pre - existing objects...
Executing common preamble SQL
Executing DDL script for schema 'sakila'
```

```
[Worker001] Executing DDL script for 'sakila'.'actor'
[Worker000] Executing DDL script for 'sakila'.'store'
[Worker002] Executing DDL script for 'sakila'.'language'
[Worker003] Executing DDL script for 'sakila'.'rental'
...
```

8.6　四种逻辑备份工具的对比

本节介绍 4 种逻辑备份工具(mysqldump、mysqlpump、mydumper 和 MySQL shell)的对比。

8.6.1　功能对比

表 8.1 列出了本章介绍的 4 种逻辑备份工具的部分功能对比。

表　8.1

工具	并行导出	大表并行导出	并行导入	默认压缩	支持 InnoDB 表的热备	支持元数据和实际数据分开
mysqldump	×	×	×	×	√	√
mysqlpump	√	×	×	×	√	×
mydumper/myloader	√	√	√	×	√	√
MySQL Shell	√	√	√	√	√	√

8.6.2　性能对比

在虚拟机的 MySQL 实例上创建一个 sbtest 数据库，准备了两个表，每个表 100 万条记录，分别使用这 4 种备份工具进行备份，结果如下。

1. mysqldump

备份：

```
$ time mysqldump --databases sbtest > sbtest
real   0m8.655s
user   0m6.725s
sys  0m1.157s
```

恢复：

```
$ time mysql < sbtest
real   4m26.065s
user   0m4.228s
sys  0m1.060s
```

2. mysqlpump

备份：

```
$ time mysqlpump --set-gtid-purged=off  --databases sbtest > sbtest
Dump progress: 0/2 tables, 250/1973754 rows
Dump progress: 0/2 tables, 621250/1973754 rows
Dump progress: 0/2 tables, 1263000/1973754 rows
Dump progress: 0/2 tables, 1804000/1973754 rows
Dump completed in 3635

real   0m4.165s
user   0m5.832s
sys   0m0.840s
-rw-r--r-- 1 root  root  382M Feb 21 06:02 sbtest
```

恢复：

```
$ time mysql < sbtest
real   4m19.480s
user   0m4.277s
sys   0m0.950s
```

3. mydumper

备份：

```
$ time mydumper -u root -p dingjia -P 3306 -h localhost -B sbtest
--chunk-filesize 20 --trx-consistency-only -o ./output

real   0m4.329s
user   0m3.248s
sys   0m0.761s
```

恢复：

```
$ time myloader  -u root -p dingjia -P 3306 -h localhost -B sbtest -d ./output
real   3m27.018s
user   0m2.074s
sys   0m0.963s
```

4. MySQL Shell

备份：

```
MySQL  localhost:33060+ ssl  Py > util.dump_schemas(["sbtest"],"datadump")
...
4 thds dumping - 80% (1.59M rows / ~1.97M rows), 645.92K rows/s, 125.20 MB/s uncompressed,
56.75 MB/s compressed
1 thds dumping - 101% (2.00M rows / ~1.97M rows), 655.80K rows/s, 127.15 MB/s uncompressed,
57.67 MB/s compressed
Duration: 00:00:03s
Schemas dumped: 1
Tables dumped: 2
Uncompressed data size: 387.78 MB
Compressed data size: 175.86 MB
Compression ratio: 2.2
```

```
Rows written: 2000000
Bytes written: 175.86 MB
Average uncompressed throughput: 123.92 MB/s
Average compressed throughput: 56.20 MB/s
```

恢复：

```
MySQL  localhost:33060+ ssl  Py > util.load_dump("datadump")
...
15 chunks (2.00M rows, 387.78 MB) for 2 tables in 1 schemas were loaded in 2 min 28 sec (avg
throughput 2.62 MB/s)
0 warnings were reported during the load.
```

由于备份环境和数据量的限制，这里的测试结果不一定具有代表性，但有一定的参考价值，对比结果如表 8.2 所示。

表 8.2

工　具	备份用时/s	恢复用时	备份集大小/MB
mysqldump	8.6	4min26s	382
mysqlpump(默认 2 线程)	4.2	4min19s	382
mydumper/myloader(并行块单位 20MB)	4.3	3min27s	384
util.dump_schemas/util.load_dump	3	2min28s	169

可以看到压缩比不压缩快，并行比不并行快，其中 MySQL Shell 的备份速度最快。

8.7 备份集的一致性

由于逻辑备份过程中 MySQL 数据库处于打开状态，数据在备份过程中可能会发生变化，因此在执行逻辑备份的过程中如何保证生成备份集的一致性就成为一个让很多人关心的问题。一个便捷的方法是在 MySQL 启动时设定选项 skip_networking 禁止远端连接，这样只要本地用户不修改数据库，数据就是一致的，但这样影响了数据库向外提供服务，因此并不常用，常见的方式是采用读锁，或者把整个备份过程放入单个事务中来保证数据的一致性。

8.7.1 读锁

对需要备份的对象加上只读锁可以保证备份集的一致性，例如，对某个表上的只读锁使用如下的命令：

```
mysql > flush tables sakila.tab_a with read lock;
```

之后如果要修改该表会被拒绝，例如：

```
mysql > insert into tab_a values(2);
ERROR 1099 (HY000): Table 'tab_a' was locked with a READ lock and can't be updated
```

如果要进行全库备份，可以使用如下命令加上全局读锁：

```
mysql > flush tables with read lock;
```

如上命令会关闭所有打开的表，对于 MyISAM 的表，会把系统缓存中的脏数据刷新到磁盘上，保证备份的数据是一致的，然后对所有的表加上了读锁。这样设置后，客户端执行修改会等待超时，超时时间受参数 lock_wait_timeout 控制，默认是 31 536 000 秒，也就是 365 天，因此应再设置下面的参数：

```
mysql > set global read_only = on;
```

这样设置后，客户端执行修改时会马上返回类似下面的错误：

```
ERROR 1290 (HY000): The MySQL server is running with the -- read - only option so it cannot
execute this statement
```

执行这两个命令后，数据库将不会发生变化，此时进行备份不需要再加锁。可以使用 mysqldump 等逻辑备份工具进行备份，但不能采用直接复制数据文件的方式进行备份，因为缓存里仍然有 InnoDB 表的已经修改了但没有被刷新到磁盘上的数据，这时备份的数据文件是不一致的。

因为 flush tables with read lock 会关闭所有打开的表，因此该锁需要等到数据库中时间运行最长的 SQL 语句完成后才能加锁成功。如果遇到长时间加锁不成功时，使用如下的 SQL 语句可以找出阻止加锁的连接：

```
mysql > select id,user,host,time from information_schema.processlist where time >(select time
from information_schema.processlist where info = 'flush tables with read lock') and info is not
null;
+----+------+-----------+------+
| id | user | host      | time |
+----+------+-----------+------+
| 10 | root | localhost | 525  |
+----+------+-----------+------+
1 row in set (0.00 sec)
```

也可以设置系统参数 max_execution_time 来限制最长执行 SQL 语句的时间，该参数的单位是毫秒，例如，如下命令为设置最长执行的 SQL 语句不能超过 10 秒：

```
mysql > set max_execution_time = 10000;
Query OK, 0 rows affected (0.00 sec)
```

执行一个需要 10 秒的 SQL 语句：

```
mysql > select 1 where sleep(10);
Empty set (10.00 sec)
```

可以成功，再执行一个需要 11 秒的 SQL 语句：

```
mysql > select 1 where  sleep(11);
ERROR 3024 (HY000): Query execution was interrupted, maximum statement execution time exceeded
```

执行时间需要 11 秒的 SQL 语句被拒绝了。

也可以使用提示指定最长的 SQL 语句执行时间，例如：

```
mysql> select /* + max_execution_time(10000) */ 1 where  sleep(11);
ERROR 3024 (HY000): Query execution was interrupted, maximum statement execution time exceeded
```

当设置了该系统参数后，用户可能需要知道哪些 SQL 语句是因此超时被拒绝执行的，这些 SQL 语句可以按照错误号是 3024 的规则在 performance_schema 数据库中的 events_statements_history 和 events_statements_history_long 视图中查询出，如下所示：

```
mysql> select * from performance_schema.events_statements_history where mysql_errno = 3024;
```

但有时生成统计报告的 SQL 语句就是需要运行很长时间，这种情况下只能避开这种 SQL 运行的时段进行加锁。

这种上读锁的方法虽然能保证数据的一致性，但毕竟影响了数据库对外提供服务，下面介绍的利用 InnoDB 引擎的事务特性在保证备份集一致性的同时还能同时对外提供读写服务。

8.7.2 单个事务

对于 InnoDB 的表，可以利用 InnoDB 的事务特性，使用--single-transaction 选项把整个导出过程放在一个事务中完成，这样既可以保证备份集的一致性，又能同时对外提供数据库的读写服务。但这种方法只能保证 InnoDB 表的数据一致性，对 MyISAM 之类的非 InnoDB 表无效。采用这种方法时，在通用查询日志里可以看到在备份开始阶段 mysqldump 执行了如下命令：

```
SET SESSION TRANSACTION ISOLATION LEVEL REPEATABLE READ
START TRANSACTION /*!40100 WITH CONSISTENT SNAPSHOT */
```

在备份开始时把事务的隔离级别设置为可重复读(repeatable read)，在导出数据之前提交一个 start transaction 语句开启事务，备份完成退出时该事务结束。这种备份方法既可以让数据库对外提供读写服务，又能得到一致性的备份集，但事情并没有想象的那么美好，维持单个事务的长期活跃也是有成本的，在高负载下有时这种成本高到让人无法接受。事务在活跃过程中主要消耗两种资源：锁和 undo 日志。

隔离级别为可重复读(这也是 MySQL 默认的隔离级别)的事务会对所有的访问对象加上元数据共享锁，这些锁在 SQL 执行完成也不会释放，直到事务执行完成才会释放，如果大量的锁得不到释放，可能会出现 ER_LOCK_TABLE_FULL 的错误，造成事务失败，在 MySQL 的错误日志里有如下记录：

```
ERROR: 1206: The total number of locks exceeds the lock table size
```

锁会占用 InnoDB 的缓存，如果占用过多，在 MySQL 的错误日志里有如下警告信息：

```
[Warning] InnoDB: Over 67 percent of the buffer pool is occupied by lock heaps or the adaptive hash index! Check that your transactions do not set too many row locks. Your buffer pool size is 8 MB. Maybe you should make the buffer pool bigger?. Starting the InnoDB Monitor to print diagnostics, including lock heap and hash index sizes.
```

长事务消耗的另外一个资源就是 undo 日志，会造成存放 undo 日志的表空间不断膨胀，在 MySQL 5.7 里 undo 日志是在 ibdata1 文件中，在 MySQL 8.0 里默认有两个独立的 undo 表空间。很多人误以为备份过程是只读，不会产生数据变更，不需要 undo 日志。但实际上隔离级别为可重复读的事务，需要在事务开始时建立一个一致性的快照，在此之后其他事务修改的数据也要在 undo 日志里保存原始的快照，这样越是数据变化快的数据库，undo 日志增长得就越快。而且 undo 日志越长，要访问原始快照的用时就越长。在备份过程中，检查 show engine innodb status 里的 transactions 部分可以得到事务的相关信息，例如：

```
------------
TRANSACTIONS
------------
Trx id counter 9270
Purge done for trx's n: o < 2567 undo n: o < 0 state: running but idle
History list length 3233
LIST OF TRANSACTIONS FOR EACH SESSION:
--- TRANSACTION 421456989354560, not started
0 lock struct(s), heap size 1136, 0 row lock(s)
--- TRANSACTION 421456989353704, not started
0 lock struct(s), heap size 1136, 0 row lock(s)
--- TRANSACTION 9269, ACTIVE (PREPARED) 0 sec
mysql tables in use 1, locked 1
...
```

其中，第一行显示的是当前的事务 id；第二行是 purge 线程清除的 undo 日志信息，小于事务 2567 的 undo 日志已经被清除；第三行的 History list length 是需要保留的 undo 日志的长度；后面是当前的事务信息，为了节约篇幅，没有全部列出。

过了一小段时间后，在备份过程中再次检查事务的信息如下：

```
------------
TRANSACTIONS
------------
Trx id counter 31117
Purge done for trx's n: o < 2567 undo n: o < 0 state: running but idle
History list length 13762
LIST OF TRANSACTIONS FOR EACH SESSION:
--- TRANSACTION 421456989354560, not started
0 lock struct(s), heap size 1136, 0 row lock(s)
--- TRANSACTION 421456989353704, not started
0 lock struct(s), heap size 1136, 0 row lock(s)
...
```

可以看到，事务 id 从 9270 增加到 31 117，但清除的 undo 日志事务号仍是 2567，History list length 从 3233 增长到 13 762。这是因为为了保证备份过程中数据的一致性，已经变化的数据的旧镜像没有被清除。

从 information_schema.innodb_metrics 视图中也能查询到 History list length，例如：

```
mysql > select name, count, status
from information_schema.innodb_metrics
where subsystem = 'transaction';
```

```
+---------------------------+-------+----------+
| name                      | count | status   |
+---------------------------+-------+----------+
| trx_rw_commits            |     0 | disabled |
| trx_ro_commits            |     0 | disabled |
| trx_nl_ro_commits         |     0 | disabled |
| trx_commits_insert_update |     0 | disabled |
| trx_rollbacks             |     0 | disabled |
| trx_rollbacks_savepoint   |     0 | disabled |
| trx_rollback_active       |     0 | disabled |
| trx_active_transactions   |     0 | disabled |
| trx_allocations           |     0 | disabled |
| trx_on_log_no_waits       |     0 | disabled |
| trx_on_log_waits          |     0 | disabled |
| trx_on_log_wait_loops     |     0 | disabled |
| trx_rseg_history_len      | 13762 | enabled  |
| trx_undo_slots_used       |     0 | disabled |
| trx_undo_slots_cached     |     0 | disabled |
| trx_rseg_current_size     |     0 | disabled |
+---------------------------+-------+----------+
16 rows in set (0.00 sec)
```

可以看到，这里只有 trx_rseg_history_len 是激活的，它的值也是 13 762，和使用 show engine innodb status 查询到的 History list length 一致。在 sys.metrics 视图里也能查询到类似的信息：

```
mysql > select Variable_name, Variable_value, Enabled from sys.metrics where Type = 'InnoDB
Metrics - transaction';
+---------------------------+----------------+---------+
| Variable_name             | Variable_value | Enabled |
+---------------------------+----------------+---------+
| trx_active_transactions   | 0              | NO      |
| trx_allocations           | 0              | NO      |
| trx_commits_insert_update | 0              | NO      |
| trx_nl_ro_commits         | 0              | NO      |
| trx_on_log_no_waits       | 0              | NO      |
| trx_on_log_wait_loops     | 0              | NO      |
| trx_on_log_waits          | 0              | NO      |
| trx_ro_commits            | 0              | NO      |
| trx_rollback_active       | 0              | NO      |
| trx_rollbacks             | 0              | NO      |
| trx_rollbacks_savepoint   | 0              | NO      |
| trx_rseg_current_size     | 0              | NO      |
| trx_rseg_history_len      | 13762          | YES     |
| trx_rw_commits            | 0              | NO      |
| trx_undo_slots_cached     | 0              | NO      |
| trx_undo_slots_used       | 0              | NO      |
```

```
+-----------------------------+-----------------+----------+
16 rows in set (0.02 sec)
```

History list length 这个值越大，对系统性能的负面影响就越大。如果把事务的隔离级别设置为提交读(read committed)，在备份过程中持有的锁和产生的 undo 日志将会少得多，但这种隔离级别无法保证备份集的一致性。

从前面的分析可以看出，采用单一事务的方式解决备份集一致性的问题对系统的性能影响也很大，要尽量在系统负载小的时候进行备份，同时备份的时间不能太长，不然对系统性能的负面影响很大。

8.8 提高恢复的速度

逻辑备份的最大缺点就是速度慢，该缺点在对 InnoDB 表进行备份时并不特别突出，因为这些备份工具通常支持对 InnoDB 表进行热备。但恢复时速度慢就让人着急了，因为恢复时数据库不能对外提高服务，恢复的时间越长，对业务的影响越大。

很多时候提高逻辑备份的恢复速度采用的方法和提高 MySQL 实例的联机交易速度的方法一样，例如，增加 InnoDB Buffer 的大小，增加 logfile 的大小，提高硬件的性能等。本节介绍除此以外的提高逻辑备份恢复速度的方法。

8.8.1 禁用日志

禁用日志可以减少导入过程中的 IO，从而提高导入的速度，这里的禁用日志包括二进制日志、联机日志和 InnoDB 的 Double Write。

1. 禁用二进制日志

MySQL 8.0 中二进制日志默认是激活的，但导出的文件通常有下面的语句禁止在导入过程中记录二进制日志：

```
SET @@SESSION.SQL_LOG_BIN= 0;
```

如果导出的文件没有这样的语句，可以在实例启动时增加--skip-log-bin 或--disable-log-bin 参数来禁止二进制日志。

如果使用 source 执行加载的 SQL 语句，可以在执行之前禁用二进制日志，执行完成后打开二进制日志，例如：

```
mysql > set SQL_LOG_BIN = 0;
mysql > source backup.sql
mysql > set SQL_LOG_BIN = 1;
```

2. 禁用 InnoDB 的 Double Write

InnoDB 的 Double Write 机制是为了保证 InnoDB 的原子写而产生的，在数据导入的过程中，为了提高效率，可以禁用 InnoDB 的 Double Write。通过设置系统参数 innodb_doublewrite 为 off 禁用 InnoDB 的 Double Write，如下所示：

```
mysql > set persist_only innodb_doublewrite = off;
```

系统参数 innodb_doublewrite 是静态参数，设置完成后需要重新启动 MySQL 实例，也可以在启动 MySQL 实例时使用参数--skip-innodb-doublewrite 禁用 InnoDB 的 Double Write。注意，在数据导入完成后要取消该设置。

3. 禁用联机日志

在 MySQL 8.0.21 中新增了禁用联机日志的功能，同时还会禁用 InnoDB 的 Double Write。禁用联机日志是实例级，不支持表级。启动这项功能的命令如下：

```
mysql > alter instance disable innodb redo_log;
```

可以使用状态参数 Innodb_redo_log_enabled 检查联机日志的使用情况，相应 SQL 语句及运行结果如下：

```
mysql > show global status like 'Innodb_redo_log_enabled';
+-------------------------+-------+
| Variable_name           | Value |
+-------------------------+-------+
| Innodb_redo_log_enabled | OFF   |
+-------------------------+-------+
```

重新激活联机日志的命令如下：

```
mysql > alter instance enable innodb redo_log;
```

禁用联机日志时可以正常地关闭和重启实例，但在异常宕机的情况下，可能会导致数据丢失或页面损坏，禁用联机日志时，异常宕机的实例可能需要废弃重建，直接重启会有如下报错：

```
[ERROR] [MY - 013578] [InnoDB] Server was killed when Innodb Redo logging was disabled. Data files could be corrupt. You can try to restart the database with innodb_force_recovery = 6.
```

表 8.3 是在一个虚拟机上导入 200 万条记录的性能对比。

表 8.3

对比项	实验 1	实验 2
二进制日志	×	×
InnoDB 的 Double Write	√	×
联机日志	√	×
用时	4min44s	55s

从表 8.3 中可以看到，禁用联机日志后的导入性能提高了 4.2 倍。

对于压缩后的备份集，不要在完成解压后再进行加载，应该在单个操作中同时完成解压和加载，这样会快很多，例如：

```
$ gzip -d backup.sql.gz|mysql
```

8.8.2 并行导入

从前面4种逻辑备份的工具对比中可以发现，并行比串行操作要快得多，但mysqldump和mysqlpump导出的文件在使用mysql导入时只能串行操作，怎么解决这个问题呢？可以使用--tab选项将数据导出成文本文件，在导入时使用mysqlimport，并使用选项--use-threads指定多个线程并行操作。但这种并行是在表之间进行的，对于一个表同时只能有一个线程进行操作。为了解决这个问题，可以使用split命令将导出的文本文件切割成若干文件，再用parallel工具并发多个mysqlimport进行导入，下面实验一下。

对一个有100万条记录的表首先使用传统的方法进行导出，命令如下：

```
$ mysqldump  sbtest sbtest1 > sbtest1.sql
```

再使用传统的方法进行导入，命令如下：

```
$  time mysql -e "source sbtest1.sql" sbtest
real  0m44.579s
user  0m1.896s
sys  0m0.386s
```

使用mysqldump的--tab选项将数据导出，命令如下：

```
$ mysqldump --tab=/var/lib/mysql-files/ sbtest sbtest1
```

再使用split将含有100万条记录的文件切割成10个文件，命令如下：

```
$ split -l 100000 /var/lib/mysql-files/sbtest1.txt sbtest1
$ ll sbtest1.*
-rw-r--r-- 1 root root  19288895 Feb 22 04:15 sbtest1.aa
-rw-r--r-- 1 root root  19400000 Feb 22 04:15 sbtest1.ab
-rw-r--r-- 1 root root  19400000 Feb 22 04:15 sbtest1.ac
-rw-r--r-- 1 root root  19400000 Feb 22 04:15 sbtest1.ad
-rw-r--r-- 1 root root  19400000 Feb 22 04:15 sbtest1.ae
-rw-r--r-- 1 root root  19400000 Feb 22 04:15 sbtest1.af
-rw-r--r-- 1 root root  19400000 Feb 22 04:15 sbtest1.ag
-rw-r--r-- 1 root root  19400000 Feb 22 04:15 sbtest1.ah
-rw-r--r-- 1 root root  19400000 Feb 22 04:15 sbtest1.ai
-rw-r--r-- 1 root root  19400001 Feb 22 04:15 sbtest1.aj
-rw-r--r-- 1 root root 199896635 Feb 22 03:40 sbtest1.sql
```

每个文件包含10万条记录。

安装并发的工具parallel，命令如下：

```
$ sudo apt install parallel
```

导入时先创建表，命令如下：

```
$ mysql sbtest </var/lib/mysql-files/sbtest1.sql
```

创建表后使用parallel并发调用mysqlimport进行导入，命令如下：

```
$ time parallel 'mysqlimport sbtest' :::/var/lib/mysql-files/sbtest1.*
Academic tradition requires you to cite works you base your article on.
When using programs that use GNU Parallel to process data for publication
please cite:

  O. Tange (2011): GNU Parallel - The Command-Line Power Tool,
  ;login: The USENIX Magazine, February 2011: 42-47.

This helps funding further development; AND IT WON'T COST YOU A CENT.
If you pay 10000 EUR you should feel free to use GNU Parallel without citing.

To silence this citation notice: run 'parallel --citation'.

sbtest.sbtest1: Records: 100000  Deleted: 0  Skipped: 0  Warnings: 0
sbtest.sbtest1: Records: 100000  Deleted: 0  Skipped: 0  Warnings: 0
sbtest.sbtest1: Records: 100000  Deleted: 0  Skipped: 0  Warnings: 0
sbtest.sbtest1: Records: 100000  Deleted: 0  Skipped: 0  Warnings: 0
sbtest.sbtest1: Records: 100000  Deleted: 0  Skipped: 0  Warnings: 0
sbtest.sbtest1: Records: 100000  Deleted: 0  Skipped: 0  Warnings: 0
sbtest.sbtest1: Records: 100000  Deleted: 0  Skipped: 0  Warnings: 0
sbtest.sbtest1: Records: 100000  Deleted: 0  Skipped: 0  Warnings: 0
sbtest.sbtest1: Records: 100000  Deleted: 0  Skipped: 0  Warnings: 0
sbtest.sbtest1: Records: 100000  Deleted: 0  Skipped: 0  Warnings: 0

real  0m18.280s
user  0m0.304s
sys  0m0.163s
```

可以看到，在并发了 10 个 mysqlimport 进程进行导入时，用时 18 秒完成，相对于之前单进程的 44 秒，速度大幅提高。

对于需要使用自增序列生成主键的表，在导入的过程中很可能会发生自增锁冲突，此时用 show engine innodb status 查看 InnoDB 引擎的状态，可以看到大量的 setting auto-inc lock 锁。这是因为并发的线程竞争自增锁产生的，可以设定 --innodb_autoinc_lock_mode=2 来消除该锁的竞争，但此时生成的自增序列可能不是连续的。

8.8.3 其他方法

导入数据的方法有两种，一种是用 insert，另外一种是 load data，通常 load data 会比 insert 快，因为省去了 SQL 语句解析的工作。

对于 InnoDB 表，导入的数据文件如果是按主键排序的，导入时速度会快很多，因为在导入的过程中不会产生页分裂。因此在用 mysqldump 导出时，最好加上--order-by-primary 选项。mysqlpump 没有这种类似的选项，因为多线程之间无法进行排序。但对于 InnoDB 的表，无论加不加这些选项，通常是按照主键导出的。该选项在将堆表(如 MyISAM 的表)导入 InnoDB 表中很有用。

其他可以加快导入速度的方法还包括：

- 把创建索引放在导入完成后进行，这也是 mysqlpump 的默认方法。

- 在导入过程中禁用外键检查，设置 foreign_key_checks 为 off。
- 在导入过程中禁用唯一值检查，设置 unique_checks 为 off。
- 把事务的隔离级别从可重复读(repeatable read)改为未提交读(read uncommitted)，使用系统参数 transaction_isolation 进行设置。

这些选项都有一定的副作用，例如，禁用外键和唯一值检查要在确定源数据符合这些规则的情况下才能启用，不然将造成数据不一致。

8.9 实验

视频演示

1. mysqldump 备份恢复工具

使用 mysqldump 进行 MySQL 的备份恢复(参考 8.2 节)。

视频演示

2. mysqlpump 备份恢复工具

使用 mysqlpump 进行 MySQL 的备份(参考 8.3 节)。

视频演示

3. MySQL Shell 中的备份恢复工具

使用 MySQL Shell 中的工具进行 MySQL 的备份恢复(参考 8.5 节)。

第9章

XtraBackup

MySQL 的备份分为逻辑备份和物理备份，第 8 章介绍了逻辑备份，第 9 章和第 10 章将分别介绍两个流行的 MySQL 的物理备份工具 XtraBackup 和 MySQL Enterprise Backup。

9.1 特点介绍

XtraBackup 是 Percona 公司 CTO Vadim 参与开发的一款基于 InnoDB 的在线热备工具，它有以下特点：

- 开源、免费。
- 支持 InnoDB 表的热备。
- 对 MyISAM 表和其他非 InnoDB 表进行备份时数据库处于只读状态。
- 支持增量备份。
- 支持加密和压缩备份。
- 通过压缩减少占用硬盘空间和网络带宽。
- 备份时不增加 MySQL 服务的负载。

MySQL 5.6 和 5.7 对应的 Percona XtraBackup 版本是 2.4，MySQL 8.0 对应的 Percona XtraBackup 版本是 8.0。和向下兼容的惯例不同，Percona XtraBackup 8.0 不能备份 MySQL 5.6 和 5.7。

Percona XtraBackup 和 MySQL 官方的物理备份工具 MySQL Enterprise Backup 对比见表 9.1。

表 9.1

特　　性	Percona XtraBackup	MySQL Enterprise Backup
收费	免费	每台服务器参考价 5000 美元
许可证	GPL	专有软件

续表

特　性	Percona XtraBackup	MySQL Enterprise Backup
源码	开源	闭源
支持 MySQL 分支	MySQL、Percona Server for MySQL、Percona XtraDB Cluster	MySQL
支持操作系统	Linux	Linux、Solaris、Windows、OSX、FreeBSD
InnoDB 表热备	√	√
非 InnoDB 表只读备份	√	√
增量备份	√	√
使用 redo log 进行增量备份	×	√
归档备份	×	√
备份时跳过二级索引	√	×
节流备份	√	√
单表备份恢复	√	√
备份历史表	√	√
备份进度表	×	√

9.2 安装

Percona XtraBackup 的下载地址是 https://www.percona.com/downloads/Percona-XtraBackup-LATEST/#。

下载界面如图 9.1 所示。

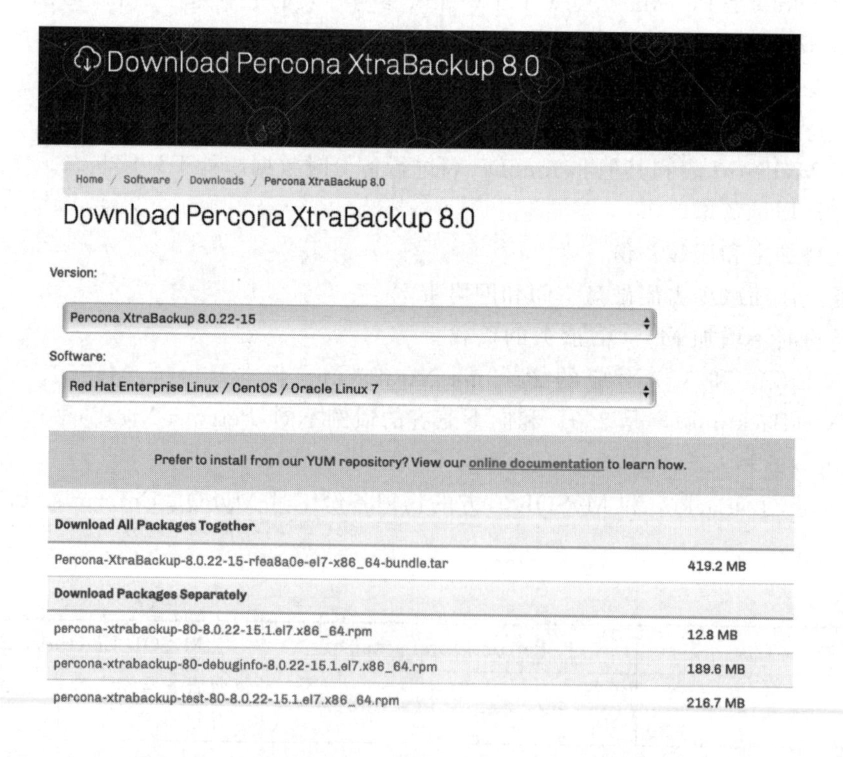

图 9.1

选择版本和操作系统后(这里以64位的Red Hat 7为例),会出现下载软件包,有如下3个rpm安装包:

(1) percona-xtrabackup-80是Percona XtraBackup的安装包。

(2) percona-xtrabackup-80-debuginfo是包括调试信息的Percona XtraBackup的安装包。

(3) percona-xtrabackup-test-80是Percona XtraBackup的测试组件。

通常只需要安装percona-xtrabackup-80,可以通过如下链接下载:https://downloads.percona.com/downloads/Percona-XtraBackup-LATEST/Percona-XtraBackup-8.0.22-15/binary/redhat/7/x86_64/percona-xtrabackup-80-8.0.22-15.1.el7.x86_64.rpm。

下面完成后使用yum localinstall进行安装,命令如下:

```
$ sudo yum localinstall percona-xtrabackup-80-8.0.22-15.1.el7.x86_64.rpm
```

9.3 工作原理

Percona XtraBackup实现了对InnoDB表的热备。Percona XtraBackup在备份InnoDB数据文件的同时,会把整个过程中重做日志的变化备份下来。在恢复之前,使用InnoDB的崩溃恢复(crash recovery)机制将重做日志的变化应用到InnoDB的数据文件,使InnoDB的数据文件达到一致的状态,这和使用二进制日志进行时间点恢复是完全不同的。Percona XtraBackup的备份恢复过程分成3个阶段,分别为:备份(backup)、准备(prepare)、恢复(restore),下面分别说明。

9.3.1 备份阶段

Percona XtraBackup备份的基本语法如下:

```
$ xtrabackup --backup --target-dir=/data/backups
```

其中,/data/backups是备份集存放的目录。

在Percona XtraBackup运行时打开通用日志,根据Percona XtraBackup运行时的提示和通用日志的记录,可分析出Percona XtraBackup全量备份MySQL的大致步骤如下。

(1) 启动一个线程读取重做日志,该线程在整个备份过程中一直运行并每秒读取一次重做日志,会不断地输出类似下面的提示:

```
210225 10:08:32 >> log scanned up to (25765184)
210225 10:08:33 >> log scanned up to (25765184)
210225 10:08:34 >> log scanned up to (25765184)
```

(2) 使用如下SQL语句找出所有的InnoDB表空间进行复制:

```
SELECT T2.PATH,          T2.NAME,          T1.SPACE_TYPE FROM
INFORMATION_SCHEMA.INNODB_TABLESPACES T1          JOIN
INFORMATION_SCHEMA.INNODB_TABLESPACES_BRIEF T2 USING (SPACE) WHERE
T1.SPACE_TYPE = 'Single' && T1.ROW_FORMAT != 'Undo'UNION SELECT T2.PATH,
```

```
SUBSTRING_INDEX(SUBSTRING_INDEX(T2.PATH, '/', -1), '.', 1) NAME,
T1.SPACE_TYPE FROM   INFORMATION_SCHEMA .INNODB_TABLESPACES T1
JOIN INFORMATION_SCHEMA .INNODB_TABLESPACES_BRIEF T2 USING (SPACE) WHERE
T1.SPACE_TYPE =  'General' && T1.ROW_FORMAT !=  'Undo';
```

备份时的类似提示如下：

```
210225 10: 08: 30 [01] Copying ./ibdata1 to /data/backups/ibdata1
210225 10: 08: 31 [01]          ...done
```

(3) 对 MySQL 上只全局读锁(flush tables with read lock)，然后备份非 InnoDB 表，注意，此过程中 InnoDB 类型的表也被锁住，提示如下：

```
210225 10: 08: 31 Executing FLUSH NO_WRITE_TO_BINLOG TABLES...
210225 10: 08: 31 Executing FLUSH TABLES WITH READ LOCK...
210225 10: 08: 31 Starting to backup non - InnoDB tables and files
210225 10: 08: 31 [01] Copying mysql/general_log_211.sdi to
/data/backups/mysql/general_log_211.sdi
210225 10: 08: 31 [01]          ...done
...
```

(4) 备份非 InnoDB 表完成后，刷新日志并记录与日志相关的信息，然后释放锁，提示如下：

```
210225 10: 08: 34 Finished backing up non - InnoDB tables and files
210225 10: 08: 34 Executing FLUSH NO_WRITE_TO_BINLOG BINARY LOGS
210225 10: 08: 34 Selecting LSN and binary log position from p_s.log_status
210225 10: 08: 34 [00] Copying /disk1/data/binlog.000003 to
/data/backups/binlog.000003 up to position 156
210225 10: 08: 34 [00]          ...done
210225 10: 08: 34 [00] Writing /data/backups/binlog.index
210225 10: 08: 34 [00]          ...done
210225 10: 08: 34 [00] Writing /data/backups/xtrabackup_binlog_info
210225 10: 08: 34 [00]          ...done
210225 10: 08: 34 Executing FLUSH NO_WRITE_TO_BINLOG ENGINE LOGS...
xtrabackup: The latest check point (for incremental): '25765184'
xtrabackup: Stopping log copying thread at LSN 25765194.
210225 10: 08: 34 >> log scanned up to (25765204)
Starting to parse redo log at lsn =  25764913

210225 10: 08: 35 Executing UNLOCK TABLES
210225 10: 08: 35 All tables unlocked
```

其中 FLUSH NO_WRITE_TO_BINLOG ENGINE LOGS 是强制将 InnoDB 的 redo 刷新到磁盘，NO_WRITE_TO_BINLOG 是指不将该语句记录到二进制日志中。

其中查询日志相关信息的 SQL 语句如下：

```
mysql >   SELECT server_uuid, local, replication, storage_engines FROM performance_schema.log_
status \G
*************************** 1. row ***************************
    server_uuid: 108d5424 - 766d - 11eb - b140 - fa163e368ff4
```

```
        local: {"gtid_executed": "", "binary_log_file": "binlog.000003", "binary_log_
position": 156}
    replication: {"channels": []}
storage_engines: {"InnoDB": {"LSN": 25765214, "LSN_checkpoint": 25765214}}
```

（5）备份之前没有备份的日志文件和元数据文件，然后成功结束，提示如下：

```
210225 10:08:35 [00] Copying ib_buffer_pool to /data/backups/ib_buffer_pool
210225 10:08:35 [00]        ...done
210225 10:08:35 Backup created in directory '/data/backups/'
MySQL binlog position: filename 'binlog.000003', position '156'
210225 10:08:35 [00] Writing /data/backups/backup-my.cnf
210225 10:08:35 [00]        ...done
210225 10:08:35 [00] Writing /data/backups/xtrabackup_info
210225 10:08:35 [00]        ...done
xtrabackup: Transaction log of lsn (25765184) to (25765214) was copied.
210225 10:08:37 completed OK!
```

因为整个备份过程没有对数据库进行任何改动，用户可以在备份过程中随时中断备份。备份完成后的目录结构如下：

```
$ ll ./xtrb/back1
total 57400
-rw-r-----. 1 root root      476 Feb 25 10:08 backup-my.cnf
-rw-r-----. 1 root root      156 Feb 25 10:08 binlog.000003
-rw-r-----. 1 root root       16 Feb 25 10:08 binlog.index
-rw-r-----. 1 root root     5502 Feb 25 10:08 ib_buffer_pool
-rw-r-----. 1 root root 12582912 Feb 25 10:08 ibdata1
drwxr-x---. 2 root root     4096 Feb 25 10:08 mysql
-rw-r-----. 1 root root 25165824 Feb 25 10:08 mysql.ibd
drwxr-x---. 2 root root     8192 Feb 25 10:08 performance_schema
drwxr-x---. 2 root root       82 Feb 25 10:08 sbtest
drwxr-x---. 2 root root       27 Feb 25 10:08 sys
-rw-r-----. 1 root root 10485760 Feb 25 10:08 undo_001
-rw-r-----. 1 root root 10485760 Feb 25 10:08 undo_002
-rw-r-----. 1 root root       18 Feb 25 10:08 xtrabackup_binlog_info
-rw-r-----. 1 root root       95 Feb 25 10:08 xtrabackup_checkpoints
-rw-r-----. 1 root root      452 Feb 25 10:08 xtrabackup_info
-rw-r-----. 1 root root     2560 Feb 25 10:08 xtrabackup_logfile
-rw-r-----. 1 root root       39 Feb 25 10:08 xtrabackup_tablespaces
```

如上目录里包括了和原 MySQL 数据目录一样的内容（但没有重做日志文件），还有如下 6 个 Percona XtraBackup 专用的文件。

（1）backup-my.cnf：在准备阶段要用到的 InnoDB 参数。

（2）xtrabackup_binlog_info：备份结束时二进制日志坐标，包括二进制文件号和位置。

（3）xtrabackup_checkpoints：备份的元数据——备份类型、LSN 等。

（4）xtrabackup_info：备份的信息，包括服务版本、开始和结束时间、二进制日志坐标等。

（5）xtrabackup_logfile：备份过程中 MySQL 产生的重做日志，在准备阶段使用。

（6）xtrabackup_tablespaces：备份的表空间信息。

9.3.2 准备阶段

由于备份的同时MySQL实例在运行着，备份的数据文件不是一致的，这样的备份集恢复到生产系统，MySQL实例将无法启动，准备阶段通过完成如下两项工作将备份集准备到可以恢复的状态：

（1）将数据文件恢复到一致的状态。

（2）生成联机日志文件。

Percona XtraBackup会使用InnoDB的崩溃恢复机制将备份过程中产生的重做日志应用到数据文件上，使数据文件达到一致的状态，此项工作通过Percona XtraBackup内置的一个类似Innodb引擎完成，因此准备阶段不需要在原服务器上完成，可以将备份集复制到任意一台服务器上完成此项工作，下面是一个例子：

```
$ xtrabackup --prepare  --target-dir=/data/backups
xtrabackup: recognized server arguments: --innodb_checksum_algorithm=crc32
--innodb_log_checksums=1 --innodb_data_file_path=ibdata1:12M:autoextend
--innodb_log_files_in_group=2 --innodb_log_file_size=104857600
--innodb_page_size=16384 --innodb_undo_directory=./
--innodb_undo_tablespaces=2 --server-id=0 --innodb_log_checksums=ON
--innodb_redo_log_encrypt=0 --innodb_undo_log_encrypt=0
xtrabackup: recognized client arguments: --prepare=1
--target-dir=/data/backups
xtrabackup version 8.0.22-15 based on MySQL server 8.0.22 Linux (x86_64)
(revision id: fea8a0e)
xtrabackup: cd to /data/backups/
xtrabackup: This target seems to be not prepared yet.
...
Starting shutdown...
Log background threads are being closed...
Shutdown completed; log sequence number 25765388
210224 19:36:18 completed OK!
```

在准备阶段完成时会有类似InnoDB关闭的提示。

Percona XtraBackup会检测一个备份集是否已经准备好，如果对一个已经完成了准备阶段的备份集再次进行准备会遇到下面的报错：

```
xtrabackup: This target seems to be already prepared.
xtrabackup: notice: xtrabackup_logfile was already used to '--prepare'.
```

在准备过程中不建议中断程序的执行，因为数据文件已经发生变化，如果中断将会破坏数据文件的一致性。

9.3.3 恢复阶段

应满足如下3个条件才能进入恢复阶段：

（1）备份集成功准备好。

（2）目标 MySQL 实例关闭。

（3）要恢复的 datadir 目录为空。

在恢复阶段，Percona XtraBackup 会读取配置文件中与目录相关的参数，如 datadir、innodb_data_home_dir、innodb_data_file_path、innodb_log_group_home_dir，并判断这些目录是否存在，因此恢复时不用指定恢复的目录，可以使用如下命令进行恢复：

```
$ xtrabackup --copy-back  --target-dir=/data/backups
```

为了节约硬盘空间，也可以使用 --move-back 选项代替--copy-back 将数据移动到 datadir 目录。还可以直接使用操作系统的命令 rsync 或 cp 将备份集直接复制过去，一个用 rsync 进行恢复的例子如下：

```
$ rsync -avrP /data/backups /var/lib/mysql/
```

恢复完成后必须保证目录和文件的属主和权限是正确的，可以使用如下命令改变文件的属主：

```
$ sudo chown -R mysql:mysql /var/lib/mysql
```

这样就恢复完成了。

9.4 典型用例

9.3 节介绍备份的 3 个阶段时已经介绍了全量备份，下面介绍其他的典型用例。

9.4.1 增量备份

Percona XtraBackup 支持增量备份，用户可以在两次全备之间创建任意次数的增量备份。一个典型的用例是每周日做一次全量备份，每周其他天做增量备份。

1. 备份阶段

MySQL 数据文件的每个页都包含着一个 LSN(log sequence number，日志序号)，LSN 是 MySQL 实例级别的系统版本号，每页的 LSN 代表该页的修改时间。增量备份时 Percona XtraBackup 扫描所有页，只备份 LSN 大于上次备份时的 LSN 的页。在备份集里有个 xtrabackup_checkpoints 文件记录着上次备份的 to_lsn，如下代码显示了该文件的内容：

```
$ cat /data/backups/base/xtrabackup_checkpoints
backup_type = full-prepared
from_lsn = 0
to_lsn = 25765184
last_lsn = 25765194
flushed_lsn = 0
```

Percona XtraBackup 根据 to_lsn 进行增量备份，一个增量备份的例子如下：

```
$ xtrabackup --backup --target-dir=/data/backups/inc1
--incremental-basedir=/data/backups/base
```

增量备份中也包括一个 xtrabackup_checkpoints 文件，其中记录的 from_lsn 恰好是上次备份的 to_lsn，如下代码显示了该文件的内容：

```
$ cat /data/backups/inc1/xtrabackup_checkpoints
backup_type = incremental
from_lsn = 25765184
to_lsn = 28531408
last_lsn = 28531418
flushed_lsn = 0
```

在增量备份集中，可以看到很多以 delta 结尾的文件，这些文件只记录原文件的变化部分，例如：

```
$ ls /data/backups/inc1
backup-my.cnf   ib_buffer_pool   mysql                 performance_schema
undo_001.delta  undo_002.meta           xtrabackup_info
binlog.000006  ibdata1.delta   mysql.ibd.delta   sbtest
undo_001.meta   xtrabackup_binlog_info   xtrabackup_logfile
binlog.index   ibdata1.meta    mysql.ibd.meta   sys
undo_002.delta  xtrabackup_checkpoints   xtrabackup_tablespaces
$ ls /data/backups/inc1/sbtest/
sbtest1_361.sdi  sbtest1.MYD  sbtest1.MYI  sbtest2.ibd.delta  sbtest2.ibd.meta
```

2. 准备阶段

增量备份的准备阶段与全量备份的准备阶段不同。在全量备份中，执行两种类型的操作以使数据库保持一致：根据日志文件向数据文件重放已提交的事务，并回滚未提交的事务。但在准备增量备份时，必须跳过未提交事务的回滚，因为备份时未提交的事务可能正在进行中，并且很可能会在下一次增量备份中提交。应该使用--apply-log-only 选项来防止未提交事务的回滚，如果没有这样做，增量备份集将无法使用。

对全量备份集进行准备的命令如下：

```
$ xtrabackup --prepare --apply-log-only --target-dir=/data/backups/base
```

全量备份集准备好后，再用如下命令将增量备份应用到全量备份上：

```
$ xtrabackup --prepare --apply-log-only --target-dir=/data/backups/base \
--incremental-dir=/data/backups/inc1
```

如上命令只是将增量部分应用到全量备份集上，在输出中可以看到类似下面的内容：

```
xtrabackup: page size for /data/backups/inc1/ibdata1.delta is 16384 bytes
Applying /data/backups/inc1//ibdata1.delta to ./ibdata1...
```

准备好的备份集在全量备份目录，而不是在增量备份目录。

如果还有第二个增量备份集，再使用如下命令进行准备：

```
$ xtrabackup --prepare --target-dir=/data/backups/base \
--incremental-dir=/data/backups/inc2
```

--apply-log-only 在最后一次应用增量备份时不用，因为此时要将未提交的事务回滚。如果没有第二个增量备份集，在应用第一个增量备份集时不要加--apply-log-only，如果加了也没有关系，可以再次准备备份集，也可以直接恢复，MySQL 在启动时会自动将未提交的事务回滚。

3. 恢复阶段

增量备份的恢复阶段和全量备份一样，此处不再赘述。

9.4.2　部分备份

Percona XtraBackup 可以只备份一部分指定的数据库和表，该功能称为部分备份(partial backups)。

指定部分表的选项如下。

- --tables=name：用正则表达式的方式指定包括的表。
- --tables-file=name：将要包括的表名放在指定的文件中。
- --tables-exclude=name：用正则表达式的方式指定排除的表，该选项比--tables 的优先级高。

指定部分数据库的选项如下。

- --databases=name
- --databases-file=name
- --databases-exclude=name

如上 3 个选项的含义和指定部分表选项的含义类似。

例如，如下命令只备份一个 sbtest.sbtest2 表：

```
$ xtrabackup --backup --datadir=/disk1/data --target-dir=/data/backups\
--tables="^sbtest[.]sbtest2"
```

如下命令对备份集进行准备：

```
$ xtrabackup --prepare --export --target-dir=/data/backups
```

准备阶段完成的工作和全量备份类似，但多产生了一个/data/backups/sbtest/sbtest2.cfg 文件，该文件保存着表的元数据。

要恢复的数据库中如果没有 sbtest.sbtest2 表，要先创建该表，然后使用如下命令抛弃表空间：

```
mysql> alter table sbtest.sbtest2 discard tablespace;
```

再把 sbtest2.cfg 和 sbtest2.ibd 文件复制到对应的目录下，然后使用如下命令导入表空间：

```
mysql> alter table sbtest.sbtest2 import tablespace;
```

这样部分恢复就完成了。这个例子是在部分备份集上完成的，实际上也可以使用全量备份集进行部分恢复。

9.4.3 压缩备份

Percona XtraBackup 支持压缩备份，压缩和解压可以在本地或通过流的方式完成，例如：

```
$ xtrabackup --backup --compress --target-dir=/data/compress
```

为了提高压缩的性能，可以使用--compress-threads 指定并行度进行压缩，如下命令启动了 8 个线程进行压缩：

```
$ xtrabackup --backup --compress --compress-threads=8
--target-dir=/data/compress
```

查看备份的文件，以.qp 结尾的文件都是压缩的文件：

```
$ ls /data/compress/
backup-my.cnf.qp  ib_buffer_pool.qp  mysql.ibd.qp          sys
xtrabackup_binlog_info.qp  xtrabackup_logfile.qp
binlog.000010.qp  ibdata1.qp          performance_schema  undo_001.qp
xtrabackup_checkpoints       xtrabackup_tablespaces.qp
binlog.index.qp   mysql                sbtest               undo_002.qp
xtrabackup_info.qp
```

解压需要安装 qpress 包，安装完成后使用如下命令进行解压：

```
$ for bf in 'find . -iname "*\.qp"'; do qpress -d $bf $(dirname $bf) && rm $bf;
done
```

也可以用 Percona XtraBackup 调用 qpress 进行解压：

```
$ xtrabackup --decompress --target-dir=/data/compress
```

Percona XtraBackup 不会自动删除压缩文件，可以使用--remove-original 选项指定删除压缩文件。解压后的准备与恢复与正常的操作一样，只是恢复时 Percona XtraBackup 不会将压缩文件复制到数据目录。

9.4.4 compact 备份

Percona XtraBackup 的 compact 备份方式在备份时会跳过二级索引，这样的备份集占用的空间会减少，这种备份方式的备份命令如下：

```
$ xtrabackup --backup --compact --target-dir=/path/to/backup
```

在准备阶段会创建二级索引，用时较长，准备阶段的命令如下：

```
$ xtrabackup --prepare --apply-log --rebuild-indexes
--target-dir=/path/to/backup Restore:
```

恢复命令和其他类型的恢复命令类似。

9.4.5 加密备份

Percona XtraBackup 可以支持本地和流式(stream)的加密备份。备份时的加密选项有如下 3 个。

- --encrypt：指定加密算法。
- --encrypt-key：指定密钥。
- --encrypt-key-file：指定存放密钥的文件。

其中--encrypt-key 和--encrypt-key-file 不能同时使用。

可以使用 openssl 生成 base64 的随机密钥，命令如下：

```
$ openssl rand -base64 24
pANfkEv0tEGznvoPWYuo4YYj73LDErev
```

一个生成加密备份集的命令如下：

```
$ xtrabackup --backup --target-dir=/data/backups --encrypt=AES256 \
--encrypt-key="pANfkEv0tEGznvoPWYuo4YYj73LDErev"
```

备份集生成后可以看到所有的文件都是以.xbcrypt 结尾的加密文件，如下所示：

```
$ ls /data/backups/
backup-my.cnf.xbcrypt    ibdata1.xbcrypt      sbtest
xtrabackup_binlog_info.xbcrypt  xtrabackup_tablespaces.xbcrypt
binlog.000011.xbcrypt    mysql                sys
xtrabackup_checkpoints
binlog.index.xbcrypt     mysql.ibd.xbcrypt    undo_001.xbcrypt
xtrabackup_info.xbcrypt
ib_buffer_pool.xbcrypt   performance_schema   undo_002.xbcrypt
xtrabackup_logfile.xbcrypt
```

可以将密钥存储到文件中，例如：

```
$ xtrabackup --backup --target-dir=/data/backups --encrypt=AES256 \
--encrypt-key-file=/path/to/key-file
```

可以通过如下两个选项提高加密过程的速度。

- --encrypt-threads：指定并行加密线程的个数，默认是 1。
- --encrypt-chunk-size：指定用于加密的缓存，默认为 64KB。

解密的例子如下：

```
$ xtrabackup --decrypt=AES256
--encrypt-key="pANfkEv0tEGznvoPWYuo4YYj73LDErev" --target-dir=/data/backups
xtrabackup: recognized server arguments: --datadir=/disk1/data
--innodb_log_file_size=100m
xtrabackup: recognized client arguments: --user=root --password=*
--decrypt=AES256 --encrypt-key=* --target-dir=/data/backups
xtrabackup version 8.0.22-15 based on MySQL server 8.0.22 Linux (x86_64)
(revision id: fea8a0e)
210226 14:13:06 [01] decrypting ./xtrabackup_logfile.xbcrypt
210226 14:13:06 [01] decrypting ./ibdata1.xbcrypt
```

```
210226 14:13:08 [01] decrypting ./sys/sys_config.ibd.xbcrypt
210226 14:13:08 [01] decrypting ./sbtest/sbtest2.ibd.xbcrypt
210226 14:13:09 [01] decrypting ./sbtest/sbtest1.MYI.xbcrypt
...
```

选项--parallel 可以在解密时用于提高解密的并行度。

和压缩备份类似，Percona XtraBackup 不会自动删除加密文件，可以使用--remove-original 选项指定删除加密文件。解密后的准备和恢复与正常的操作一样，只是恢复时 Percona XtraBackup 不会将加密文件复制到数据目录。

9.4.6 流备份

Percona XtraBackup 支持以流的方式将备份文件输出到 UNIX 的标准输出，这样可以在备份过程的同时对备份集进行后续处理(如加密或压缩等)或传输到远端。

指定选项--stream=xbstream 可以启动流式备份，如下命令在备份的同时将备份集传输到远端的机器：

```
$ xtrabackup --backup --stream=xbstream --target-dir=/data/backup | ssh
root@192.168.88.88 "cat - > /tmp/backup.tar"
```

如下命令在备份时进行压缩：

```
$ xtrabackup --backup --stream=xbstream --target-dir=/data/backup | gzip - > backup.
stream.gz
```

xbsteam 是 Percona 提供的处理流的工具，如下命令将流备份的文件解开到本地：

```
$ xbstream -x < backup.tar
```

9.5 高级功能

9.5.1 节流

虽然 Percona XtraBackup 不会阻止数据库的操作，但备份本身会占用大量的硬盘 IO 带宽，为了减少对联机业务的影响，可以通过--throttle 选项设置每秒备份的 chunk 个数，一个 chunk 的大小是 10MB。图 9.2 是当--throttle 设置为 1 时的硬盘 IO 情况。

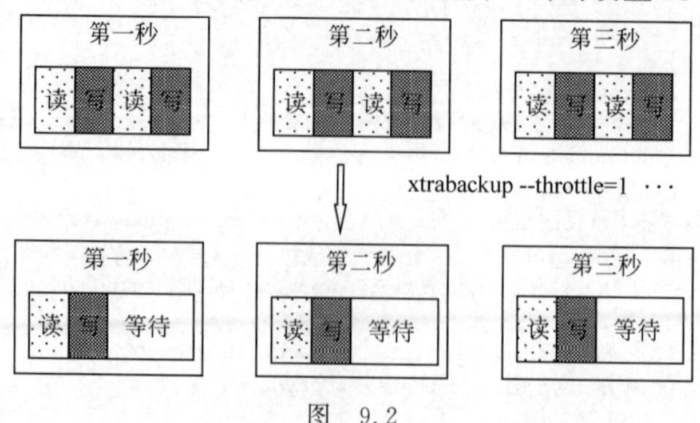

图 9.2

当 Percona XtraBackup 在一秒内完成了--throttle 指定的 chunk 个数后，就进入等待状态。这里--throttle 指定 chunk 个数的同时包括读和写，对于增量备份的只读过程，chunk 个数只包括读。

9.5.2 在数据库中记录备份历史

Percona XtraBackup 可以将备份信息记录在数据库中，例如，使用--history 指定备份名称为 full_sunday 进行全量备份，命令如下：

```
$ xtrabackup --backup --history=full_sunday --target-dir=/data/backup
```

检查数据库中表 percona_schema. xtrabackup_history 中的记录，代码如下：

```
mysql> select * from percona_schema.xtrabackup_history \G
*************************** 1. row ***************************
             uuid: ed4b8315-796a-11eb-932c-fa163e368ff4
             name: full_sunday
        tool_name: xtrabackup
     tool_command: --backup --history=full_sunday --target-dir=/data/backup
     tool_version: 8.0.22-15
  ibbackup_version: 8.0.22-15
    server_version: 8.0.22
        start_time: 2021-02-28 10:15:57
          end_time: 2021-02-28 10:16:11
         lock_time: 10
        binlog_pos: filename 'binlog.000016', position '156'
innodb_from_lsn: 0
    innodb_to_lsn: 28627193
           partial: N
       incremental: N
            format: file
           compact: N
        compressed: N
         encrypted: N
1 row in set (0.00 sec)
```

以后在该全量备份基础上进行增量备份可以使用--incremental-history-name 或--incremental-history-uuid 指定备份的基础，percona_schema. xtrabackup_history 会根据这两个参数指定的条件在表 percona_schema. xtrabackup_history 中找到最大的 innodb_to_lsn 作为增量备份的起点。这两个参数不能同时指定，指定了这两个参数也不能再指定--incremental-basedir 或--incremental-lsn，例如，如下命令在上次全量备份的基础上再进行增量备份：

```
$ xtrabackup --backup --history --target-dir=/data/incr1
--incremental-history-name=full_sunday
```

备份完成后，再次检查 percona_schema. xtrabackup_history 表，代码如下：

```
$ mysql -e "select * from percona_schema.xtrabackup_history\G"
```

```
*************************** 1. row ***************************
              uuid: 8b5c8f9e-796d-11eb-932c-fa163e368ff4
              name: NULL
         tool_name: xtrabackup
      tool_command: --backup --history --target-dir=/data/incr1
--incremental-history-name=full_sunday
      tool_version: 8.0.22-15
 ibbackup_version: 8.0.22-15
    server_version: 8.0.22
        start_time: 2021-02-28 10:34:48
          end_time: 2021-02-28 10:34:56
         lock_time: 5
        binlog_pos: filename 'binlog.000017', position '156'
  innodb_from_lsn: 28627193
     innodb_to_lsn: 28628303
           partial: N
       incremental: Y
            format: file
           compact: N
        compressed: N
         encrypted: N
*************************** 2. row ***************************
              uuid: ed4b8315-796a-11eb-932c-fa163e368ff4
              name: full_sunday
         tool_name: xtrabackup
      tool_command: --backup --history=full_sunday --target-dir=/data/backup
      tool_version: 8.0.22-15
 ibbackup_version: 8.0.22-15
    server_version: 8.0.22
        start_time: 2021-02-28 10:15:57
          end_time: 2021-02-28 10:16:11
         lock_time: 10
        binlog_pos: filename 'binlog.000016', position '156'
   innodb_from_lsn: 0
     innodb_to_lsn: 28627193
           partial: N
       incremental: N
            format: file
           compact: N
        compressed: N
         encrypted: N
```

发现 percona_schema.xtrabackup_history 表里有两条记录，分别记录了全量备份和增量备份的信息。

9.6 实验

XtraBackup 备份恢复工具

(1) 使用 XtraBackup 进行全量备份和恢复。

(2) 使用 XtraBackup 进行增量备份和恢复。

视频演示

第10章

MySQL Enterprise Backup

第 9 章介绍了一个流行的 MySQL 的物理备份工具 XtraBackup,本章介绍另一个流行的 MySQL 的物理备份工具 MySQL Enterprise Backup(简称 MEB)。

10.1 简介

MySQL Enterprise Backup 是 Oracle 官方提供的多平台、高性能的 MySQL 物理备份工具,它有以下功能:

- InnoDB 表的热备。
- MyISAM 表和其他非 InnoDB 表的温备。
- 增量备份和差异备份。
- 加密和压缩备份。
- 选择性备份和恢复。
- 云存储备份和恢复。

从 MySQL 8.0.18 版本以后,在对非 InnoDB 表进行备份时,只将非 InnoDB 表置于只读状态,而不会将 InnoDB 表也置于只读状态,InnoDB 表在这个过程中可以进行 DML 操作,但不能进行 DDL 操作。为了减少对 MySQL 实例的影响,也可以通过--only-innodb 选项只备份 InnoDB 表。

MySQL Enterprise Backup 和 MySQL 的版本兼容关系见表 10.1。

表 10.1

MySQL Enterprise Backup 版本	MySQL 5.6	MySQL 5.7	MySQL8.0
3.12	√	×	×
4.1	×	√	×
8.0	×	×	√

可以看到，和向下兼容的惯例不同，MySQL Enterprise Backup 8.0 不能备份 MySQL 5.6 和 MySQL 5.7。

所有 MySQL Enterprise Backup 的功能都使用 mysqlbackup 客户端执行。它用于执行不同类型的备份和恢复操作，以及备份压缩、解压缩、验证等其他相关任务。

10.2 工作原理

备份恢复大致分为如下 3 个步骤。

(1) 备份(backup)：复制源数据库的相关文件，生成原始(raw)备份集。

(2) 应用日志(apply-log)：把备份过程中日志的变化应用到备份集中的数据文件，使原始备份集转换成一致的备份集。

(3) 复制(copy-back)：将转换好的备份集复制到目标数据库的对应目录。

前面两步可以使用 backup-and-apply-log 命令合并在一起执行，后面两步也可以使用 copy-back-and-apply-log 命令合并在一起执行。

10.2.1 备份步骤

MySQL Enterprise Backup 备份 MySQL 的大致步骤和相关输出信息如下。

(1) 复制 InnoDB 数据文件、重做日志文件、二进制日志文件、中继(relay)日志文件(正在使用的日志文件除外)，在此过程中数据库正常对外提供服务，数据和数据结构也可以发生变化。此过程中备份工具的输出如下：

```
210307 08:28:03 RDR1    INFO: Starting to copy all innodb files...
210307 08:28:03 RDR1    INFO: Copying /disk1/data/ibdata1.
...
```

(2) 对 MySQL 实例上备份锁(lock instance for backup)，该锁会阻止 DDL 操作，不会阻止 DML 操作。然后 mysqlbackup 将会扫描第一步中备份的数据文件，将此过程中 DDL 语句修改的数据应用到备份中，此过程中绝大部分对 MySQL 实例的读写都不会受到影响。此过程中备份工具的输出如下：

```
210307 08:28:04 RDR1    INFO: Starting to lock instance for backup...
210307 08:28:04 RDR1    INFO: The server instance is locked for backup.
```

(3) 对所有用户创建的非 InnoDB 表上读锁(flush tables tbl_name [,tbl_name] ... with read lock)，然后备份这些表。如果没有用户创建的非 InnoDB 表，跳过这一步。此过程中备份工具的输出如下：

```
210307 08:28:05 RDR1    INFO: Starting to read - lock tables...
210307 08:28:05 RDR1    INFO: 1 tables are read - locked.
210307 08:28:05 RDR1    INFO: Opening backup source directory '/disk1/data'
210307 08:28:05 RDR1    INFO: Starting to copy non - innodb files in subdirs of '/disk1/data'
...
210307 08:28:09 RDR1    INFO: Completing the copy of all non - innodb files.
```

(4) 短暂中断日志记录，记录与日志相关的信息，包括InnoDB的LSN、二进制日志坐标、GTID和复制状态。此过程中备份工具的输出如下：

```
210307 08:28:09 RDR1     INFO: Requesting consistency information...
210307 08:28:09 RDR1     INFO: Locked the consistency point for 999 microseconds.
210307 08:28:09 RDR1     INFO: Consistency point server_uuid '108d5424-766d-11eb-b140-
fa163e368ff4'.
210307 08:28:09 RDR1     INFO: Consistency point gtid_executed ''.
210307 08:28:09 RDR1     INFO: Consistency point binary_log_file 'binlog.000022'.
210307 08:28:09 RDR1     INFO: Consistency point binary_log_position 156.
210307 08:28:09 RDR1     INFO: Consistency point InnoDB lsn 52840491.
210307 08:28:09 RDR1     INFO: Consistency point InnoDB lsn_checkpoint 52840491.
210307 08:28:09 RDR1     INFO: Requesting completion of redo log copy after LSN 52840491.
```

(5) 释放非InnoDB表的读锁。此过程中备份工具的输出如下：

```
210307 08:28:09 RDR1     INFO: Read-locked tables are unlocked.
```

(6) 根据第(4)步记录的二进制日志信息，复制当前重做日志和二进制日志，把从第一步开始的所有数据库的变化都备份下来，以后可以将这些变化应用到备份集上，使备份集达到一致的状态。此过程中备份工具的输出如下：

```
210307 08:28:09 RLW1     INFO: A copied database page was modified at 52836375. (This is the
highest lsn found on a page)
210307 08:28:09 RLW1     INFO: Scanned log up to lsn 52840491.
210307 08:28:09 RLW1     INFO: Was able to parse the log up to lsn 52840491.
210307 08:28:09 RLW1     INFO: Copied redo log
                                log_start_lsn      52835328
                                start_checkpoint 52835405
                                start_lsn          52835405
                                last_checkpoint   52840491
                                consistency_lsn   52840491
                                log_end_lsn        52840491
210307 08:28:09 RLR1     INFO: Redo log reader waited 161 times for a total of 805.00 ms for
logs to generate.
210307 08:28:09 RDR1     INFO: Truncating binary log index '/data/backups/datadir/binlog.
index' to 352.
210307 08:28:09 RDR1     INFO: Truncating binary log to '/data/backups/datadir/binlog.
000022': 156.
210307 08:28:09 RDR1     INFO: Copying /disk1/data/binlog.000022.
210307 08:28:09 RDR1     INFO: Completed the copy of binlog files...
```

(7) 释放MySQL实例上的备份锁，数据库恢复到正常状态。此过程中备份工具的输出如下：

```
210307 08:28:09 RDR1     INFO: The server instance is unlocked after 4.198 seconds.
```

(8) 备份之前没有备份的日志文件和元数据文件。此过程中备份工具的输出如下：

```
210307 08:28:09 RDR1     INFO: Reading all global variables from the server.
210307 08:28:09 RDR1     INFO: Completed reading of all 586 global variables from the server.
210307 08:28:09 RDR1     INFO: Writing server defaults files 'server-my.cnf' and 'server-
```

```
all.cnf' for server '8.0.23' in '/data/backups'.
210307 08:28:09 RDR1     INFO: Copying meta file /data/backups/meta/backup_variables.txt.
210307 08:28:09 RDR1     INFO: Copying meta file /data/backups/datadir/ibbackup_logfile.
210307 08:28:09 RDR1     INFO: Copying meta file /data/backups/server-all.cnf.
210307 08:28:09 RDR1     INFO: Copying meta file /data/backups/server-my.cnf.
210307 08:28:09 RDR1     INFO: Copying meta file /data/backups/meta/backup_content.xml.
210307 08:28:09 RDR1     INFO: Copying meta file /data/backups/meta/image_files.xml.
```

(9) 备份完成,mysqlbackup 返回成功。此过程中备份工具的输出如下:

```
210307 08:28:09 MAIN     INFO: Full Image Backup operation completed successfully.
210307 08:28:09 MAIN     INFO: Backup image created successfully.
210307 08:28:09 MAIN     INFO: Image Path = /data/backups/bk0.mbi
210307 08:28:09 MAIN     INFO: MySQL binlog position: filename binlog.000022, position 156
-------------------------------------------------------------------
   Parameters SuMySQLary
-------------------------------------------------------------------
   Start LSN                  : 52835328
   Last Checkpoint LSN        : 52840491
   End LSN                    : 52840491
-------------------------------------------------------------------

mysqlbackup completed OK!
```

10.2.2 恢复步骤

MySQL Enterprise Backup 恢复 MySQL 的大致步骤如下。

(1) 关闭 MySQL 实例。

(2) 删除--datadir、--datadir--innodb_data_home_dir、--innodb_log_group_home_dir 和--innodb_undo_directory 目录下的所有文件和目录。

(3) 把备份过程中日志的变化应用到备份集中的数据文件,使原始(raw)备份集转换成一致的备份集。

(4) 将转换好的备份集复制到目标数据库的对应目录。

最后两步经常用 copy-back-and-apply-log 命令合并在一起执行。

10.2.3 备份锁

为了保证备份过程中的数据一致性,在备份非 InnoDB 表时,通常需要对 MySQL 实例应用全局读锁(flush tables with read lock)。该锁的强度比较大,会阻止 DML 和 DDL 操作。当 MySQL 实例上该锁处于激活状态时,在 performance_schema.metadata_locks 表中可以查到,其 lock_duration 是 explicit,object_type 是 COMMIT 表示加锁成功,如下所示:

```
mysql> select owner_thread_id,object_type,lock_type,
lock_duration,lock_status
from   performance_schema.metadata_locks
```

```
where lock_duration = 'explicit';
+------------------+-------------+-----------+-------------+-----------+
| owner_thread_id  | object_type | lock_type | lock_duration | lock_status |
+------------------+-------------+-----------+-------------+-----------+
|               52 | GLOBAL      | SHARED    | EXPLICIT    | GRANTED   |
|               52 | COMMIT      | SHARED    | EXPLICIT    | GRANTED   |
+------------------+-------------+-----------+-------------+-----------+
2 rows in set (0.00 sec)
```

从 MySQL Enterprise Backup 8.0.16 后，MySQL Enterprise Backup 使用备份锁(lock instance for backup)代替全局读锁，备份锁会阻止 DDL 操作，但不会阻止 DML 操作，对 MySQL 的影响也就小很多。

备份锁在表 performance_schema.metadata_locks 中的 lock_duration 字段也是 explicit，如下所示：

```
mysql > select owner_thread_id,object_type,
lock_type,lock_duration,lock_status
from   performance_schema.metadata_locks
where   lock_duration = 'explicit';
+------------------+-------------+-----------+-------------+-----------+
| owner_thread_id  | object_type | lock_type | lock_duration | lock_status |
+------------------+-------------+-----------+-------------+-----------+
|               52 | BACKUP LOCK | SHARED    | EXPLICIT    | GRANTED   |
+------------------+-------------+-----------+-------------+-----------+
1 row in set (0.00 sec)
```

从图 10.1 中可以看出全局读锁和备份锁对 SQL 语句的作用。

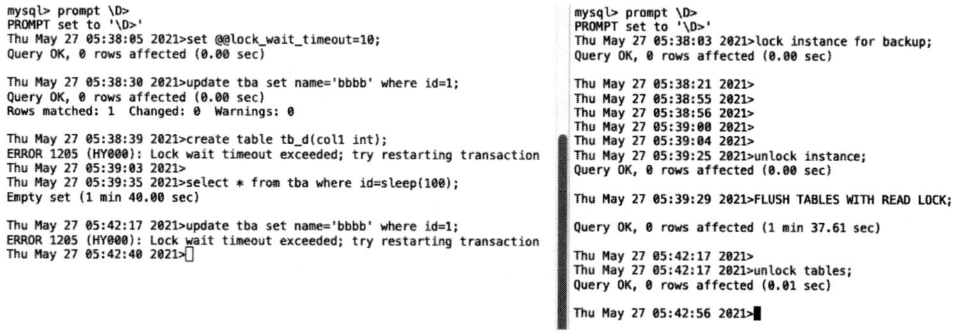

```
mysql> prompt \D>
PROMPT set to '\D>'
Thu May 27 05:38:05 2021>set @@lock_wait_timeout=10;
Query OK, 0 rows affected (0.00 sec)

Thu May 27 05:38:30 2021>update tba set name='bbbb' where id=1;
Query OK, 0 rows affected (0.00 sec)
Rows matched: 1  Changed: 0  Warnings: 0

Thu May 27 05:38:39 2021>create table tb_d(col1 int);
ERROR 1205 (HY000): Lock wait timeout exceeded; try restarting transaction
Thu May 27 05:39:03 2021>
Thu May 27 05:39:35 2021>select * from tba where id=sleep(100);
Empty set (1 min 40.00 sec)

Thu May 27 05:42:17 2021>update tba set name='bbbb' where id=1;
ERROR 1205 (HY000): Lock wait timeout exceeded; try restarting transaction
Thu May 27 05:42:40 2021>
```

```
mysql> prompt \D>
PROMPT set to '\D>'
Thu May 27 05:38:03 2021>lock instance for backup;
Query OK, 0 rows affected (0.00 sec)

Thu May 27 05:38:21 2021>
Thu May 27 05:38:55 2021>
Thu May 27 05:38:56 2021>
Thu May 27 05:39:00 2021>
Thu May 27 05:39:04 2021>
Thu May 27 05:39:25 2021>unlock instance;
Query OK, 0 rows affected (0.00 sec)

Thu May 27 05:39:29 2021>FLUSH TABLES WITH READ LOCK;

Query OK, 0 rows affected (1 min 37.61 sec)

Thu May 27 05:42:17 2021>
Thu May 27 05:42:17 2021>unlock tables;
Query OK, 0 rows affected (0.01 sec)

Thu May 27 05:42:56 2021>
```

图 10.1

仔细观察这些 SQL 语句的执行顺序可以看出：

(1) 备份锁不阻止 DML 语句。

(2) 备份锁阻止 DDL 语句。

(3) 当 MySQL 实例中有长时间的查询语句时，在查询语句执行完成之前，全局读锁无法加锁成功。

(4) 全局读锁阻止 DML 语句。

为了防止应用中有长时间执行的 SQL，可以设置系统参数 max_execution_time 来控制 SQL 语句的最长执行时间，该方法已在第 8.7.1 节中介绍过。

10.3 典型用例

10.3.1 全量备份和恢复

mysqlbackup 的 backup 命令将会在--backup-dir 指定的目录下生成全量备份集，例如：

```
$ mysqlbackup --backup-dir=/data/backups backup
```

可以使用--with-timestamp 在指定目录下创建一个以时间标签为目录名的子目录存放全量备份集：

```
$ mysqlbackup --backup-dir=/data/backups --with-timestamp  backup
...
210308 16:04:22 MAIN     INFO: Backup directory created: '/data/backups/2021-03-08_16-04-22'
...
```

将备份结果生成一个单一文件的备份集便于管理和传输，如下命令会生成一个单一文件的备份集：

```
$ mysqlbackup --backup-image=/data/bk0.mbi --backup-dir=/data/backups backup-to-image
```

这里的单一的备份集是/data/bk0.mbi，备份过程中产生的临时文件存放在/data/backups 目录，该目录可以看成 mysqlbackup 的工作目录。

可以使用 validate 命令对备份集进行验证，如下所示：

```
$ mysqlbackup --backup-image=/data/bk0.mbi  validate
```

也可以使用 list-image 列出单一备份集中的文件清单，如下所示：

```
$ mysqlbackup  --backup-image=/data/bk0.mbi list-image
```

使用 image-to-backup-dir 可以将单一的备份集展开，如下所示：

```
$ mysqlbackup --backup-dir=/data/backup --backup-image=/data/bk0.mbi image-to-backup-dir
```

恢复时首先关闭 MySQL 实例，然后删除数据目录下的所有文件，包括参数 innodb_data_home_dir、innodb_log_group_home_dir 和 innodb_undo_directory 指定的目录下的所有文件，使用 copy-back-and-apply-log 进行恢复，命令如下：

```
$ mysqlbackup  --backup-image=/data/bk0.mbi --backup-dir=/data/backups_dir copy-back-and-apply-log
```

其中--backup-dir 指定存放恢复过程中临时文件的目录，恢复完成后要检查文件属主，通常需要将文件属主改成 mysql 用户。

10.3.2 差异备份和增量备份

差异备份(differential backup)指备份自上次全量备份后的所有的变化数据，恢复时要恢复全量备份集和本次差异备份集。

增量备份(incremental backup)指备份自上次备份后的所有的变化数据，上次备份不一定是全量备份，这种备份集通常比差异备份小，但恢复时要恢复与增量备份相关的全量备份集和所有增量或差异备份集。

可以使用--incremental-base 指定上次的备份集目录，在此基础上进行增量备份，命令如下：

```
$ mysqlbackup --incremental --incremental-base=dir:/data/backup
--backup-dir=/data/temp_dir --backup-image=incremental_image1.bi
backup-to-image
```

mysqlbackup 会从指定的目录中读取上次备份集结束的 lsn(end_lsn)，end_lsn 在--backup-dir 指定的目录下的 meta/backup_variables.txt 文件中，也可以使用--start-lsn 直接指定这次备份开始的 lsn。注意，--start-lsn 等于上次的 end_lsn，而不是 end_lsn+1。

增量备份在恢复时有两种方法，第一种方法是把全量备份恢复到目标路径，再用 copy-back-and-apply-log 命令恢复增量备份，例如：

```
$ mysqlbackup --defaults-file=<my.cnf> -uroot --backup-image=<inc_image_name>
--backup-dir=<incBackupTmpDir> --datadir=<restoreDir> --incremental
copy-back-and-apply-log
```

其中<inc_image_name>是存放在<restoreDir>目录的增量备份集名，--backup-dir 是恢复过程中用到的临时目录，存放临时文件和元数据。如果还有其他的增量备份集，继续采用这种方法进行增量恢复。对于 MySQL Enterprise Backup 8.0.21 及更高版本，恢复增量备份时不再需要--incremental 选项。

第二种方法是在全量备份集的基础上进行恢复，先用 apply-log 命令准备全量备份集，例如：

```
$ mysqlbackup --backup-dir=/data/backup apply-log
...
210307 20:33:49 MAIN     INFO: Apply-log operation completed successfully.
210307 20:33:49 MAIN     INFO: Full backup prepared for recovery successfully.

mysqlbackup completed OK!
```

然后将增量备份集应用到全量备份集上，如下命令先将增量备份集展开到/data/temp_dir2 目录：

```
$ mysqlbackup --backup-dir=/data/temp_dir2
--backup-image=/data/temp_dir/incremental_image1.bi image-to-backup-dir
...
210307 20:45:18 MAIN     INFO: Extract operation completed successfully.
210307 20:45:18 MAIN     INFO: Requested contents of backup image extracted successfully.
```

```
210307 20:45:18 MAIN      INFO: Source Image Path = /data/temp_dir/incremental_image1.bi

mysqlbackup completed OK!
```

然后将展开的增量备份集用apply-incremental-backup命令应用到全量备份集目录：

```
$ mysqlbackup --incremental-backup-dir=/data/temp_dir2
--backup-dir=/data/backup apply-incremental-backup
...
210307 20:49:22 MAIN      INFO: Apply-log operation completed successfully.
210307 20:49:22 MAIN      INFO: Full backup prepared for recovery successfully.

mysqlbackup completed OK!
```

如果还有增量备份集就继续采用这种方式应用到全量备份集上,这样全量备份集就准备好了。后续还可以继续进行增量备份,再继续将增量备份集应用到全量备份集上,这样全量备份集就一直处于准备好的状态,恢复时用copy-back一步即可恢复。这种方法在恢复时比第一种方法更加便捷,减少了出错的可能性。

10.3.3 部分备份

MySQL Enterprise Backup默认会备份MySQL数据目录下的所有文件,但可以通过设定参数进行选择性备份,创建一个部分备份集(partial backup)。例如,可以通过--include-tables或--exclude-tables包含或排除一些表,这两个参数支持正则表达式。例如,下面的命令备份了sbtest数据库中除sbtest1表之外的所有表：

```
$ mysqlbackup --backup-dir=/data/partial1 --include-tables="^sbtest\."
--exclude-tables="^sbtest\.sbtest1" backup
```

备份完成后每个表多了一个以.cfg结尾的文件用于部分恢复时的校验,例如：

```
$ ls /data/partial1/datadir/sbtest/
sbtest2.cfg  sbtest2.ibd
```

如下命令在上面命令的基础上增加了一个--use-tts选项,tts是传输表空间(transportable tablespace)的缩写：

```
$ mysqlbackup --backup-dir=/data/partial2 --include-tables="^sbtest\."
--exclude-tables="^sbtest\.sbtest1" --use-tts backup
```

备份完成后,每个表又多了一个.sql,用于创建表,命令如下：

```
$ ls /data/partial2/datadir/sbtest/
sbtest2.cfg  sbtest2.ibd  sbtest2.sql
```

查询该sbtest2.sql文件的内容如下：

```
$ cat /data/partial2/datadir/sbtest/sbtest2.sql
CREATE TABLE 'sbtest2' (
  'id' int NOT NULL AUTO_INCREMENT,
```

```
  'k' int NOT NULL DEFAULT '0',
  'c' char(120) NOT NULL DEFAULT '',
  'pad' char(60) NOT NULL DEFAULT '',
  PRIMARY KEY ('id'),
  KEY 'k_2' ('k')
) ENGINE = InnoDB AUTO_INCREMENT = 10001 DEFAULT CHARSET = utf8mb4 COLLATE = utf8mb4_0900_ai_ci
```

对于部分备份集，恢复时数据库应处于运行状态，例如，可以使用如下命令恢复前面创建的部分备份集：

```
$ mysqlbackup --backup-dir=/data/partial1  copy-back-and-apply-log
```

mysqlbackup 会自动判断这是一个部分备份集，对于非 tts 备份集，需要恢复的表必须在目标数据库上已经存在，并且表结构和备份时一样。对于 tts 备份集，需要恢复的表不能存在于目标数据库上，恢复时可以使用--rename 选项修改表名。

选择性备份还有如下两个选项。

- --only-innodb：只备份 InnoDB 的表。
- --only-known-file-types：只备份 MySQL 的相关文件，用户在 MySQL 数据目录中保存的其他文件将不会被备份。

10.3.4 流备份

为了节约本地硬盘空间，可以在备份时使用 --backup-to-image，并指定--backup-image=-，将备份集定向到标准输出，同时用管道线将标准输出定向到其他机器，例如：

```
$ mysqlbackup --defaults-file=~/my_backup.cnf --backup-image=-
--backup-dir=/tmp backup-to-image | ssh <user name>@<remote host name> 'cat > ~/
backups/my_backup.img'
```

其中 ssh 可以用其他通信协议，如 ftp 代替，cat 也可以用其他命令，如 dd 或 tar 代替。

如下命令可以一边备份，一边将备份集恢复到一个远端的 MySQL 数据库：

```
$ mysqlbackup  --backup-dir=backup --backup-image=-  --compress backup-to-
image | ssh <user name>@<remote host name> 'mysqlbackup
--backup-dir=backup_tmp --datadir=/data --innodb_log_group_home_dir=.
--uncompress --backup-image=- copy-back-and-apply-log'
```

这样做节省了硬盘空间。

10.3.5 压缩备份

为了节省磁盘空间，可以使用 mysqlbackup 的--compress 选项压缩 InnoDB 备份数据文件。通过压缩，可以保留更多备份数据集，或者在将备份数据发送到另一台服务器时节省传输时间。此外，由于 IO 减少，压缩通常会加快备份速度。

例如，如下命令可创建一个压缩的备份集：

```
$ mysqlbackup --defaults-file=/home/dbadmin/my.cnf --compress
```

```
--backup-image=backup.img backup-to-image
```

可以使用--compressmethod选项指定压缩算法，也可以使用--compresslevel选项指定压缩级别。在创建压缩备份集的同时指定压缩算法和压缩级别的代码如下：

```
$ mysqlbackup --defaults-file=/home/dbadmin/my.cnf --compress-method=zlib
--compress-level=5 backup
```

恢复时可以使用--uncompress进行解压，例如：

```
$ mysqlbackup --defaults-file=<my.cnf> -uroot --backup-image=<image_name>\
  --backup-dir=<backupTmpDir> --datadir=<restoreDir> --uncompress copy-back-and
-apply-log
```

从MySQL Enterprise Backup 8.0.21版本之后，--uncompress选项不再需要。

10.3.6 加密备份

为了提高备份集的安全性，MySQL Enterprise Backup支持加密备份，加密和解密使用同一个密钥，密钥只支持十六进制的数字，如果采用其他形式的密钥，会出现类似下面的报错：

```
ERROR: Invalid encryption key given. It should be a string of 64 hexadecimal digits.
```

可以使用openssl生成十六进制的随机密钥，命令如下：

```
$ openssl rand -hex 32
e9b8c5170dca44a516961ed78a869bf4ce60989d9a84755008ca7b5a8989f670
```

如下是创建一个加密备份集的例子：

```
$ mysqlbackup --backup-image=/data/bk0.enc --encrypt
--key=e9b8c5170dca44a516961ed78a869bf4ce60989d9a84755008ca7b5a8989f670
--backup-dir=/data/backups backup-to-image
```

也可以将密钥保存到文件里，例如：

```
$ openssl rand  -hex 32 >keyfile
$ cat keyfile
d7b0eb8bcdfc1f796682cbb7984306763ac8832e4d62f586568ea55e4ba04b8d
```

再使用--key-file指定密钥文件进行加密，命令如下：

```
$ mysqlbackup --backup-image=/data/bk0.enc --encrypt --key-file=keyfile
--backup-dir=/data/backups backup-to-image
```

解密使用--decrypt选项，密钥使用同一个密钥，命令如下：

```
$ mysqlbackup --backup-image=/data/bk0.enc --decrypt --key-file=keyfile
--backup-dir=/data/backups/extract-dir  extract
```

10.4 高级功能

10.4.1 重做日志归档

为了保证数据的一致性，MySQL Enterprise Backup 会把备份过程中产生的所有重做日志都备份下来，当 MySQL 短时间内产生的重做日志太多时，前面的重做日志可能还没有备份就被覆盖了。为了避免这个问题，MySQL Enterprise Backup 从 8.0.17 版本后推出了重做日志归档的功能(redo log archiving)的功能。

启用重做日志归档需要为 innodb_redo_log_archive_dirs 系统变量设置一个值，该值是以分号分隔的带标签的重做日志归档目录列表，标签和目录由冒号分隔。其语法如下：

```
mysql > set global
innodb_redo_log_archive_dirs = 'label1: directory_path1[; label2: directory_path2; … ]';
```

例如，如下命令指定/data 为第一个归档目录：

```
mysql > set global innodb_redo_log_archive_dirs = 'archive_redo1:/data';
```

当设置了 innodb_redo_log_archive_dirs 系统变量后，mysqlbackup 在备份开始时会启动重做日志归档，结束时会终止重做日志归档，在 mysqlbackup 的输出中会看到下面的提示：

```
...
210308 16: 27: 42 MAIN      INFO: Started redo log archiving.
...
210308 16: 28: 02 RLR1      INFO: Stopping redo log archiving.
...
```

如果使用 root 用户调用 mysqlbackup，会输出下面的错误：

```
210308 16: 22: 11 MAIN  WARNING: MySQL query 'DO
innodb_redo_log_archive_start('archive_redo1','16151917317537689'); ': 3847,
Cannot create redo log archive file '/data/16151917317537689/archive.199d4998 - 7f1e - 11eb -
a5c7 - fa163e368ff4.000001.log' (OS errno: 13  -  Permission denied)
```

这是因为 mysqlbackup 会用 root 用户在/data 目录下创建的子目录 16151917317537689，而 mysql 用户没有权限向里面写归档文件，如果 mysql 用户不能登录，可以使用如下 sudo 的命令将其转换成 mysql 用户执行备份：

```
$ sudo  - u mysql mysqlbackup  -- backup - dir = /data/backups  -- with - timestamp  backup
```

不使用 mysqlbackup 也可以启动重做日志归档，其语法如下：

```
mysql > select innodb_redo_log_archive_start('label', 'subdir');
```

或：

```
mysql > do innodb_redo_log_archive_start('label', 'subdir');
```

如下命令为启动重做日志归档，归档目录是/data/redo1：

```
mysql >  DO innodb_redo_log_archive_start('archive_redo1','redo1');
Query OK, 0 rows affected (0.07 sec)
```

启动后生成的归档日志文件名的格式是：

```
directory_identified_by_label/[subdir/]archive.serverUUID.000001.log
```

例如：

```
/data/redo1/archive.199d4998-7f1e-11eb-a5c7-fa163e368ff4.000001.log
```

终止重做日志归档的语法如下：

```
mysql > select innodb_redo_log_archive_stop();
```

或：

```
mysql > do innodb_redo_log_archive_stop();
```

10.4.2 节流

虽然 MySQL Enterprise Backup 在备份过程中不会阻止数据库的绝大部分操作，但备份本身会占用大量的硬盘 IO 带宽，为了减少对联机业务的影响，可以通过--sleep 选项指定从 InnoDB 表复制一个数据块后睡眠的毫秒数。每个数据块有 1024 个 InnoDB 数据页，通常总计 16MB。该选项的默认值是 0，也就是不睡眠。

10.4.3 数据库中的历史记录

mysqlbackup 在运行过程中会将进度信息写入 mysql.backup_progress 表，当完成备份操作时，它将状态信息记录在 mysql.backup_history 表中。可以通过查询这些表来监视正在进行的备份作业，查看各个阶段所用的时间，并检查是否发生了任何错误。在增量备份时指定--incremental-base=history：last_backup，也是查询 mysql.backup_history 表找出最后一次全量备份时结束的 LSN。

10.4.4 优化备份

优化备份（optimistic backup）是一种提高备份和恢复大型数据库性能的功能，在这些数据库中，只有少量的表经常被修改。在对一个繁忙的大型数据库进行备份的过程中，可能产生大量的重做日志，这些日志会延长备份和应用日志的时间，增加了出错的可能。为了减少需要备份的重做日志，优化备份将备份分成如下两个阶段。

（1）优化阶段，也是第一阶段：备份不太可能修改的表（也称为“非活跃表”），不锁定 MySQL 实例。而且在该阶段，mysqlbackup 不会备份重做日志、撤销日志和系统表空间。

（2）正常阶段，也是第二阶段：备份第一阶段未备份的表（也称为“繁忙表”），备份方式

与普通备份中的处理方式类似，首先复制 InnoDB 文件，然后复制或处理其他相关文件，并在不同时间对数据库应用各种锁。重做日志、撤销日志和系统表空间也在此阶段备份。

后备份繁忙表，可以减少需要备份的重做日志。可以使用选项--optimistic-time 或--optimistic-busy-tables 启动优化备份。在下面的例子中，自 2021 年 3 月 16 日中午以来修改的表将被视为繁忙表，在正常阶段进行备份，所有其他表在优化阶段进行备份：

```
$ mysqlbackup --defaults-file=/home/dbadmin/my.cnf
--optimistic-time=210316120000  --backup-image=<image-name>
--backup-dir=<temp-dir>  backup-to-image
```

当 optimistic-time=now 时，所有的表都被视为非活跃表。

在下面的例子中，mydatabase 中以 mytables 为前缀的表将被视为繁忙表，在正常阶段进行备份，所有其他表在优化阶段进行备份：

```
$ mysqlbackup --defaults-file=/home/dbadmin/my.cnf
--optimistic-busy-tables="^mydatabase\.mytables-.*"
--backup-image=<image-name> --backup-dir=<temp-dir> backup
```

优化备份可以和部分备份、优化增量备份一起使用，进一步提高备份的效率。

10.4.5 使用页跟踪的增量备份

对于 MySQL Enterprise Backup 8.0.18 及更高版本，MySQL Enterprise Backup 支持使用 MySQL 的页面跟踪功能创建增量备份，mysqlbackup 在 InnoDB 数据文件中查找自上次备份以来修改过的页面，然后进行复制。如果数据库中的大部分数据没有被修改，那么使用页跟踪的增量备份比其他类型的增量备份速度要快。

使用此功能需要先安装相应组件，命令如下：

```
mysql> install component "file://component_mysqlbackup";
Query OK, 0 rows affected (0.02 sec)
```

安装完成后，使用下面的函数激活页跟踪，返回值是当前的 LSN：

```
mysql> select mysqlbackup_page_track_set(true);
+-----------------------------------+
| mysqlbackup_page_track_set(true)  |
+-----------------------------------+
|                        94414952   |
+-----------------------------------+
1 row in set (0.14 sec)
```

使用下面的函数可以查询记录的起始 LSN：

```
mysql> select mysqlbackup_page_track_get_start_lsn();
+-----------------------------------------+
| mysqlbackup_page_track_get_start_lsn()  |
+-----------------------------------------+
|                              94414952   |
+-----------------------------------------+
```

```
1 row in set (0.00 sec)
```

使用下面的函数终止页跟踪，返回值是当前的 LSN：

```
mysql > select mysqlbackup_page_track_set(false);
+--------------------------------------+
| mysqlbackup_page_track_set(false)    |
+--------------------------------------+
|                         94429122     |
+--------------------------------------+
1 row in set (0.15 sec)
```

如果使用--incremental 选项时未指定任何值，mysqlbackup 将使用页跟踪功能执行增量备份。用户还可以指定--incremental＝page-track 使 mysqlbackup 使用页面跟踪功能。但使用页跟踪功能进行增量备份有如下两个前提条件：

（1）页跟踪功能正常运行，并且在创建基础备份之前已启用。

（2）更改页数少于总页数的 50％。

以上两个条件如果没有满足，当--incremental＝page-track 时，mysqlbackup 会抛出错误，当--incremental 未设定时，执行完全扫描增量备份（full-scan incremental backup）。

10.4.6 优化增量备份

优化增量备份（optimistic incremental backup）根据 InnoDB 数据文件的修改时间来确定自上次备份后被修改的表，只备份修改过的表。通过指定--incremental＝optimistic，可以执行优化增量备份，在备份过程中会提示哪些表自上次备份后被修改过，需要包括在此次增量备份中，例如：

```
210309 11: 18: 22 RDR1      INFO: Innodb table 'sbtest'.'sbtest2' has been modified  since
previous backup. It should be copied. consistency time : 1615258503, InnoDB table mtime
: 1615259872
```

优化增量备份可能会缩短备份时间，但它有以下限制：

（1）服务器上的系统时间和数据目录的位置不能变化。否则，备份可能会失败，或者生成不一致的增量备份。

（2）不能使用--incremental-with-redo-log-only 进行增量备份。

（3）当指定了--start-lsn 时不能进行优化备份，只能进行完全扫描增量备份。

--incremental＝optimistic 和--incremental＝page-track 的区别是前者是基于文件的修改时间进行增量备份，后者是基于页的 LSN 进行增量备份。

10.4.7 只用重做日志进行增量备份

mysqlbackup 可以通过指定--incremental-with-redo-log-only 来实现只用重做日志进行增量备份。这种备份方式只备份重做日志文件中的内容，它在数据库中的表比较大，但修改的数据比较少时比--incremental 的方式效率高。

实施这个备份方式要先计算重做日志文件的大小，查询重做日志文件的相关参数：

```
mysql > show variables like 'innodb_log_file%';
+---------------------------+-----------+
| Variable_name             | Value     |
+---------------------------+-----------+
| innodb_log_file_size      | 104857600 |
| innodb_log_files_in_group | 2         |
+---------------------------+-----------+
2 rows in set (0.01 sec)
```

将 innodb_log_file_size 和 innodb_log_files_in_group 相乘即可得到重做日志的大小。

再计算 MySQL 实例的日志生成速度，InnoDB 的 LSN 对应写入重做日志中的字节数，下面是查询 10 秒内重做日志的生成量的例子：

```
mysql > pager grep sequence
PAGER set to 'grep sequence'

mysql > SHOW ENGINE INNODB STATUS;
Log sequence number           94429160
1 row in set (0.00 sec)

mysql > select sleep(10);
1 row in set (10.00 sec)

mysql > SHOW ENGINE INNODB STATUS;
Log sequence number           99653686
1 row in set (0.01 sec)
```

两次查询到的 Log sequence number 的差就是这段时间产生的重做日志量。根据重做日志的产生速度和重做日志文件的大小配置备份策略。例如，如果一天产生 1GB 的重做日志，而重做日志文件的大小是 7GB，那每个星期必须进行一次只用重做日志的增量备份，实际上，为了防止突然产生大量的重做日志，可能每天都进行这种增量备份。如果--start-lsn 指定的 LSN 在重做日志文件中已经被覆盖，这种增量备份就会失败，备份脚本应该能捕捉到这种报错，而改用--incremental 的方式进行增量备份。

10.5　实验

mysqlbackup 备份恢复工具

（1）使用 mysqlbackup 进行全量备份和恢复。

（2）使用 mysqlbackup 进行增量备份和恢复。

视频演示

第11章

数 据 救 援

第 8～10 章介绍了 MySQL 的逻辑备份和物理备份，但有时 MySQL 实例无法启动并且没有备份，那么此时是不是一定会面临数据的丢失呢？本章将介绍遇到这种情况时进行数据救援的方法。

11.1 InnoDB 强制恢复

MySQL 实例启动时会进行 InnoDB 崩溃恢复(crash recovery)，很多时候 MySQL 实例启动失败是由于 InnoDB 崩溃恢复失败造成的。此时在错误日志中有类似下面的提示：

```
InnoDB: We intentionally generate a memory trap.
InnoDB: Submit a detailed bug report to http://bugs.mysql.com.
InnoDB: If you get repeated assertion failures or crashes, even
InnoDB: immediately after the mysqld startup, there may be
InnoDB: corruption in the InnoDB tablespace. Please refer to
InnoDB: http://dev.mysql.com/doc/refman/8.0/en/forcing-innodb-recovery.html
InnoDB: about forcing recovery.
...
```

系统参数 innodb_force_recovery 控制着 InnoDB 崩溃恢复的操作，该参数的取值范围为 0～6，默认是 0，数值每增大 1，InnoDB 的崩溃恢复就会跳过一部分操作，从而提高操作成功的可能性，下面说明 innodb_force_recovery 的值对应的操作。

(1) 0：执行完整的 InnoDB 的崩溃恢复步骤。

(2) 1(SRV_FORCE_IGNORE_CORRUPT)：忽略检查到的损坏的页。

(3) 2(SRV_FORCE_NO_BACKGROUND)：如果在执行 purge 操作时会退出，将阻止主线程和 purge 线程的运行。

(4) 3(SRV_FORCE_NO_TRX_UNDO)：不执行事务回滚操作。

(5) 4(SRV_FORCE_NO_IBUF_MERGE)：如果会造成崩溃就不执行插入缓冲

(insert buffer)的合并操作。

(6) 5(SRV_FORCE_NO_UNDO_LOG_SCAN)：不查询撤销日志(undo log)，InnoDB存储引擎会将未提交的事务视为已提交。

(7) 6(SRV_FORCE_NO_LOG_REDO)：不执行事务前滚的操作。

不同 innodb_force_recovery 的值会对 InnoDB 有不同的影响，例如，限制部分操作或将 InnoDB 置于只读状态。表 11.1 说明了不同 innodb_force_recovery 的值与允许的操作和对数据的影响。

表　11.1

操作类型	innodb_force_recovery 的值		
	0	1、2、3	4、5、6
可以执行 INSERT、UPDATE 和 DELETE 操作	√	×	×
可以执行 DROP 或 CREATE 表操作	√	√	×
InnoDB 未处于只读状态	√	√	×
不会造成数据永久损坏	√	√	×

通过增大 innodb_force_recovery 的值可以增加 MySQL 实例成功启动的可能性，但应从 0 开始，每次只增加 1，当 innodb_force_recovery 的值大于 3 后，可能造成数据永久损坏。

当以跳过 InnoDB 部分崩溃恢复的方法启动 MySQL 实例后，应尽快将数据导出，此时不能使用物理备份，因为物理备份会把损坏的页原样备份出来，应使用逻辑备份将所有数据导出，类似命令如下：

```
$ mysqldump --force --all-databases --routines --events --flush-privileges > all_databases.sql
```

接着执行如下步骤恢复数据。

(1) 关闭 MySQL 实例。

(2) 将系统参数 innodb_force_recovery 恢复到默认值 0。

(3) 将 datadir 目录下的所有文件进行备份后删除。

(4) 对数据库进行初始化，代码如下：

```
$ sudo /usr/sbin/mysqld --initialize-insecure --user=mysql
```

(5) 重新启动 MySQL 后把刚才备份的数据导入，代码如下：

```
$ mysql -uroot -p < all_databases.sql
Enter password:
```

因为对数据库进行初始化时使用了--initialize-insecure 参数，生成的用户 root@localhost 是空密码，错误日志里有下面的信息：

```
6 [Warning] [MY-010453] [Server] root@localhost is created with an empty password ! Please consider switching off the --initialize-insecure option.
```

所以在恢复数据时，并不需要输入密码，在“Enter password：”后面直接按 Enter 键即可开始进行恢复。

因为在备份数据时加入了--flush-privileges选项，所以在恢复数据的最后会执行flush privileges语句将用户权限表刷新到内存中，恢复完成后，原来的用户、密码和权限即刻生效，用户、密码和权限保持不变，不用重新创建用户和赋予权限，也不用重新启动。

11.2 迁移MyISAM表

MyISAM引擎因为不支持事务，因此对MyISAM表进行备份时应上只读锁，或者在MySQL关闭的情况下进行，恢复的方法是直接从操作系统上使用文件复制的方法进行恢复。当没有备份时，可以使用表迁移的方法减少数据的损失。MyISAM表的迁移比InnoDB表要容易得多，在MySQL 8.0中对MyISAM表的元数据保存发生了一些变化，因此迁移方法也有所不同，下面分别进行说明。

11.2.1 MySQL 8.0之前

每个MyISAM表在MySQL 8.0之前版本上有如下3个对应的文件：

(1) 以.FRM结尾的文件储存表结构。

(2) 以.MYI结尾的文件储存索引。

(3) 以.MYD结尾的文件储存数据。

MyISAM表的迁移非常简单，直接复制到对应的目录即可，如下例子是把mysql数据库中的tba表复制到performance_schema数据库中。

首先查询mysql数据库里的tba表如下：

```
root:/var/lib/mysql# mysql -e "select * from mysql.tba";
+----+---------+
| id | name    |
+----+---------+
|  1 | aaaaaaa |
|  2 | bbbbbb  |
+----+---------+
```

再检查performance_schema数据库中的tba表如下：

```
root:/var/lib/mysql# mysql -e "select * from performance_schema.tba";
ERROR 29 (HY000) at line 1: File './performance_schema/tba.MYD' not found (Errcode: 2 - No such
file or directory)
```

发现mysql数据库里有tba表，而performance_schema数据库里没有该表。

把tba开头的文件从mysql目录复制到performance_schema目录的命令如下：

```
root:/var/lib/mysql# sudo -u mysql cp mysql/tba.* performance_schema
```

如上命令中使用sudo -u mysql进行复制是为了使新建文件的属主是mysql。

最后检查performance_schema数据库中的tba表如下：

```
root:/var/lib/mysql# mysql -e "select * from performance_schema.tba";
+----+---------+
```

```
| id  | name      |
+----+----------+
|  1 | aaaaaaa   |
|  2 | bbbbbb    |
+----+----------+
```

可以看到,把 tba 表相关的 3 个文件从 mysql 目录复制到 performance_schema 目录中,迁移就完成了。用这种方法还可以实现跨 MySQL 实例的 MyISAM 表的迁移。

11.2.2 在 MySQL 8.0 中

在 MySQL 8.0 中,MyISAM 表以.FRM 结尾的文件消失了,多了一个以.sdi 结尾的文件,该文件以 JSON 格式存储表结构。

下面以把一个 test 数据库中的 tba 表迁移到 sakila 数据库中为例进行说明。

首先把以.MYD 和.MYI 结尾的文件从 test 目录复制到 sakila 目录,命令如下:

```
root:/var/lib/mysql# sudo -u mysql cp test/tba.MY{D,I} sakila
```

再把表的定义文件 tba_3077.sdi 复制到/tmp 目录,命令如下:

```
root:/var/lib/mysql# sudo -u mysql cp test/tba_3077.sdi /tmp
```

然后把 sdi 文件中的数据库名从 test 改成 sakila,命令如下:

```
root:/var/lib/mysql# sed -e 's/test/sakila/' -i /tmp/tba_3077.sdi
```

最后使用 import 命令把 tba 表的定义导入 MySQL 中,命令如下:

```
root:/var/lib/mysql# mysql sakila -e "import table from '/tmp/tba_3077.sdi'"
```

检查 tba 表对应的文件,命令如下:

```
root:/var/lib/mysql#  ls  sakila/tba*
sakila/tba_3078.sdi  sakila/tba.MYD  sakila/tba.MYI
```

可以看到 import 命令生成了一个 tba_3078.sdi 的表定义文件。

至此,tba 表迁移的工作就完成了,在不同 MySQL 实例之间进行表迁移的方法也相同。

11.3 只有表空间文件时批量恢复 InnoDB 表

当通过设置系统参数 innodb_force_recovery 仍然不能启动 MySQL 实例时,也不一定意味着应用的数据丢失了,此时可能只是系统表空间等其他文件损坏,或者遇到 MySQL 的 Bug。如果应用表对应的数据文件没有损坏,数据还是有效的。对于 InnoDB 引擎的表空间可以采用传输表空间的方式把数据文件传输到新的实例上进行恢复。下面是一个批量恢复 InnoDB 表的例子,分成两步,第一步恢复表结构,第二步恢复数据文件。

11.3.1 恢复表结构

恢复表结构的方法很多,例如,可以从同一个应用的其他 MySQL 实例中将创建表的语

句导出来,使用如下命令将 yy_db 数据库中所有表的建表语句导出:

```
$ mysqldump --no-data --compact yy_db > createtb.sql
```

官方提供的 MySQL 的工具集(mysql-utilities)中有个 mysqlfrm 工具,可以从.frm 文件中找回建表语句。从 t1.frm 文件生成建表语句的命令如下:

```
$ mysqlfrm --diagnostic t1.frm
```

把一个目录下的全部.frm 文件转换成建表语句的命令如下:

```
$ mysqlfrm --diagnostic /var/lib/mysql/yy_db/bk/ > createtb.sql
$ grep "^CREATE TABLE" createtb.sql | wc -l
124
```

可以看到共生成了 124 个建表语句。有了建表语句,就可以在一个正常的 MySQL 实例上生成空表,如下所示:

```
mysql > create database yy_db;
mysql > use yy_db
Database changed
mysql > source createtb.sql
Query OK, 0 rows affected (0.07 sec)
```

这样表结构就恢复完成了。

11.3.2 恢复数据文件

生成空表的同时已经生成了数据文件,经检查,以.ibd 结尾的数据文件也是 124 个:

```
$ ll *.ibd|wc
    124     1116     7941
```

使用如下命令将所有.ibd 文件抛弃:

```
$ mysql -e "show tables from yy_db" |grep -v  Tables_in_yy_db |while read a;
do mysql -e "ALTER TABLE yy_db.$a DISCARD TABLESPACE"; done
```

再次检查以.ibd 结尾的数据文件:

```
$ ls *.ibd|wc
ls: cannot access '*.ibd': No such file or directory
```

可以看到,所有的.idb 文件都已经被抛弃了。然后把旧的有数据的.ibd 文件复制到 yy_db 目录下,注意需要把属主改成 mysql,再把这些数据文件导入数据库中,命令如下:

```
$ mysql -e "show tables from yy_db" |grep -v  Tables_in_yy_db |while read a;
do mysql -e "ALTER TABLE yy_db.$a import TABLESPACE"; done
```

导入完成后使用 mysqlcheck 对数据库 yy_db 下的所有表进行检查:

```
$ mysqlcheck -c yy_db
yy_db.cdp_backup_point                          OK
...
```

经过验证,所有表均导入成功。

11.4　使用 ibd2sdi 恢复表结构

在 MySQL 8.0 中,11.3.2 节介绍的导入.ibd 的功能仍然可以使用,但.frm 文件已经取消,无法从.frm 文件中导出表结构,如果 MySQL 实例无法启动,如何导出表结构呢? MySQL 8.0 中,表空间(临时表空间和 undo 表空间除外)中会保存着 SDI(serialized dictionary information,序列化词典信息),该 SDI 作为数据字典的冗余以 JSON 格式保存着数据库对象的元信息,可以使用 MySQL 自带的工具 ibd2sdi 导出表结构。

导出 sakila 数据库中 actor 表结构的代码如下:

```
$ ibd2sdi /var/lib/mysql/sakila/actor.ibd
["ibd2sdi"
,
{
    "type": 1,
    "id": 2448,
    "object":
        {
    "mysqld_version_id": 80022,
    "dd_version": 80022,
    "sdi_version": 80019,
    "dd_object_type": "Table",
    "dd_object": {
        "name": "actor",
        "mysql_version_id": 80022,
        "created": 20210221073024,
        "last_altered": 20210221073024,
        "hidden": 1,
        "options": "avg_row_length = 0; encrypt_type = N; key_block_size = 0; keys_disabled =
0; pack_record = 1; stats_auto_recalc = 0; stats_sample_pages = 0; ",
        "columns": [
            {
                "name": "actor_id",
                "type": 3,
...
```

11.5　TwinDB 数据恢复工具

TwinDB 是一款专门用于 InnoDB 数据恢复的工具,它还有一个名字叫 undrop for InnoDB。它可以在 MySQL 实例无法启动的情况下,直接访问 InnoDB 表空间,并尽量多地恢复数据。

11.5.1　安装

使用如下命令进行下载:

```
$ wget https://github.com/chhabhaiya/undrop-for-innodb/archive/master.zip
```

或者：

```
$ git clone https://github.com/twindb/undrop-for-innodb.git
```

在编译前需要安装如下依赖包：

```
$ sudo apt install make gcc flex bison
```

使用 make 生成执行文件：

```
$ sudo make
```

目前的 TwinDB 只能用于 MySQL 5.7，不能用于 MySQL 8.0。

11.5.2 恢复表结构

当 MySQL 无法启动时，TwinDB 工具集可以从系统表空间文件 ibdata1 中直接恢复表结构，相应步骤如下。

TwinDB 工具集安装完成后，dictionary 目录下有创建 4 个数据字典表的 SQL 语句，如下所示：

```
$ ls dictionary/
SYS_COLUMNS.sql   SYS_FIELDS.sql   SYS_INDEXES.sql   SYS_TABLES.sql
```

创建一个 data_recovered 数据库，命令如下：

```
$ mysqladmin create data_recovered
```

在 data_recovered 数据库中生成 4 个数据字典表，命令如下：

```
$ cat dictionary/SYS_* | mysql data_recovered
```

检查生成的数据，命令及其运行结果如下：

```
$ mysqlshow -vv data_recovered
Database: data_recovered
+-------------+----------+------------+
|  Tables     | Columns  | Total Rows |
+-------------+----------+------------+
| SYS_COLUMNS |        7 |          0 |
| SYS_FIELDS  |        3 |          0 |
| SYS_INDEXES |        7 |          0 |
| SYS_TABLES  |        8 |          0 |
+-------------+----------+------------+
4 rows in set.
```

使用工具集中的 stream_parser 对系统表空间文件 ibdata1 进行解析，代码如下：

```
$ ./stream_parser -f /var/lib/mysql/ibdata1
Opening file:/var/lib/mysql/ibdata1
File information:
```

```
ID of device containing file:          64768
inode number:                         29252470
protection:                              100640 (regular filc)
number of hard links:                       1
user ID of owner:                          27
group ID of owner:                         27
device ID (if special file):                0
blocksize for filesystem I/O:            4096
number of blocks allocated:             24576
Opening file:/var/lib/mysql/ibdata1
...
```

解析完成后将解析的数据放在当前目录下的 pages-ibdata1 目录下，代码如下：

```
$ ls pages - ibdata1/ *
pages - ibdata1/FIL_PAGE_INDEX:
0000000000000001.page  0000000000000011.page  0000000000000016.page
0000000000000026.page  0000000000000043.page
0000000000000002.page  0000000000000012.page  0000000000000017.page
0000000000000032.page  0000000000000044.page
0000000000000003.page  0000000000000013.page  0000000000000020.page
0000000000000033.page  0000000000000045.page
0000000000000004.page  0000000000000014.page  0000000000000024.page
0000000000000041.page  18446744069414584320.page
0000000000000005.page  0000000000000015.page  0000000000000025.page
0000000000000042.page

pages - ibdata1/FIL_PAGE_TYPE_BLOB:
```

根据 MySQL 的源码，这 4 个数据字典表的数据分别位于前面 4 个页中，使用如下 4 条语句从这 4 个页中读出数据：

```
$ ./c_parser  - 4f pages - ibdata1/FIL_PAGE_INDEX/0000000000000001.page  - t \
dictionary/SYS_TABLES.sql > dumps/default/SYS_TABLES \
2 > dumps/default/SYS_TABLES.sql

$ ./c_parser  - 4f pages - ibdata1/FIL_PAGE_INDEX/0000000000000002.page  - t \
dictionary/SYS_COLUMNS.sql > dumps/default/SYS_COLUMNS \
2 > dumps/default/SYS_COLUMNS.sql

$ ./c_parser  - 4f pages - ibdata1/FIL_PAGE_INDEX/0000000000000003.page  - t \
dictionary/SYS_INDEXES.sql > dumps/default/SYS_INDEXES \
2 > dumps/default/SYS_INDEXES.sql

$ ./c_parser  - 4f pages - ibdata1/FIL_PAGE_INDEX/0000000000000004.page  - t \
dictionary/SYS_FIELDS.sql > dumps/default/SYS_FIELDS \
2 > dumps/default/SYS_FIELDS.sql
```

使用如下命令将数据导入这 4 个表中：

```
$ cat dumps/default/ * .sql |  mysql data_recovered
```

导出完成后再检查一下，命令及其运行结果如下：

```
$   mysqlshow -vv data_recovered
Database: data_recovered
+-------------+----------+------------+
|  Tables     | Columns  | Total Rows |
+-------------+----------+------------+
| SYS_COLUMNS |        7 |        157 |
| SYS_FIELDS  |        3 |         50 |
| SYS_INDEXES |        7 |         35 |
| SYS_TABLES  |        8 |         28 |
+-------------+----------+------------+
4 rows in set.
```

这样所有表的数据字典信息都恢复到这 4 个表中了，下面编译生成 sys_parser 工具：

```
$ sudo make sys_parser
/bin/mysql_config
cc -o sys_parser sys_parser.c 'mysql_config --cflags' 'mysql_config --libs'
```

使用 sys_parser 工具可以从数据字典表中生成任意表的建表语句，用法如下：

```
$ ./sys_parser
sys_parser [-h <host>] [-u <user>] [-p <passowrd>] [-d <db>] databases/table
```

例如，使用如下命令生成 world.city 表的建表语句：

```
$ ./sys_parser -u root -p dingjia  -d data_recovered world/city
CREATE TABLE 'city'(
    'ID' INT NOT NULL,
    'Name' CHAR(35) CHARACTER SET 'utf8mb4' COLLATE 'utf8mb4_general_ci' NOT NULL,
    'CountryCode' CHAR(3) CHARACTER SET 'utf8mb4' COLLATE 'utf8mb4_general_ci' NOT NULL,
    'District' CHAR(20) CHARACTER SET 'utf8mb4' COLLATE 'utf8mb4_general_ci' NOT NULL,
    'Population' INT NOT NULL,
    PRIMARY KEY ('ID')
) ENGINE = InnoDB;
```

11.5.3 恢复被删除的表

TwinDB 工具集可以用于恢复刚刚被误删除的表，如果误删除了 world.city 表，该表对应的独立表空间文件/var/lib/mysql/world/city.ibd 也被删除，使用 11.5.2 节介绍的方法可以恢复这个表的结构，然后使用如下命令恢复 InnoDB 的页：

```
$ ./stream_parser -f /dev/sda2 -t 59.5G
```

这里的/dev/sda2 是 world.city 表所在的硬盘分区，59.5G 是该分区的大小，如果该表是放在通用表空间的，那么删除该表时不会同时删除该表空间，应将上面命令中的/dev/sda2 替换成对应的表空间文件名。stream_parser 会扫描该分区，将所有的 InnoDB 页都解析出来放到如下目录中：

```
$ ls pages-sda2/FIL_PAGE_INDEX
0000000000000001.page  0000000000000033.page  0000000000000062.page
0000000000000092.page  0000000000000125.page
0000000000000002.page  0000000000000034.page  0000000000000063.page
0000000000000093.page  0000000000000126.page
0000000000000003.page  0000000000000035.page  0000000000000064.page
0000000000000094.page  0000000000000127.page
...
```

根据之前生成的数据字典信息，查询 world.city 表 ID 的命令及其运行结果如下：

```
mysql> select * from data_rrecovered.SYS_TABLES where name = 'world/city';
+------------+----+--------+-------+--------+----------+-------------+-------+
| NAME       | ID | N_COLS | TYPE  | MIX_ID | MIX_LEN  | CLUSTER_NAME| SPACE |
+------------+----+--------+-------+--------+----------+-------------+-------+
| world/city | 40 |      5 |  33   |      0 |     80   |             |    23 |
+------------+----+--------+-------+--------+----------+-------------+-------+
1 row in set (0.00 sec)
```

知道 world.city 的表 ID 为 40 后，再查询主键的 ID，其命令及其运行结果如下：

```
mysql> select * from data_recovered.SYS_INDEXES where table_id = 40;
+----------+----+-------------+----------+------+-------+---------+
| TABLE_ID | ID | NAME        | N_FIELDS | TYPE | SPACE | PAGE_NO |
+----------+----+-------------+----------+------+-------+---------+
|       40 | 41 | PRIMARY     |        1 |    3 |    23 |       3 |
|       40 | 42 | CountryCode |        1 |    0 |    23 |       4 |
+----------+----+-------------+----------+------+-------+---------+
2 rows in set (0.00 sec)
```

知道主键的 ID 为 41，因为 InnoDB 都是索引组织表，数据按主键进行存放，因此 world.city 表的数据就在第 41 页，可以使用 c_parser 解析第 41 页，将数据输出到文件中，代码如下：

```
$ ./c_parser -6f pages-city.ibd/FIL_PAGE_INDEX/0000000000000041.page -t city.sql \
>> dumps/default/city \
> 2> dumps/default/city.sql
```

查看生成的 SQL 文件，代码如下：

```
$ cat dumps/default/city.sql
SET FOREIGN_KEY_CHECKS = 0;
LOAD DATA LOCAL INFILE
'/root/install/undrop-for-innodb-master/dumps/default/city' REPLACE INTO
TABLE 'city' FIELDS TERMINATED BY '\t' OPTIONALLY ENCLOSED BY '"' LINES STARTING
BY 'city\t' ('ID', 'Name', 'CountryCode', 'District', 'Population');
```

执行该 SQL 文件将数据加载到数据库中，命令如下：

```
$ cat dumps/default/city.sql | mysql world
```

至此，数据恢复成功。

11.5.4 从损坏的表空间中挽救数据

造成表空间损坏的原因有很多，如硬盘坏道、多个进程同时访问数据文件、突然掉电等。遇到表空间损坏时，TwinDB 工具集可以跳过损坏的部分，尽量读取更多的数据。

表空间损坏后可能造成 MySQL 实例无法启动，或者客户端在查询损坏的表时崩溃，在 MySQL 的错误日志中可能会有如下提示：

```
2021-02-23T11:32:01.326823Z 0 [ERROR] InnoDB: Database page corruption on disk or a failed
file read of page [page id: space=24, page number=4]. You may have to recover from a backup.
```

使用 innochecksum 检查损坏的表可能会有如下提示：

```
$ innochecksum  /var/lib/mysql/world/country.ibd
Fail: page 0 invalid
Exceeded the maximum allowed checksum mismatch count::0
```

这时可以使用 TwinDB 工具集将对应的表空间分成页，命令如下：

```
$ ./stream_parser -f /var/lib/mysql/world/country.ibd
```

如上命令执行后会生成一个页，如下所示：

```
$ ls pages-country.ibd/FIL_PAGE_INDEX/
0000000000000043.page
```

使用如下命令从第 43 页中导出数据：

```
$ ./c_parser -6f pages-country.ibd/FIL_PAGE_INDEX/0000000000000043.page \
-t country.sql > dumps/default/country 2> dumps/default/country.sql
```

对导出的文件 dumps/default/city 进行检查，确认无误后再导入重新创建的表中，命令如下：

```
$ cat dumps/default/city.sql | mysql world
```

11.6 实验

MyISAM 表的迁移

分别在 MySQL 5.7 和 MySQL 8.0 下进行 MyISAM 表的迁移。

视频演示

第四部分

高 可 用

第12章

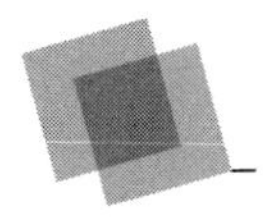

MySQL Shell

MySQL 用户使用最多的客户端工具应该是 mysql,但 MySQL 官方新推出的客户端工具 MySQL Shell 功能更加强大,本章介绍这种工具的相关内容,包括 MySQL Shell 的通用命令、客户化 MySQL Shell、全局对象和报告架构等。

后续两章都会用到 MySQL Shell 的管理 API。

12.1 简介

MySQL Shell 是 MySQL 官方于 2017 年推出的新一代命令行客户端工具。相对于传统的 mysql 客户端工具,它的功能非常强大,它不仅支持传统的关系数据库,还支持 NoSQL 文档型数据库。它同时支持 JavaScript、Python 和 SQL 三种语言,其设计目标如下:

- 成为大多数场景下的客户端工具。
- 同时提供给开发人员和 DBA 用。
- 将复杂的操作简单化。
- 成为管理 InnoDB cluster 和 InnoDB ReplicaSet 的工具。
- 开源和可扩展。

现在看来,MySQL Shell 的设计目标基本都实现了。它可以在 64 位的 Microsoft Windows、Linux 和 macOS 平台上运行,并且在这些平台上的操作方法都一样,例如,按 Ctrl+A 可以回到命令行的行首,按 Ctrl+D 组合键退出 MySQL Shell。可以到 MySQL 的官方网站(https://dev.mysql.com)上下载它的安装版本,其安装方法和 MySQL 的其他工具,如 workbench 类似。

启动 MySQL Shell 的方法是在 Linux 中调用 mysqlsh,例如:

```
$ mysqlsh
MySQL Shell 8.0.23
```

```
Copyright (c) 2016, 2021, Oracle and/or its affiliates.
Oracle is a registered trademark of Oracle Corporation and/or its affiliates.
Other names may be trademarks of their respective owners.

Type '\help' or '\?' for help; '\quit' to exit.
MySQL  JS >
```

在 Windows 中是运行 mysqlsh. exe 程序。

12.2 通用命令

MySQL Shell 提供以“\”开头的通用命令可以在 JavaScript、Python 和 SQL 三种语言下执行。

12.2.1 命令清单

表 12.1 是 MySQL Shell 的通用命令的清单。

表 12.1

命令	快捷方式	说明
\help	\h 或\?	打印有关 MySQL Shell 的帮助，或搜索联机帮助
\quit	\q 或\exit	退出 MySQL Shell
\	—	在 SQL 模式下，开始多行模式，缓存输入的代码，遇到空行时执行
\status	\s	显示当前的 MySQL Shell 状态
\js	—	将执行语言模式切换为 JavaScript
\py	—	将执行语言模式切换为 Python
\sql	—	将执行语言模式切换为 SQL
\connect	\c	连接到 MySQL 实例
\reconnect	—	重新连接到同一个 MySQL 实例
\use	\u	指定默认的数据库
\source	\.	使用当前活动的语言模式执行脚本文件
\warnings	\W	显示语句生成的任何警告
\nowarnings	\w	不显示语句生成的任何警告
\history	—	查看和编辑命令行历史记录
\rehash	—	手动更新自动完成名称缓存
\option	—	查询和更改 MySQL Shell 配置选项
\show	—	使用提供的选项和参数运行指定的报告
\watch	—	使用提供的选项和参数运行指定的报告，并定期刷新结果
\edit	\e	使用默认的系统编辑器编辑当前命令
\pager	\P	指定用来处理输出的命令(pager)
\nopager	—	取消指定用来处理输出的命令(pager)
\system	\!	执行操作系统的命令，并将结果输出到 MySQL Shell

12.2.2 连接到 MySQL 实例

在 MySQL Shell 中，一个简单的连接到 MySQL 实例的命令如下：

```
MySQL  SQL > \c root@localhost
Creating a session to 'root@localhost'
Fetching schema names for autocompletion... Press ^C to stop.
Your MySQL connection id is 227 (X protocol)
Server version: 8.0.22 MySQL Community Server - GPL
No default schema selected; type \use <schema> to set one.
```

连接 MySQL 的语法如下：

```
\connect [<TYPE>] <URI>  或    \c [<TYPE>] <URI>
```

其中，TYPE 有两种，分别对应如下两种协议：

(1) --mc,--mysql——经典的 MySQL 协议，默认端口是 3306，只支持关系数据库。

(2) --mx,--mysqlx——新增的 X 协议，默认端口是 33060，既支持关系数据库，也支持 NoSQL 文档型数据库。

URI 的格式如下：

```
[user[:password]@]hostname[:port]
```

也可以在 MySQL Shell 调用时直接指定要连接的 MySQL 实例，可以使用 uri 的格式指定，例如：

```
$ mysqlsh  --mysqlx --uri root@localhost:33060
```

也可以使用键值的方法指定，例如：

```
$ mysqlsh --mysqlx -u root -h localhost -P 33060
```

在 SQL 语言模式中只有一个连接，该连接称为全局连接，它可以在 3 种语言模式切换过程中共享，但在 Python 和 JavaScript 中可以有多个连接，例如，如下 Python 小程序中有到两个不同数据库的连接：

```
$ cat ex1.py
session1 = mysql.get_session("root:dingjia@localhost:3306")
session2 = mysql.get_session("root:dingjia@192.168.17.149:3306")

result = session1.run_sql(" select * from sakila.actor; ")
row = result.fetch_all()
print(row[1])
session2.run_sql("select @@hostname, @@port")
```

在 MySQL Shell 的 Python 语言模式中执行该小程序的输出如下：

```
MySQL  localhost:33060+ ssl  sakila  Py > \source ~/ex1.py
[
    2,
```

```
    "NICK",
    "WAHLBERG",
    "2006-02-15 16:04:33"
]
+------------+--------+
| @@hostname | @@port |
+------------+--------+
| scutech    |  3306  |
+------------+--------+
1 row in set (0.0004 sec)
```

可以看到，该 Python 小程序可以同时操作两个连接。

\reconnect 命令没有任何参数，它使用当前的参数重新连接到 MySQL 实例，如下所示：

```
MySQL  localhost:33060+ ssl  SQL > \reconnect
Attempting to reconnect to 'mysqlx://root@localhost:33060'..
The global session was successfully reconnected.
```

\disconnect 也没有任何参数，如下命令为终止和当前 MySQL 的连接，如下所示：

```
MySQL  localhost:33060+ ssl  SQL > \disconnect
MySQL  SQL >
```

12.2.3 帮助信息

MySQL Shell 的功能非常强大，为了便于使用，该工具中也内置了丰富的帮助信息，下面分类说明。

（1）输入\? 或\h 后不接任何参数可以得到总的帮助信息，如下所示：

```
MySQL  JS > \?
The Shell Help is organized in categories and topics. To get help for a
specific category or topic use: \? <pattern>

The <pattern> argument should be the name of a category or a topic.

The pattern is a filter to identify topics for which help is required, it can
use the following wildcards:

- ? matches any single character.
- * matches any character sequence.

The following are the main help categories:
...
```

其他子类帮助信息的查询使用\? 或\h 后接要查询的命令即可。

（2）查询通用命令的帮助信息，使用\? 或\h 后接要查询的命令即可得到它的帮助信息，例如，查询\pager 命令的帮助信息及其输出结果如下：

```
MySQL  localhost:33060+  ssl  SQL>\? \pager
NAME
      \pager - Sets the current pager.

SYNTAX
      \pager [command]
      \P [command]

DESCRIPTION
      The current pager will be automatically used to:

       - display results of statements executed in SQL mode,
       - display text output of \help command,
       - display text output in scripting mode, after shell.enablePager() has
         been called,

      Pager is going to be used only if shell is running in interactive mode.

EXAMPLES
      \pager
            With no parameters this command restores the initial pager.

      \pager ""
            Restores the initial pager.

      \pager more
            Sets pager to "more".

      \pager "more -10"
            Sets pager to "more -10".

      \pager more -10
            Sets pager to "more -10".
```

(3) 查询SQL语言的帮助信息,使用\?或\h后接要查询的SQL命令即可得到它的帮助信息,例如,查询select命令的帮助信息及其输出结果如下:

```
MySQL  localhost:33060+  ssl  SQL>\h select
Syntax:
SELECT
    [ALL | DISTINCT | DISTINCTROW ]
    [HIGH_PRIORITY]
    [STRAIGHT_JOIN]
    [SQL_SMALL_RESULT] [SQL_BIG_RESULT] [SQL_BUFFER_RESULT]
    [SQL_NO_CACHE] [SQL_CALC_FOUND_ROWS]
    select_expr [, select_expr] ...
    [into_option]
    [FROM table_references
    ...
```

(4) 查询全局对象的帮助信息,使用\?或\h后接要查询的全局对象名(全局对象名可以在总的帮助信息中查询）即可得到它的帮助信息，例如，查询 shell 对象的帮助信息及其输出结果如下：

```
MySQL  localhost:33060+ ssl  Py > \? shell
NAME
      shell - Gives access to general purpose functions and properties.

DESCRIPTION
      Gives access to general purpose functions and properties.

PROPERTIES
      options
            Gives access to options impacting shell behavior.

      reports
            Gives access to built-in and user-defined reports.

FUNCTIONS
      add_extension_object_member(object, name, member[, definition])
            Adds a member to an extension object.

      connect(connectionData[, password])
            Establishes the shell global session.
...
```

（5）查询全局对象中函数帮助信息的方法有两种，一种是使用\? 或\h 后接要查询的全局对象中的函数名，例如，查询 util.export_table 函数的语句及其输出结果如下：

```
MySQL  localhost:33060+ ssl  Py > \h util.export_table
NAME
      export_table - Exports the specified table to the data dump file.

SYNTAX
      util.export_table(table, outputUrl[, options])

WHERE
      table: Name of the table to be exported.
      outputUrl: Target file to store the data.
      options: Dictionary with the export options.

DESCRIPTION
      The value of table parameter should be in form of table or schema.table,
...
```

另一种方法是使用全局对象内置的一个 help()函数输出该函数的使用方法，例如：

```
MySQL  Py > util.help('export_table')
```

这两种方法查询的结果完全一样。

（6）每个全局对象都内置的一个 help()函数输出该对象的使用方法，例如，dba.help()的输出结果如下：

```
MySQL  localhost:33060+ ssl  sakila  JS > dba.help()
NAME
      dba - InnoDB cluster and replicaset management functions.

DESCRIPTION
      Entry point for AdminAPI functions, including InnoDB clusters and replica
      sets.
...
```

help()函数还可以输出全局对象里函数的使用方法,这种方法前面已经介绍过,此处不再赘述。

12.2.4 命令示例

MySQL Shell 中可以通过\sql、\py、\js 三个命令在 SQL、Python 和 JavaScript 三种语言模式中进行切换,登录时默认是 JavaScript 语言模式,可以通过如下命令设置登录后默认的语言模式为 SQL:

```
MySQL  localhost:33060+ ssl  SQL > \option --persist defaultMode sql
```

设置完成,可以用如下命令查询:

```
MySQL  localhost:33060+ ssl  SQL > \option defaultMode
Sql
```

取消设置,恢复到默认语言模式的方法如下:

```
MySQL  localhost:33060+ ssl  SQL > \option --unset --persist defaultMode
```

全局对象 shell 中的 options 属性保存着 MySQL Shell 的所有设置,是字典类型的数据:

```
MySQL  localhost:33060+ ssl  Py > shell.options
{
    "autocomplete.nameCache": true,
    "batchContinueOnError": false,
    "credentialStore.excludeFilters": [],
    "credentialStore.helper": "default",
    "credentialStore.savePasswords": "prompt",
    "dba.gtidWaitTimeout": 60,
    "dba.logSql": 0,
    "dba.restartWaitTimeout": 60,
    "defaultCompress": false,
    "defaultMode": "sql",
    "devapi.dbObjectHandles": true,
    "history.autoSave": true,
    "history.maxSize": 1000,
    "history.sql.ignorePattern": "*IDENTIFIED*:*PASSWORD*",
    "interactive": true,
    "logLevel": 5,
```

```
    "oci.configFile": "/root/.oci/config",
    "oci.profile": "DEFAULT",
    "outputFormat": "table",
    "pager": "",
    "passwordsFromStdin": false,
    "resultFormat": "table",
    "sandboxDir": "/root/mysql-sandboxes",
    "showColumnTypeInfo": false,
    "showWarnings": true,
    "useWizards": true,
    "verbose": 0
}
```

查询其中的默认语言模式，命令如下：

```
MySQL  localhost:33060+ ssl  Py > shell.options['defaultMode']
sql
```

在 MySQL Shell 中可以通过上下箭头调出历史的命令，但退出时默认不会保存，可以通过如下方法把属性 history.autoSave 设置成 1，这样退出时会保存历史命令：

```
MySQL  localhost:33060+ ssl  Py >
shell.options.set_persist('history.autoSave',1)
```

设置完成后查询属性 history.autoSave 的值如下：

```
MySQL  localhost:33060+ ssl  Py > shell.options['history.autoSave']
True
```

命令 status 可以查询当前的 MySQL Shell 的状态，如下所示：

```
MySQL  localhost:33060+ ssl  Py > \status
MySQL Shell version 8.0.23

Connection Id:              235
Default schema:
Current schema:
Current user:               root@localhost
SSL:                        Cipher in use: ECDHE-RSA-AES128-GCM-SHA256
TLSv1.2
Using delimiter:            ;
Server version:             8.0.22 MySQL Community Server - GPL
Protocol version:           X protocol
Client library:             8.0.23
Connection:                 localhost via TCP/IP
TCP port:                   33060
Server characterset:        utf8mb4
Schema characterset:        utf8mb4
Client characterset:        utf8mb4
Conn. characterset:         utf8mb4
Result characterset:        utf8mb4
Compression:                Enabled (DEFLATE_STREAM)
```

```
Uptime:                           4 days 7 hours 35 min 44.0000 sec
```

命令 use 可以设置当前的数据库，如下所示：

```
MySQL  localhost: 33060 + ssl  Py > \use sakila
Default schema 'sakila' accessible through db.
MySQL  localhost: 33060 + ssl  sakila  Py >
```

命令 system（或\!）可以执行操作系统的命令，例如：

```
MySQL  localhost: 33060 + ssl  sakila  Py > \! uname - r
3.10.0 - 123.el7.x86_64
```

命令 edit 会使用默认的系统编辑器编辑当前命令，在 Linux 系统上是 vi，该命令也可以通过按 Ctrl+X 组合键和 Ctrl+E 组合键调用。

12.3 客户化 MySQL Shell

12.3.1 启动脚本

当 MySQL Shell 第一次启动到或切换到 Python 或 JavaScript 语言模式时，会自动调用启动脚本。可以使用启动脚本完成 Python 或 JavaScript 语言的运行环境设置工作，Python 语言模式的启动脚本是 mysqlshrc.py，JavaScript 语言模式的启动脚本是 mysqlshrc.js。MySQL Shell 搜索启动脚本的路径如下。

（1）系统的标准路径。

- Windows 系统：%PROGRAMDATA%\MySQL\mysqlsh\mysqlshrc.[js|py]。
- UNIX/Linux 系统：/etc/mysql/mysqlsh/mysqlshrc.[js|py]。

（2）MySQL Shell 家目录下的 share/mysqlsh 子目录，MySQL Shell 家目录由环境变量 MYSQLSH_HOME 决定，如果没有定义该环境变量，则 MySQL Shell 家目录在执行文件 mysqlsh 路径 bin 的上一级目录。

- Windows 系统：%MYSQLSH_HOME%\share\mysqlsh\mysqlshrc.[js|py]。
- Unix/Linux 系统：$MYSQLSH_HOME/share/mysqlsh/mysqlshrc.[js|py]。

（3）如果 MySQL Shell 家目录在第 2 项中没有找到，则在执行文件 mysqlsh 所在的目录。

- Windows 系统：执行文件 mysqlsh 所在的目录\mysqlshrc.[js|py]。
- UNIX/Linux 系统：执行文件 mysqlsh 所在的目录/mysqlshrc.[js|py]。

（4）环境变量 MYSQLSH_USER_CONFIG_HOME 指定的路径。

- Windows 系统：%MYSQLSH_USER_CONFIG_HOME%\mysqlshrc.[js|py]。
- UNIX/Linux 系统：$MYSQLSH_USER_CONFIG_HOME/mysqlshrc.[js|py]。

（5）当第 4 项指定的环境变量没有设置时，则在当前用户环境决定的目录。

- Windows 系统：%APPDATA%\MySQL\mysqlsh\mysqlshrc.[js|py]。
- UNIX/Linux 系统：$HOME/.mysqlsh/mysqlshrc.[js|py]。

启动脚本是可选的，如果在上面的路径中找到了两个启动脚本，则选择后面的脚本执行。启动脚本的一个典型用例是增加 Python 或 JavaScript 语言加载模块的搜索路径，例如，如下 Python 启动脚本：

```
$ cat ~/.mysqlsh/mysqlshrc.py
import sys
sys.path.append('~/custom')
```

12.3.2 配置提示符

一个默认的提示符的例子如图 12.1 所示。

MySQL localhost:33060+ ssl sakila Py >

图 12.1

图 12.1 所示默认提示符的开头是 MySQL 字符串，接着是主机名和端口号，"+"表示 X 协议，ssl 表示协议，sakila 是默认的数据库，Py 是语言模式，为蓝色；如果是 SQL，即为橘黄色；JavaScript 是黄色。

提示符的定义由一个 JSON 文件完成，该文件在 Windows 平台上位于目录 C:\Program Files\MySQL\中，在 Linux 平台上位于如下目录中：

```
$ ll /usr/share/mysqlsh/prompt/
total 48
-rw-r--r--. 1 root root 1245 Dec 11 16:36 prompt_16.json
-rw-r--r--. 1 root root 1622 Dec 11 16:36 prompt_256inv.json
-rw-r--r--. 1 root root 2137 Dec 11 16:36 prompt_256.json
-rw-r--r--. 1 root root 2179 Dec 11 16:36 prompt_256pl+aw.json
-rw-r--r--. 1 root root 1921 Dec 11 16:36 prompt_256pl.json
-rw-r--r--. 1 root root  183 Dec 11 16:36 prompt_classic.json
-rw-r--r--. 1 root root 2172 Dec 11 16:36 prompt_dbl_256.json
-rw-r--r--. 1 root root 2250 Dec 11 16:36 prompt_dbl_256pl+aw.json
-rw-r--r--. 1 root root 1992 Dec 11 16:36 prompt_dbl_256pl.json
-rw-r--r--. 1 root root 1205 Dec 11 16:36 prompt_nocolor.json
-rw-r--r--. 1 root root 6197 Dec 11 16:36 README.prompt
```

每个 JSON 文件定义一种提示符的配置方式，默认的配置文件是 prompt_256.json，最后一个 README.prompt 文件说明了如何编写 JSON 文件，如果要编写自己的 JSON 文件，可以将类似的 JSON 文件进行修改后复制到如下目录。

- Linux 平台：~/.mysqlsh/prompt.json。
- Windows 平台：%AppData%\MySQL\mysqlsh\prompt.json。

在说明文件 README.prompt 中有一个定义变量的例子如下：

```
"variables" : {
# set the is_production variable to "production"
# if the %host% variable is contained in the PRODUCTION_SERVERS
# environment variable, which is a list of hosts separated by ;
"is_production": {
    "match" : {
```

```
            "pattern": " * ; % host % ; * ",
            "value": "; % env: PRODUCTION_SERVERS % "
        },
        "if_truc" : "production",
        "if_false" : ""
    }
}
```

如上变量表示如果连接到环境变量 PRODUCTION_SERVERS 中定义的主机名，提示符前面将会增加红色的 PRODUCTION 提示，表示已经连接到生产系统，进行变更时要谨慎，使用如下命令设置本机为生产环境：

```
$  export PRODUCTION_SERVERS = localhost
```

设置后的提示符如图 12.2 所示。

```
PRODUCTION  MySQL  localhost:33060+ ssl  SQL >
```

图　12.2

12.3.3　配置输出格式

MySQL Shell 的输出模式有以下 4 种，通过选项 resultFormat 进行控制。

(1) 表格模式(table)：输出结果放在已格式化的表格中，格式美观。例如：

```
MySQL  localhost: 33060 + ssl  SQL > \option resultFormat = table
MySQL  localhost: 33060 + ssl  SQL > select * from sakila.actor where last_name = 'WOOD';
+----------+------------+-----------+---------------------+
| actor_id | first_name | last_name | last_update         |
+----------+------------+-----------+---------------------+
|       13 | UMA        | WOOD      | 2006 - 02 - 15 04: 34: 33 |
|      156 | FAY        | WOOD      | 2006 - 02 - 15 04: 34: 33 |
+----------+------------+-----------+---------------------+
2 rows in set (0.0015 sec)
```

(2) 制表符分隔模式(tabbed)：输出结果通过制表符进行分隔。例如：

```
MySQL  localhost: 33060 + ssl  SQL > \option resultFormat = tabbed
MySQL  localhost: 33060 + ssl  SQL > select * from sakila.actor where last_name = 'WOOD';
actor_id  first_name  last_name  last_update
13  UMA  WOOD  2006 - 02 - 15 04: 34: 33
156  FAY  WOOD  2006 - 02 - 15 04: 34: 33
2 rows in set (0.0037 sec)
```

(3) 垂直模式(vertical)：输出结果通过垂直的方式而不是水平的方式进行显示，和 SQL 语句使用\G 结束符的效果一样，在结果集中有较长字段时，这种显示方式的效果更好。例如：

```
MySQL  localhost: 33060 + ssl  SQL > \option resultFormat = vertical
MySQL  localhost: 33060 + ssl  SQL > select * from sakila.actor where last_name = 'WOOD';
```

```
*************************** 1. row ***************************
   actor_id: 13
 first_name: UMA
  last_name: WOOD
last_update: 2006-02-15 04:34:33
*************************** 2. row ***************************
   actor_id: 156
 first_name: FAY
  last_name: WOOD
last_update: 2006-02-15 04:34:33
2 rows in set (0.0049 sec)
```

(4) JSON模式(json或json/pretty)：通过JSON格式显示输出结果。例如：

```
MySQL  localhost:33060+ ssl  SQL > \option resultFormat=json/pretty
MySQL  localhost:33060+ ssl  SQL > select * from sakila.actor where last_name='WOOD';
{
    "actor_id": 13,
    "first_name": "UMA",
    "last_name": "WOOD",
    "last_update": "2006-02-15 04:34:33"
}
{
    "actor_id": 156,
    "first_name": "FAY",
    "last_name": "WOOD",
    "last_update": "2006-02-15 04:34:33"
}
2 rows in set (0.0048 sec)
```

(5) RAW JSON模式(json/raw或ndjson)：通过原始JSON格式显示输出结果。例如：

```
MySQL  localhost:33060+ ssl  SQL > \option resultFormat=json/raw
MySQL  localhost:33060+ ssl  SQL > select * from sakila.actor where last_name='WOOD';
{"actor_id":13,"first_name":"UMA","last_name":"WOOD","last_update":"2006-02-15 04:34:33"}
{"actor_id":156,"first_name":"FAY","last_name":"WOOD","last_update":"2006-02-15 04:34:33"}
2 rows in set (0.0019 sec)
```

(6) 数组JSON模式(json/array)：通过放在数组中的原始JSON格式显示输出结果。例如：

```
MySQL  localhost:33060+ ssl  SQL > \option resultFormat=json/array
MySQL  localhost:33060+ ssl  SQL > select * from sakila.actor where last_name='WOOD';
[
{"actor_id":13,"first_name":"UMA","last_name":"WOOD","last_update":"2006-02-15 04:34:33"},
{"actor_id":156,"first_name":"FAY","last_name":"WOOD","last_update":"2006-02-15 04:34:33"}
]
```

```
2 rows in set (0.0028 sec)
```

在交互式模式下，默认的输出模式是表格模式；在批处理模式下，默认的输出模式是制表符分隔模式。

在 Python 或 JavaScript 语言模式中，输出模式和在 SQL 语言模式中一样，如下命令为在 Python 中设置为表格模式的输出格式：

```
MySQL  localhost: 33060 + ssl  Py > shell.options.set('resultFormat','table')
```

设置完成后的输出例子如下：

```
MySQL  localhost: 33060 + ssl  Py > session.sql("select * from sakila.actor where last_name
= 'WOOD'")
+----------+------------+-----------+---------------------+
| actor_id | first_name | last_name | last_update         |
+----------+------------+-----------+---------------------+
|       13 | UMA        | WOOD      | 2006-02-15 04:34:33 |
|      156 | FAY        | WOOD      | 2006-02-15 04:34:33 |
+----------+------------+-----------+---------------------+
2 rows in set (0.0047 sec)
```

12.3.4　保存登录凭证

MySQL Shell 内置保存密码的功能，当第一次登录某个服务的 URL 时，会提示用户是否保存密码，如下所示：

```
$ mysqlsh --mysqlx -u root -h localhost -P 33060
Please provide the password for 'root@localhost: 33060': *******
Save password for 'root@localhost: 33060'? [Y]es/[N]o/Ne[v]er (default No):
```

MySQL Shell 内置的保存密码的功能支持如下 3 种接口：

- Mysql login-path(myql_config_editor)
- macOS keychain
- Windows Credentials Management API

可以使用 shell. listCredentialHelpers()查询当前支持的保存密码的接口，命令如下：

```
MySQL  localhost: 33060 + ssl  JS > shell.listCredentialHelpers()
[
    "login-path"
]
```

可以使用 shell. listCredentials()查询当前保存的密码，命令如下：

```
MySQL  localhost: 33060 + ssl  JS > shell.listCredentials();
[
    "root@localhost: 33060"
]
```

使用工具 mysql_config_editor 也能查询当前保存的密码，命令如下：

```
$ mysql_config_editor print --all
[root@localhost: 33060]
user = root
password = *****
host = localhost
port = 33060
```

这两种查询方法的结果是一致的。

和保存密码相关的函数如下：

(1) shell.options.credentialStore.savePasswords = "value"

设置保存密码的规则，value 可以设置成：

- Always——一直保存密码。
- Never——一直不保存密码。
- Prompt——提示用户后，根据用户的选择决定保存或不保存密码，这也是默认设置。

(2) shell.options.credentialStore.excludeFilters = ["*@myserver.com:*"];

指定不保存密码的 MySQL 服务的 URL，可以使用“*”和“?”进行模式匹配，“*”匹配任意长度的字符，“?”匹配单个字符。

(3) var list = shell.listCredentialHelpers();

返回当前平台保存密码的接口。

(4) shell.storeCredential(url[,password]);

保存 URL 和对应的密码，如果没有输入密码参数，提示用户手工输入。

(5) shell.deleteCredential(url);

删除保存的 URL 和对应的密码。

(6) shell.deleteAllCredentials();

删除所有保存的密码。

(7) var list = shell.listCredentials();

列出当前保存的密码。

12.4 全局对象

MySQL Shell 通过内置的全局对象将功能分组，用户也可以自己添加全局对象，MySQL Shell 中已经内置了以下全局对象，这些全局对象在 Python 和 JavaScript 中都存在。

(1) session：当存在全局连接时，该对象代表全局连接。例如：

```
MySQL  localhost: 33060+ ssl  sakila  Py > session
<Session: root@localhost: 33060>
```

(2) db：当使用 X 协议连接到 MySQL 时，该对象代表当前默认数据库，数据库的表成了这个对象的属性。如下命令设置当前数据库为 sakila：

```
MySQL  localhost: 33060+ ssl  Py > \use sakila
```

```
Default schema 'sakila' accessible through db.
```

这时，输入全局对象 db 会显示当前的默认数据库，如下所示：

```
MySQL  localhost:33060+ ssl  sakila  Py > db
<Schema:sakila>
```

按 Tab 键会输出当前数据库下的表和 db 的函数：

```
MySQL  localhost:33060+ ssl  sakila  Py > db.
actor                        customer_list                get_collection()
help()                       sales_by_film_category
actor_info                   drop_collection()            get_collection_as_table()
inventory                    sales_by_store
address                      exists_in_database()         get_collections()
language                     schema
category                     film                         get_name()
modify_collection()          session
city                         film_actor                   get_schema()
name                         staff
country                      film_category                get_session()
nicer_but_slower_film_list   staff_list
create_collection()          film_list                    get_table()
payment                      store
customer                     film_text                    get_tables()
rental
```

(3) dba：通过 AdminAPI 提供管理 InnoDB 集群的接口。

(4) cluster：代表 InnoDB 集群的全局对象，只有当使用--cluster 选项启动 mysqlsh 时才会生成该对象。

(5) rs：代表 InnoDB 的复制集(replicaset)的全局对象，只有当使用--replicaset 选项启动 mysqlsh 时才会生成该对象。

(6) shell：包含各种通用的方法和属性的全局对象。

(7) mysql：采用经典协议连接到 MySQL 的全局对象。

(8) mysqlx：采用新的 X 协议连接到 MySQL 的全局对象。

(9) util：包含 MySQL Shell 各种工具的全局对象，例如，升级检查工具、数据导入导出工具等。

Python 和 JavaScript 两种语言的方法名和函数名很类似，只是拼写形式不同，Python 是蛇形(snake case)拼写，JavaScript 是驼形(camel case)拼写，例如，打开连接的函数在 Python 中是 shell.open_session()，在 JavaScript 中是 shell.openSession()。

全局对象提供的 API 接口还可以在 MySQL Shell 的命令行中直接调用，语法如下：

```
$ mysqlsh [options] -- shell_object object_method [arguments]
```

MySQL Shell 会把“--”后面的字符串当作 MySQL Shell 中的命令进行调用，其中：

- shell_object 是全局对象。
- object_method 是全局对象的方法。
- arguments 是传递给全局对象方法的参数。

例如，如下命令为设置自动保存历史：

```
$ mysqlsh -- shell.options set_persist history.autoSave true
```

如上命令和如下 MySQL Shell 里的命令是等价的：

```
mysql-js> shell.options.set_persist('history.autoSave', true);
```

再例如，如下命令为对 MySQL 实例进行升级前的检查：

```
$ mysqlsh -- util check-for-server-upgrade { --user=root --host=localhost --port
=3306 } --password='password' --config-path=/etc/my.cnf
```

如上命令和如下 MySQL Shell 里的命令是等价的：

```
mysql-js> util.checkForServerUpgrade({user: 'root', host: 'localhost', port: 3306},
{password: 'password', configPath: '/etc/my.cnf'})
```

12.5 报告架构

12.5.1 简介

MySQL Shell 中有一套报告架构，支持内置的报告和用户开发的报告，内置了 query、thread 和 threads 三种类型的报告。

执行报告的命令有如下两个。

- \show：单次执行报告。
- \watch：重复执行报告。
- \watch 有如下两个选项决定报告输出的时间和形式。
- --interval=float，-i float：以秒为单位规定每次执行报告之间的时间间隔，范围是 0.1～86 400 秒，默认为 2 秒。
- --nocls：执行报告时不清屏，下次执行报告的输出接在上次输出的后面。

可以按 Ctrl+C 组合键终止\watch 命令的执行。

12.5.2 query 报告

query 类型的报告执行作为参数的单条 SQL 语句，并使用报告架构输出执行结果。

如下命令为每隔 0.5 秒执行一次查询当前正在执行 SQL 语句：

```
MySQL  localhost:33060+ ssl  JS> \watch query --interval=0.5 --nocls  -E select conn_
id, current_statement from sys.session where command = 'query' and conn_id <> connection_id()
*************************** 1. row ***************************
          conn_id: 247
current_statement: INSERT INTO t1 VALUES (2038455 ... 09BiuT0L44jsvrZ2GKczwyAS6dv')
*************************** 1. row ***************************
          conn_id: 247
current_statement: INSERT INTO t1 VALUES (2038455 ... 09BiuT0L44jsvrZ2GKczwyAS6dv')
```

```
Report returned no data.
```

其中,选项--vertical(或-E)表示以垂直方式显示输出结果。

12.5.3　thread 报告

thread 类型的报告返回单个连接的线程信息。这类报告支持如下选项。

- --tid,--cid：指定线程 ID 或连接 ID。
- --general：当没有指定下面的选项时,默认输出线程的基本信息。
- --brief：在一行中显示线程的简短信息。
- --client：显示客户端的连接和会话信息。
- --innodb：显示当前存在的 InnoDB 事务信息。
- --locks：显示与线程相关的锁信息。
- --prep-stmts：显示分配到线程的准备 SQL 语句。
- --status：显示会话的状态参数,可以通过前缀只显示部分状态参数。
- --vars：显示会话的系统参数,可以通过前缀只显示部分系统参数。
- --user-vars：显示会话的用户定义参数,可以通过前缀只显示部分用户定义参数。
- --all：显示除简短信息之外的全部信息。

如下命令为显示线程 ID 为 276 的线程的通用信息、客户端信息和锁信息：

```
MySQL  localhost:33060+ ssl  JS > \show thread --tid 276  --general --client --locks
GENERAL
Thread ID:                  276
Connection ID:              247
Thread type:                FOREGROUND
Program name:               mysqlslap
User:                       root
Host:                       localhost
Database:                   mysqlslap
Command:                    Query
Time:                       00:00:00
State:                      waiting for handler commit
Transaction state:          NULL
Prepared statements:        0
Bytes received:             5064279
Bytes sent:                 133774
Info:                       INSERT INTO t1 VALUES (2038455 ... 09BiuT0L44jsvrZ2GKczwyAS6dv')
Previous statement:         NULL

CLIENT
_client_name:               libmysql
_client_version:            8.0.26
_os:                        Linux
_pid:                       4350
_platform:                  x86_64
program_name:               mysqlslap
Protocol:                   classic
Socket IP:
Socket port:                0
```

```
Socket state:                ACTIVE
Socket stats:                NULL
Compression:                 OFF
SSL cipher:
SSL version:

LOCKS
Waiting for InnoDB locks
N/A

Waiting for metadata locks
N/A

Blocking InnoDB locks
N/A

Blocking metadata locks
N/A
```

thread 类型的报告不支持选项--vertical（或-E）以垂直方式显示输出结果，因为这类报告的输出中已经同时有垂直和表格两种方式。

12.5.4 threads 报告

threads 类型的报告返回所有线程信息，它有如下参数。

- --foreground，--background，--all：这三个参数指定输出线程的类型，分别对应前台、后台和全部线程。
- --format（-o）：通过指定一组用逗号分隔的字段名自定义输出的字段。
- --where，--order-by，--desc，--limit：处理输出的结果，和 SQL 语句中的 where、order by、desc、limit 的功能一样。

也可以使用选项--vertical（或-E）以垂直方式显示输出结果。如下报告输出了前台线程和相关信息，包括线程 ID、用户名、主机名、客户端程序名、线程正在执行的命令、线程占用的内存等。

```
mysql-js> \show threads --foreground -o tid,user,host,progname,command,memory
+-----+-----------------+-----------+----------+---------+-----------+
| tid | user            | host      | progname | command | memory    |
+-----+-----------------+-----------+----------+---------+-----------+
| 26  | root            | localhost | mysqlsh  | Query   | 1.40 MiB  |
| 42  | event_scheduler | localhost | NULL     | Daemon  | 16.18 KiB |
| 46  | NULL            | NULL      | NULL     | Daemon  | 19.57 KiB |
+-----+-----------------+-----------+----------+---------+-----------+
```

视频演示

12.6 实验

MySQL Shell 客户端工具

练习 MySQL Shell 客户端的常用操作。

第13章

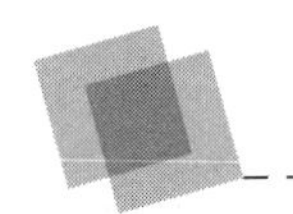

复　制

第3章介绍了 MySQL 的二进制日志，它的一个重要功能是可以用于 MySQL 的复制，本章介绍 MySQL 复制的相关内容。建立复制前，首先要根据源创建一个副本，创建副本的方法很多，这里介绍使用克隆插件创建副本的方法。

本章还介绍了两种配置复制的方法，一种是使用传统的二进制日志文件坐标，另一种是使用 GTID；此外，还介绍了如何排除复制过程中出现的错误，并举了一个常见的案例；以及使用 MySQL Shell 的管理 API 创建和管理复制的方法。

13.1　简介

复制是 MySQL 的一项重要功能，它可以把数据的更改从一个实例复制到另一个实例。这两个实例之前被称为主库（master）和从库（slave），现在通常被称为源（source）和副本（replica）。复制的实现原理是：源将所有数据和结构的更改写入二进制日志，副本向源请求二进制日志，并在本地应用其内容。

MySQL 复制要求所有的表都有主键，对于没有定义主键的 InnoDB 表，MySQL 会生成一个隐含的聚簇主键，但这样的主键并不能满足复制的要求。对于没有自定义主键的表，在副本的表上进行更新时会进行全表扫描。从操作系统层可以看到副本的磁盘读远远大于源，更新也会落后源。没有主键还无法阻止在副本上把一个记录重复插入。因此在建立复制前应采用第5章里的脚本找出没有主键的表，给所有的表创建主键，同时将系统参数 sql_require_primary_key 设置为 on，阻止创建表或更改表结构时不设主键。

MySQL 复制有如下应用场景。

- 水平扩展：将只读的交易迁移到副本上执行，通过增加副本，实现应用的水平扩展，但只能实现读扩展，不能实现写扩展。在副本上进行读需要注意的是，副本的数据和源数据之间并不完全同步，可能存在很短时间的数据更新滞后，如果应用不能接受读到陈旧的数据，则应该采用组复制或 InnoDB 集群。

- 商业智能和分析：在副本上运行对资源消耗大的报告和分析，让源专注于生产应用。在实施中可以在源和副本之间加上数据过滤规则，只复制需要的数据，而不是源的数据全量，这样会让副本的运行效率更高。
- 远距离数据分布：通过复制将远距离的数据迁移到本地，使本地应用程序就近访问数据，减少网络延迟，给客户更好的使用体验。
- 高可用性：通过多个副本提供冗余，并实现受控地切换或滚动升级。
- 备份：由于在副本上进行备份不会影响源端的应用，因此在备份过程中不需要担心应用性能的下降、锁占用和备份数据对缓冲池的污染，甚至可以把副本关闭进行冷备份。

MySQL 复制的功能主要由如下三个线程完成，一个在源，两个在副本。

(1) 二进制日志转储线程(binary log dump thread)：当副本连接到源时，源创建这个线程将二进制日志内容发送到副本。在源上使用 SHOW PROCESSLIST 查看时，该线程被标识为 Binlog Dump 线程。

(2) 复制 I/O 线程(replication I/O thread)：当在副本上发出 START REPLICA 语句时，副本创建一个 I/O 线程，该线程连接到源，并要求它发送其二进制日志中记录的更新。复制 I/O 线程接收到源的二进制日志的更新，并将其保存到副本的中继日志文件中。在 SHOW SLAVE STATUS 的输出中，该线程的状态显示为 Slave_IO_running。

(3) 复制 SQL 线程(replication SQL thread)：副本创建一个 SQL 线程来读取复制 I/O 线程写入的中继日志，并执行其中包含的事务。把系统参数 replica_parallel_workers 设置为大于 0 时，会有指定相同数量的工作(worker)线程并行执行 SQL 线程的工作和一个协调(coordinator)线程协调这些工作线程。

13.2 克隆插件

创建 MySQL 复制的第一步是创建一个新的副本，通常是把源的一个全量备份集恢复到新的副本上，可以采用的工具包括：mysqldump、Xtrabackup、mysqlbackup 等。从 MySQL 8.0.17 开始，MySQL 推出了克隆插件，可以方便地在本地或从远程 MySQL 实例克隆数据。该插件也可以用于创建新的副本。

13.2.1 加载克隆插件

加载克隆插件的方法有两种，一种是在 MySQL 启动时使用 plugin-load 或 plugin-load-add 选项加载插件，例如，在配置文件中加入如下内容：

```
[mysqld]
plugin-load=mysql_clone.so
```

这样每次 MySQL 启动时都会加载克隆插件。如果要阻止 MySQL 实例在没有加载克隆组件的情况下启动，可以设置 clone=FORCE_PLUS_PERMANENT 或 clone=FORCE。

另一种方法是在 MySQL 运行时使用 install plugin 进行加载，命令如下：

```
mysql > install plugin clone soname 'mysql_clone.so';
```

这样，插件会被注册到 mysql. plugin 表中，以后每次启动时都会加载。检查插件状态的命令及其运行结果如下：

```
mysql > select plugin_name, plugin_status from information_schema.plugins where plugin_name =
'clone';
+-------------+---------------+
| plugin_name | plugin_status |
+-------------+---------------+
| clone       | ACTIVE        |
+-------------+---------------+
1 row in set (0.03 sec)
```

13.2.2　本地克隆

执行本地克隆的用户应该具有 BACKUP_ADMIN 的权限，该权限可以让用户传输数据和阻止克隆期间的 DDL 操作，可以使用如下命令进行赋权：

```
mysql > grant backup_admin on *.* to 'clone_user';
```

运行如下命令进行本地克隆：

```
mysql > clone local data directory = '/path/to/clone_dir';
```

其中，/path/to/clone_dir 是新的 MySQL 的数据目录，mysql 用户应有权限创建该目录，该目录应是绝对路径，目录/path/to 应已经存在，而 clone_dir 应不存在。

新克隆的数据目录可以用于启动一个新的 MySQL 实例，例如：

```
$ sudo mysqld --datadir = '/path/to/clone_dir' --user = mysql
```

注意，还要配置监听端口和 socket 文件之类的参数，如果同时启动这两个实例，注意这些参数不能冲突。

本地克隆操作不支持对驻留在数据目录之外的用户创建的表或表空间进行克隆，试图克隆这类表空间会遇到下面的错误提示：

```
ERROR 1086 (HY000): File '/path/to/tablespace_name.ibd' already exist.
```

所有用户创建的 InnoDB 表和表空间、InnoDB 系统表空间、redo logs 和 undo 表空间都被克隆到指定的目录中。

13.2.3　远程克隆

远程克隆由目标端发起，发起的用户需要具有 CLONE_ADMIN 的权限，该权限可以让用户复制数据，阻止克隆期间的 DDL 操作和自动重新启动 MySQL。可以使用如下命令赋予用户权限：

```
mysql> grant clone_admin on *.* to 'clone_user';
```

目标端的用户应该具有 BACKUP_ADMIN 的权限。可以使用如下命令赋予用户权限：

```
mysql> grant backup_admin on *.* to 'donor_user'@'recipient_host';
```

在克隆开始之前先设置全局变量 clone_valid_donor_list，使其指向有效的源，例如：

```
mysql> set global clone_valid_donor_list = '192.168.17.103:3306';
```

执行类似如下的命令开始克隆：

```
mysql> clone instance from donor_user@'192.168.17.103':3306 identified by 'dingjia';
```

在克隆的过程中，如果设置 log_error_verbosity=3，可以在错误日志中看到进度，类似的提示如下：

```
2021-04-03T06:45:14.655970Z 10 [Note] [MY-013458] [InnoDB] Clone State BEGIN FILE COPY
2021-04-03T06:45:14.656638Z 10 [Note] [MY-011845] [InnoDB] Clone Start PAGE ARCH : start
LSN : 101373841, checkpoint LSN : 101373841
2021-04-03T06:45:14.657456Z 10 [Note] [MY-013458] [InnoDB] Clone State FILE COPY : 12
chunks,  chunk size : 64 M
2021-04-03T06:45:14.775361Z 10 [Note] [MY-013458] [InnoDB] Stage progress: 25%
completed.
2021-04-03T06:45:14.887643Z 10 [Note] [MY-013458] [InnoDB] Stage progress: 50%
completed.
2021-04-03T06:45:14.961622Z 10 [Note] [MY-013458] [InnoDB] Stage progress: 75%
completed.
2021-04-03T06:45:15.783160Z 10 [Note] [MY-013458] [InnoDB] Stage progress: 100%
completed.
2021-04-03T06:45:15.817921Z 11 [Note] [MY-013458] [InnoDB] Clone set state change ACK: 2
```

克隆完成后，查询二进制日志坐标的命令和运行结果如下：

```
mysql> select binlog_file, binlog_position from performance_schema.clone_status;
+---------------+-----------------+
| binlog_file   | binlog_position |
+---------------+-----------------+
| binlog.000006 |             926 |
+---------------+-----------------+
1 row in set (0.00 sec)
```

如上二进制的坐标就是后续进行复制的开始位置。

13.3 配置复制

13.3.1 源配置

源与复制相关的参数如下。

- server_id：使用一个整数代表该 MySQL 数据库，这个整数在整个复制架构里应是唯一的。
- log_bin：是一个布尔值，代表是否激活二进制日志，在 MySQL 8.0 里面默认为ON，在之前的版本中默认为 OFF。如果要使用复制功能，必须激活二进制日志。
- binlog_format：指定二进制日志的格式，有语句(STATEMENT)、混合(MIXED)和行(ROW)三种格式，通常使用行格式。

还要在源上创建一个用户，用于从副本连接到源，命令如下：

```
mysql > create user user@slave_hostname identified by 'password';
```

并给该用户赋予 REPLICATION SLAVE 的权限，命令如下：

```
mysql > grant replication slave on * . * to user@slave_hostname;
```

13.3.2 副本配置

副本与复制相关的系统参数如下。

- server_id：和源的 server_id 类似，使用一个整数代表该 MySQL 数据库。
- logs-slave-updates：布尔值，决定副本是否将其 SQL 线程执行的更新记入自己的二进制日志，默认为 OFF。为了使该选项生效，还必须用--logs-bin 选项启动副本以启用二进制日志。
- relay_log_recovery：布尔值，当为 ON 时，副本启动时会自动放弃所有未执行的中继日志，重新生成一个中继日志，将副本的 I/O 线程的位置重新指向新的中继日志；并将 SQL 线程的位置退回到与 I/O 线程的位置保持一致，重新开始同步，这样在副本中，事务不会丢失。该参数的默认值是 OFF，建议设置成 ON。
- read-only：布尔值，当该参数激活时，禁止客户端对库进行修改，除非客户端的用户有 CONNECTION_ADMIN 或 SUPER 权限，该参数的默认值是 OFF，在副本上建议设置成 ON。
- skip-slave-start：布尔值，当该参数激活时，复制进程就不会随着数据库的启动而启动，该参数的默认值是 OFF，建议设置成 ON，让数据库启动后留一些时间让用户进行检查，然后再决定是否手工启动复制，这些做更加稳妥。

13.3.3 启动复制

在副本上使用 change replication source to…配置到源的连接：

```
mysql > change replication source to source_host = 'host_name', source_user = 'user_name', source
_password = 'password',
source_log_file = 'master_log_name', source_log_pos = master_log_pos;
```

其中的 source_log_file 和 source_log_pos 指定了复制开始的二进制日志坐标，这个坐标一定要设置准确，不然会出现丢失更新或重复更新数据的情况。

使用如下命令启动副本：

```
mysql> start replica;
```

13.4 GTID

GTID是global transaction identifiers的缩写，它在整个复制架构中唯一标识一个事务。它由两部分组成，一部分是MySQL服务的UUID，它保存在MySQL数据目录的auto.cnf文件中，这一部分是不会变的；另外一部分就是事务ID了，随着事务的增加，值依次递增。两部分之间用冒号分开，例如：

```
199d4998-7f1e-11eb-a5c7-fa163e368ff4:1
```

每个MySQL的UUID都不同，在复制结构中，如果副本是从源直接复制过来的，那么主从中的auto.cnf内容是一样的。在启动复制功能时会报错："Fatal error: The slave I/O thread stops because master and slave have equal MySQL server UUIDs; these UUIDs must be different for replication to work."这时可以删除auto.cnf，然后重启MySQL，MySQL会自动生产新的UUID。UUID可以从系统变量中查到，如下所示：

```
mysql> SELECT @@server_uuid\G
*************************** 1. row ***************************
@@server_uuid: 199d4998-7f1e-11eb-a5c7-fa163e368ff4
1 row in set (0.01 sec)
```

传统的复制方式使用二进制日志文件的坐标进行源和副本之间的数据同步，新的复制方式使用GTID来保证源和副本数据的一致性。由于每个事务的GTID都有唯一性，新的复制方式不再需要二进制日志文件的坐标，这样大大简化了复制操作。

副本在执行源传输过来的事务时不会改变GTID，查询该事务的GTID就可以知道这个事务来自哪个源。MySQL在执行事务时，会把该事务的GTID记录到gtid_executed系统参数中，gtid_executed是一个只读参数，用户不能修改它。

当二进制日志文件被清除后，被清除的二进制日志文件中包含的事务GTID集被记录到gtid_purged系统参数中。gtid_purged是gtid_executed的一个子集，用户可以修改该参数。

要启动GTID，需要在源和副本的配置文件中都加入如下内容：

```
[mysqld]
gtid-mode=ON
enforce-gtid-consistency
```

重新启动后，在副本上执行如下命令切换到基于GTID的复制。

```
mysql> change replication source to source_auto_position=1;
```

13.5 排错

13.5.1 监控复制的状态

监控复制的状态主要有两种方法，一种是在副本上执行 show replica status 命令，另一种是查询 performance_schema 数据库中 replication_开头的视图。

在副本上执行 show replica status 命令检查复制的状态，如下所示：

```
mysql > show replica status \G
*************************** 1. row ***************************
             Replica_IO_State: Waiting for master to send event
                  Source_Host: 192.168.17.103
                  Source_User: repl
                  Source_Port: 3306
                Connect_Retry: 60
              Source_Log_File: binlog.000006
          Read_Source_Log_Pos: 2217
               Relay_Log_File: redhat7 - 2 - relay - bin.000002
                Relay_Log_Pos: 1612
        Relay_Source_Log_File: binlog.000006
           Replica_IO_Running: Yes
          Replica_SQL_Running: Yes
              Replicate_Do_DB:
          Replicate_Ignore_DB:
           Replicate_Do_Table:
       Replicate_Ignore_Table:
      Replicate_Wild_Do_Table:
  Replicate_Wild_Ignore_Table:
                   Last_Errno: 0
                   Last_Error:
                 Skip_Counter: 0
          Exec_Source_Log_Pos: 2217
              Relay_Log_Space: 1825
              Until_Condition: None
               Until_Log_File:
                Until_Log_Pos: 0
           Source_SSL_Allowed: No
           Source_SSL_CA_File:
           Source_SSL_CA_Path:
              Source_SSL_Cert:
            Source_SSL_Cipher:
               Source_SSL_Key:
        Seconds_Behind_Source: 0
Source_SSL_Verify_Server_Cert: No
                Last_IO_Errno: 0
                Last_IO_Error:
               Last_SQL_Errno: 0
               Last_SQL_Error:
```

```
Replicate_Ignore_Server_Ids:
           Source_Server_Id: 1
                Source_UUID: 199d4998-7f1e-11eb-a5c7-fa163e368ff4
           Source_Info_File: mysql.slave_master_info
                  SQL_Delay: 0
        SQL_Remaining_Delay: NULL
  Replica_SQL_Running_State: Slave has read all relay log; waiting for more updates
         Source_Retry_Count: 86400
                Source_Bind:
    Last_IO_Error_Timestamp:
   Last_SQL_Error_Timestamp:
             Source_SSL_Crl:
         Source_SSL_Crlpath:
         Retrieved_Gtid_Set:
          Executed_Gtid_Set:
              Auto_Position: 0
       Replicate_Rewrite_DB:
               Channel_Name:
         Source_TLS_Version:
     Source_public_key_path:
      Get_Source_public_key: 0
          Network_Namespace:
1 row in set (0.00 sec)
```

需要关注如下状态参数。

- Replica_IO_Running：复制 I/O 线程的运行状态，正常是 Yes。
- Replica_SQL_Running：复制 SQL 线程的运行状态，正常是 Yes。
- Seconds_Behind_Source：以秒为单位显示副本应用 SQL 落后与源的时间差，该值如果过大则需要关注。实质上该值是 SQL 线程和 I/O 线程之间的时间差，如果源和副本之间的通信延迟比较大，该值并不能反映它们之间的应用 SQL 的时间差。
- Source_Log_File 和 Read_Source_Log_Pos：这两个参数是 I/O 线程读取源二进制日志的坐标，可以用 show master status 命令查询源的当前二进制日志的坐标，如果相差太大，需要关注。
- Last_IO_Errno、Last_IO_Error、Last_SQL_Errno、Last_SQL_Error：当复制出现错误时，需要关注这 4 个状态参数，因为它们记录了出错的信息。

源的配置信息默认保存在 mysql. slave_master_info 表中，保存在文件中的方式从 MySQL 8.0.23 版本后已经弃用。查询该表的例子如下：

```
mysql> select * from mysql.slave_master_info \G
*************************** 1. row ***************************
                Number_of_lines: 32
                Master_log_name: binlog.000006
                 Master_log_pos: 2949
                           Host: 192.168.17.103
                      User_name: repl
                  User_password: dingjia
                           Port: 3306
```

```
                    Connect_retry: 60
                      Enabled_ssl: 0
                           Ssl_ca:
                       Ssl_capath:
                         Ssl_cert:
                       Ssl_cipher:
                          Ssl_key:
           Ssl_verify_server_cert: 0
                        Heartbeat: 30
                             Bind:
               Ignored_server_ids: 0
                             Uuid:
                      Retry_count: 86400
                          Ssl_crl:
                      Ssl_crlpath:
           Enabled_auto_position: 0
                     Channel_name:
                      Tls_version:
                  Public_key_path:
                   Get_public_key: 0
                Network_namespace:
    Master_compression_algorithm: uncompressed
   Master_zstd_compression_level: 3
                 Tls_ciphersuites: NULL
Source_connection_auto_failover: 0
1 row in set (0.00 sec)
```

副本的中继日志信息默认保存在 mysql. lave_relay_log_info 表中，保存在文件中的方式从 MySQL 8.0.23 版本后已经弃用。查询该表的例子如下：

```
mysql> select * from mysql.slave_relay_log_info \G
*************************** 1. row ***************************
                             Number_of_lines: 14
                              Relay_log_name: ./redhat7-2-relay-bin.000002
                               Relay_log_pos: 1039
                             Master_log_name: binlog.000006
                              Master_log_pos: 3667
                                   Sql_delay: 0
                           Number_of_workers: 0
                                          Id: 1
                                Channel_name:
                   Privilege_checks_username: NULL
                   Privilege_checks_hostname: NULL
                          Require_row_format: 0
             Require_table_primary_key_check: STREAM
Assign_gtids_to_anonymous_transactions_type: OFF
Assign_gtids_to_anonymous_transactions_value:
1 row in set (0.01 sec)
```

在 performance_schema 数据库中还有以 replication_开头的视图用于监控复制的状态，如下所示：

```
$ mysqlshow performance_schema|grep replication
| replication_applier_configuration                    |
| replication_applier_filters                          |
| replication_applier_global_filters                   |
| replication_applier_status                           |
| replication_applier_status_by_coordinator            |
| replication_applier_status_by_worker                 |
| replication_asynchronous_connection_failover         |
| replication_connection_configuration                 |
| replication_connection_status                        |
| replication_group_member_stats                       |
| replication_group_members                            |
```

这些性能视图包含的内容类似于 show replica status 输出,但更加详细,而且查询起来也更加方便,其中以 replication_connection 开头的视图存放着复制 I/O 线程的相关信息,以 replication_applier 开头的视图存放着复制 SQL 线程的相关信息。

除了以 replication_开头的视图,如下视图保存着日志相关的信息:

```
mysql > select * from performance_schema.log_status\G
*************************** 1. row ***************************
    SERVER_UUID: 3b1b59d2 - f44b - 11eb - 95c4 - fa163ed3d7ad
          LOCAL: {"gtid_executed": "44d8538f - f1db - 11eb - 990a - fa163ed3d7ad: 1 - 38",
"binary_log_file": "localhost - bin.000003", "binary_log_position": 196}
    REPLICATION: {"channels": [{"channel_name": "", "relay_log_file": "localhost - relay -
bin.000004", "relay_log_position": 449}]}
STORAGE_ENGINES: {"InnoDB": {"LSN": 379782374, "LSN_checkpoint": 379782374}}
1 row in set (0.00 sec)
```

13.5.2 排错案例

日常工作中对复制状态的监控首先通过 show replica status 命令进行,发现异常后,再根据症状查询相应的 performance_schema 数据库中的视图进行进一步的剖析。

下面是一个排错的例子。首先检查复制状态,如下所示:

```
mysql > show replica status\G
*************************** 1. row ***************************
             Replica_IO_State: Waiting for source to send event
                  Source_Host: localhost.localdomain
                  Source_User: mysql_innodb_rs_1349945754
                  Source_Port: 3306
                Connect_Retry: 60
              Source_Log_File: binlog.000003
          Read_Source_Log_Pos: 26628
               Relay_Log_File: localhost - relay - bin.000006
                Relay_Log_Pos: 744
        Relay_Source_Log_File: binlog.000003
           Replica_IO_Running: Yes
          Replica_SQL_Running: No
```

```
             Replicate_Do_DB:
          Replicate_Ignore_DB:
           Replicate_Do_Table:
       Replicate_Ignore_Table:
      Replicate_Wild_Do_Table:
  Replicate_Wild_Ignore_Table:
                   Last_Errno: 1062
                   Last_Error: Coordinator stopped because there were error(s) in the worker
(s). The most recent failure being: Worker 1 failed executing transaction '44d8538f-f1db-
11eb-990a-fa163ed3d7ad:43' at master log binlog.000003, end_log_pos 26597. See error log
and/or performance_schema.replication_applier_status_by_worker table for more details about
this failure or others, if any.
                 Skip_Counter: 0
          Exec_Source_Log_Pos: 26333
              Relay_Log_Space: 2322
              Until_Condition: None
               Until_Log_File:
                Until_Log_Pos: 0
           Source_SSL_Allowed: No
           Source_SSL_CA_File:
           Source_SSL_CA_Path:
              Source_SSL_Cert:
            Source_SSL_Cipher:
               Source_SSL_Key:
        Seconds_Behind_Source: NULL
Source_SSL_Verify_Server_Cert: No
                Last_IO_Errno: 0
                Last_IO_Error:
               Last_SQL_Errno: 1062
               Last_SQL_Error: Coordinator stopped because there were error(s) in the worker
(s). The most recent failure being: Worker 1 failed executing transaction '44d8538f-f1db-
11eb-990a-fa163ed3d7ad:43' at master log binlog.000003, end_log_pos 26597. See error log
and/or performance_schema.replication_applier_status_by_worker table for more details about
this failure or others, if any.
  Replicate_Ignore_Server_Ids:
             Source_Server_Id: 3940373588
                  Source_UUID: 44d8538f-f1db-11eb-990a-fa163ed3d7ad
             Source_Info_File: mysql.slave_master_info
                    SQL_Delay: 0
          SQL_Remaining_Delay: NULL
    Replica_SQL_Running_State:
           Source_Retry_Count: 86400
                  Source_Bind:
      Last_IO_Error_Timestamp:
     Last_SQL_Error_Timestamp: 210804 13:30:53
               Source_SSL_Crl:
           Source_SSL_Crlpath:
           Retrieved_Gtid_Set: 44d8538f-f1db-11eb-990a-fa163ed3d7ad:37-43
            Executed_Gtid_Set: 3b1b59d2-f44b-11eb-95c4-fa163ed3d7ad:1-2,
44d8538f-f1db-11eb-990a-fa163ed3d7ad:1-42
                Auto_Position: 1
```

```
        Replicate_Rewrite_DB:
                Channel_Name:
          Source_TLS_Version:
      Source_public_key_path:
       Get_Source_public_key: 1
           Network_Namespace:
1 row in set (0.00 sec)
```

从复制状态中可以看到如下一些出错信息：

- Last_Error 和 Last_SQL_Error 说明 SQL 线程的出错位置。
- Replica_SQL_Running 说明 SQL 线程已经停止运行。
- Retrieved_Gtid_Set 和 Executed_Gtid_Set 说明从源传输过来的事务中，GTID 最后为 43 的事务没有执行。
- Last_SQL_Error_Timestamp 是出错的时间戳。

根据提示检查视图可以得到更加详细的出错信息，如下所示：

```
mysql > select last_error_message from performance_schema.replication_applier_status_by_
worker limit 1\G
*************************** 1. row ***************************
LAST_ERROR_MESSAGE: Worker 1 failed executing transaction '44d8538f - f1db - 11eb - 990a -
fa163ed3d7ad: 43' at master log binlog.000003, end_log_pos 26597; Could not execute Write_rows
event on table sakila.actor; Duplicate entry '205' for key 'actor.PRIMARY', Error_code: 1062;
handler error HA_ERR_FOUND_DUPP_KEY; the event's master log binlog.000003, end_log_pos 26597
1 row in set (0.00 sec)
```

从出错信息中可以看到，向表 sakila.actor 中插入记录时遇到了重复的主键，主键值是 205。由此，可以分析出解决这个问题有两种方法，一是删除副本中 sakila.actor 表中主键为 205 的记录，命令如下：

```
mysql > delete from sakila.actor where actor_id = 205;
```

另一种方法是在副本上跳过该事务。由于同一个 GTID 在副本上只能执行一次，可以通过在副本上执行一个和该 GTID 相同的空事务来屏蔽这个事务的执行，方法如下。

（1）停止复制，命令如下：

```
mysql > stop replica;
```

（2）设置 GTID_NEXT 为需要跳过的事务的 GTID，因为 GTID_NEXT 默认是使用副本的 UUID 顺序产生 GTID，命令如下：

```
mysql > SET GTID_NEXT = "44d8538f - f1db - 11eb - 990a - fa163ed3d7ad: 43";
```

（3）执行一个空事务，命令如下：

```
mysql > BEGIN; COMMIT;
```

（4）恢复 GTID_NEXT 的默认设置，命令如下：

```
mysql > SET GTID_NEXT = "AUTOMATIC";
```

（5）启动复制，命令如下：

```
mysql > start replica;
```

重新启动复制后发现状态恢复正常。

13.6　使用 MySQL Shell 的 AdminAPI 管理 InnoDB 复制

MySQL Shell 8.0.19 以后的版本中，可以使用 MySQL Shell 的 AdminAPI 管理 InnoDB 复制。

13.6.1　创建复制

使用 mysqlsh 的 AdminAPI 的函数 dba.createReplicaSet()创建复制的命令及其输出结果如下：

```
MySQL  localhost:33060+ ssl  JS > var rs = dba.createReplicaSet("my")
A new replicaset with instance '192.168.17.149:3306' will be created.

* Checking MySQL instance at 192.168.17.149:3306

This instance reports its own address as 192.168.17.149:3306

NOTE: Some configuration options need to be fixed:
+-------------+-----------+------------+--------------------------------------------+
|Variable      | Current Value | Required Value |Note                                 |
+-------------+-----------+------------+--------------------------------------------+
|gtid_mode     | OFF       | ON          |Update read-only variable and restart the server |
|log_slave_updates | OFF   | ON          |Update read-only variable and restart the server |
+-------------+-----------+------------+--------------------------------------------+
Some variables need to be changed, but cannot be done dynamically on the server.
ERROR: 192.168.17.149:3306: Instance must be configured and validated with dba.
configureReplicaSetInstance() before it can be used in a replicaset.
Dba.createReplicaSet: Instance check failed (MYSQLSH 51150)
```

如上输出结果提示有两个系统参数配置不正确，对于这两个系统参数的修改可以通过函数 dba.configureReplicaSetInstance()自动完成，该函数将使用 set persist 命令设置这两个系统参数，如下命令还会创建一个账号 scott@localhost 用于管理副本集：

```
MySQL  localhost:33060+ ssl  JS >
dba.configureReplicaSetInstance('root@localhost:3306', {clusterAdmin: "'scott'@'localhost'"});
Configuring local MySQL instance listening at port 3306 for use in an InnoDB ReplicaSet...

This instance reports its own address as 192.168.17.149:3306
Password for new account: *******
Confirm password: *******

NOTE: Some configuration options need to be fixed:
+-------------+-----------+------------+--------------------------------------------+
|Variable      | Current Value | Required Value |Note                                 |
```

```
+-------------+-----------+------------+------------------------------------+
|gtid_mode         | OFF        | ON         |Update read-only variable and restart the server  |
|log_slave_updates | OFF        | ON         |Update read-only variable and restart the server  |
+-------------+-----------+------------+------------------------------------+
Some variables need to be changed, but cannot be done dynamically on the server.
Do you want to perform the required configuration changes? [y/n]: y
Do you want to restart the instance after configuring it? [y/n]: y
Cluster admin user 'scott'@'localhost' created.
Configuring instance...
The instance '192.168.17.149:3306' was configured to be used in an InnoDB ReplicaSet.
Restarting MySQL...
NOTE: MySQL server at 192.168.17.149:3306 was restarted.
```

实例配置好后，可以使用dba.createReplicaSet()函数创建复制，先检查当前的数据库，命令及其运行结果如下：

```
MySQL  localhost:33060+ ssl  JS > \sql show databases;
+--------------------+
| Database           |
+--------------------+
| information_schema |
| mysql              |
| performance_schema |
| sakila             |
| sys                |
| test               |
+--------------------+
6 rows in set (0.0038 sec)
```

使用dba.createReplicaSet()函数创建复制的命令及其输出结果如下：

```
MySQL  localhost:33060+ ssl  JS > var rs = dba.createReplicaSet("my")
A new replicaset with instance '192.168.17.149:3306' will be created.

* Checking MySQL instance at 192.168.17.149:3306

This instance reports its own address as 192.168.17.149:3306
192.168.17.149:3306: Instance configuration is suitable.

* Updating metadata...

ReplicaSet object successfully created for 192.168.17.149:3306.
Use rs.addInstance() to add more asynchronously replicated instances to this replicaset and rs.
status() to check its status.
```

再次检查数据库，命令及其运行结果如下：

```
MySQL  localhost:33060+ ssl  JS > \sql show databases;
+-------------------------------+
| Database                      |
+-------------------------------+
| information_schema            |
```

```
| mysql                              |
| mysql_innodb_cluster_metadata      |
| performance_schema                 |
| sakila                             |
| sys                                |
| test                               |
+------------------------------+
7 rows in set (0.0037 sec)
```

可以看到，创建复制后增加了一个数据库 mysql_innodb_cluster_metadata，用于保存元数据。复制创建好后，使用 status()函数检查复制状态的命令及其输出结果如下：

```
MySQL  localhost: 33060 + ssl  JS > rs.status()
{
    "replicaSet": {
        "name": "my",
        "primary": "192.168.17.149: 3306",
        "status": "AVAILABLE",
        "statusText": "All instances available.",
        "topology": {
            "192.168.17.149: 3306": {
                "address": "192.168.17.149: 3306",
                "instanceRole": "PRIMARY",
                "mode": "R/W",
                "status": "ONLINE"
            }
        },
        "type": "ASYNC"
}
}
```

这时复制环境中只有一个主库，先用 dba.deploySandboxInstance(3310)创建一个端口是 3310 的沙箱实例，再使用 addInstance()函数把该沙箱实例加入复制集，相应的命令及其输出如下：

```
MySQL  localhost: 33060 + ssl  JS > rs.addInstance('root@localhost: 3310')
Adding instance to the replicaset...


* Performing validation checks


This instance reports its own address as 127.0.0.1: 3310
127.0.0.1: 3310: Instance configuration is suitable.


* Checking async replication topology...


* Checking transaction state of the instance...


NOTE: The target instance '127.0.0.1: 3310' has not been pre - provisioned (GTID set is empty).
The Shell is unable to decide whether replication can completely recover its state.
```

```
The safest and most convenient way to provision a new instance is through automatic clone
provisioning, which will completely overwrite the state of '127.0.0.1: 3310' with a physical
snapshot from an existing replicaset member. To use this method by default, set the
'recoveryMethod' option to 'clone'.

WARNING: It should be safe to rely on replication to incrementally recover the state of the new
instance if you are sure all updates ever executed in the replicaset were done with GTIDs
enabled, there are no purged transactions and the new instance contains the same GTID set as the
replicaset or a subset of it. To use this method by default, set the 'recoveryMethod' option to
'incremental'.


Please select a recovery method [C]lone/[I]ncremental recovery/[A]bort (default Clone):
* Updating topology
Waiting for clone process of the new member to complete. Press ^C to abort the operation.
* Waiting for clone to finish...
NOTE: 127.0.0.1: 3310 is being cloned from 192.168.17.149: 3306
** Stage DROP DATA: Completed
** Clone Transfer
    FILE COPY  ############################################
############################  100%  Completed
    PAGE COPY  ############################################
############################  100%  Completed
    REDO COPY  ############################################
############################  100%  Completed

NOTE: 127.0.0.1: 3310 is shutting down...

* Waiting for server restart... ready
* 127.0.0.1: 3310 has restarted, waiting for clone to finish...
** Stage RESTART: Completed
* Clone process has finished: 97.14 MB transferred in about 1 second (~97.14 MB/s)

** Configuring 127.0.0.1: 3310 to replicate from 192.168.17.149: 3306
** Waiting for new instance to synchronize with PRIMARY...

The instance '127.0.0.1: 3310' was added to the replicaset and is replicating from 192.168.17.
149: 3306.
```

增加了一个端口是3310的沙箱实例后，再次检查复制状态的命令及其输出结果如下：

```
MySQL  localhost: 33060+ ssl  JS > rs.status()
{
    "replicaSet": {
        "name": "my",
        "primary": "192.168.17.149: 3306",
        "status": "AVAILABLE",
        "statusText": "All instances available.",
        "topology": {
            "127.0.0.1: 3310": {
                "address": "127.0.0.1: 3310",
```

```
                "instanceRole": "SECONDARY",
                "mode": "R/O",
                "replication": {
                    "applierStatus": "APPLIED_ALL",
                    "applierThreadState": "Slave has read all relay log; waiting for more
updates",
                    "receiverStatus": "ON",
                    "receiverThreadState": "Waiting for master to send event",
                    "replicationLag": null
                },
                "status": "ONLINE"
            },
            "192.168.17.149:3306": {
                "address": "192.168.17.149:3306",
                "instanceRole": "PRIMARY",
                "mode": "R/W",
                "status": "ONLINE"
            }
        },
        "type": "ASYNC"
    }
}
```

可以看到，新增的实例是从库，处于只读状态。

使用 removeInstance()函数可以删除副本，例如，删除刚刚增加副本的命令及其输出结果如下：

```
MySQL  localhost:33060+ ssl  JS > rs.removeInstance('root@localhost:3310')
The instance '127.0.0.1:3310' was removed from the replicaset.
```

删除成功后再次检查复制状态的命令及其输出结果如下：

```
MySQL  localhost:33060+ ssl  JS > rs.status()
{
    "replicaSet": {
        "name": "my",
        "primary": "192.168.17.149:3306",
        "status": "AVAILABLE",
        "statusText": "All instances available.",
        "topology": {
            "192.168.17.149:3306": {
                "address": "192.168.17.149:3306",
                "instanceRole": "PRIMARY",
                "mode": "R/W",
                "status": "ONLINE"
            }
        },
        "type": "ASYNC"
    }
}
```

发现只剩下一个主库了。

13.6.2 主库和从库的切换

MySQL复制不能自动进行主库和从库的切换，可以使用ReplicaSet.setPrimaryInstance()进行主库切换，相应命令及其输出结果如下：

```
MySQL  localhost:33060+ ssl  JS>
rs.setPrimaryInstance('root@localhost:3310')
127.0.0.1:3310 will be promoted to PRIMARY of 'my'.
The current PRIMARY is 192.168.17.149:3306.

* Connecting to replicaset instances
** Connecting to 192.168.17.149:3306
** Connecting to 127.0.0.1:3310
** Connecting to 192.168.17.149:3306
** Connecting to 127.0.0.1:3310

* Performing validation checks
** Checking async replication topology...
** Checking transaction state of the instance...

* Synchronizing transaction backlog at 127.0.0.1:3310

* Updating metadata

* Acquiring locks in replicaset instances
** Pre-synchronizing SECONDARIES
** Acquiring global lock at PRIMARY
** Acquiring global lock at SECONDARIES

* Updating replication topology
** Configuring 192.168.17.149:3306 to replicate from 127.0.0.1:3310

127.0.0.1:3310 was promoted to PRIMARY.
```

切换完成后，再次检查复制状态的命令及其输出结果如下：

```
MySQL  localhost:33060+ ssl  JS> rs.status()
{
    "replicaSet": {
        "name": "my",
        "primary": "127.0.0.1:3310",
        "status": "AVAILABLE",
        "statusText": "All instances available.",
        "topology": {
            "127.0.0.1:3310": {
                "address": "127.0.0.1:3310",
                "instanceRole": "PRIMARY",
                "mode": "R/W",
```

```
                "status": "ONLINE"
            },
            "192.168.17.149: 3306": {
                "address": "192.168.17.149: 3306",
                "instanceRole": "SECONDARY",
                "mode": "R/O",
                "replication": {
                    "applierStatus": "APPLIED_ALL",
                    "applierThreadState": "Slave has read all relay log; waiting for more
updates",
                    "receiverStatus": "ON",
                    "receiverThreadState": "Waiting for master to send event",
                    "replicationLag": null
                },
                "status": "ONLINE"
            }
        },
        "type": "ASYNC"
    }
}
```

可以看到,端口是3310的从库变成了主库,原来的主库变成了从库。

当主库出现故障,不能连接到主库时,使用ReplicaSet.setPrimaryInstance()进行主库切换会出错,相应命令及其输出结果如下:

```
MySQL  localhost: 3310 ssl  JS >
rs.setPrimaryInstance('root@localhost: 3310')
ERROR: Unable to connect to the PRIMARY of the replicaset my: MySQL Error 2003: Could not open
connection to '192.168.17.149: 3306': Can't connect to MySQL server on '192.168.17.149' (111)
Cluster change operations will not be possible unless the PRIMARY can be reached.
If the PRIMARY is unavailable, you must either repair it or perform a forced failover.
See \help forcePrimaryInstance for more information.
ReplicaSet.setPrimaryInstance: PRIMARY instance is unavailable (MYSQLSH 51118)
```

这时可以使用ReplicaSet.forcePrimaryInstance()将从库强制提升为主库,相应命令及其输出结果如下:

```
MySQL  localhost: 3310 ssl  JS >
rs.forcePrimaryInstance('root@localhost: 3310')
 * Connecting to replicaset instances
 ** Connecting to 127.0.0.1: 3310

 * Waiting for all received transactions to be applied
 ** Waiting for received transactions to be applied at 127.0.0.1: 3310
127.0.0.1: 3310 will be promoted to PRIMARY of the replicaset and the former PRIMARY will be
invalidated.

 * Checking status of last known PRIMARY
NOTE: 192.168.17.149: 3306 is UNREACHABLE
 * Checking status of promoted instance
NOTE: 127.0.0.1: 3310 has status ERROR
 * Checking transaction set status
```

```
* Promoting 127.0.0.1:3310 to a PRIMARY...

* Updating metadata...

127.0.0.1:3310 was force-promoted to PRIMARY.
NOTE: Former PRIMARY 192.168.17.149:3306 is now invalidated and must be removed from the
replicaset.
* Updating source of remaining SECONDARY instances

Failover finished successfully.
```

操作完成后，再次检查复制状态的命令及其输出结果如下：

```
MySQL  localhost:3310 ssl  JS > rs.status()
{
    "replicaSet": {
        "name": "my",
        "primary": "127.0.0.1:3310",
        "status": "AVAILABLE_PARTIAL",
        "statusText": "The PRIMARY instance is available, but one or more SECONDARY instances
are not.",
        "topology": {
            "127.0.0.1:3310": {
                "address": "127.0.0.1:3310",
                "instanceRole": "PRIMARY",
                "mode": "R/W",
                "status": "ONLINE"
            },
            "192.168.17.149:3306": {
                "address": "192.168.17.149:3306",
                "connectError": "Could not open connection to '192.168.17.149:3306': Can't
connect to MySQL server on '192.168.17.149' (111)",
                "fenced": null,
                "instanceRole": null,
                "mode": null,
                "status": "INVALIDATED"
            }
        },
        "type": "ASYNC"
    }
}
```

可以看到，原来的从库已经被提升为主库，原来的主库现在不能访问。

13.7 实验

视频演示

1. 创建复制

(1) 使用克隆插件创建一个新的副本。

(2) 启动复制。

(3) 查询复制状态。

2. 使用 MySQL Shell 创建复制

(1) 创建复制集。

(2) 创建一个沙箱实例。

(3) 将该沙箱实例作为一个副本增加到复制集中。

3. 使用 MySQL Shell 进行复制切换

(1) 检查当前复制集的状态。

(2) 将从库切换到主库。

(3) 当主库宕机时,将从库强制提升为主库。

第14章

InnoDB集群

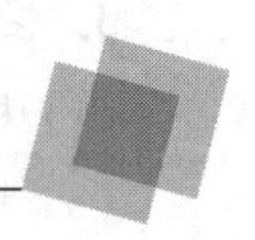

第 13 章介绍了 MySQL 复制，把 MySQL 复制的副本切换成源可以一定程度上实现 MySQL 的高可用性，但 InnoDB 集群才是成熟的 MySQL 高可用解决方案，本章介绍 InnoDB 集群的相关内容，包括 InnoDB 集群的架构、组复制、MySQL Router 和管理 InnoDB 集群等。

14.1 架构

InnoDB 集群提供了一套完整的、可扩展的 MySQL 高可用解决方案，InnoDB 集群的架构如图 14.1 所示。

InnoDB 集群基于 MySQL 组复制，组复制是一个 MySQL 服务器插件，它安装在集群中的每个服务器实例上。当集群中的服务器离线时，集群可以自动重新配置自己。必须至少有三台服务器才能组成一个可以提供高可用性的组。组可以在单一主模式下运行，其中一次只有一台服务器接受更新，也可以在多主模式下运行，其中的所有服务器都可以接受更新，即使更新是同时发布的。

组复制插件是在 MySQL 5.7 中引入的，但是当时使用起来很困难。InnoDB 集群引入了两个新的组件，使组复制更易于设置和管理。一个是 MySQL Router，它位于应用程序和集群之间，将应用的流量分配到集群中合适的实例上，MySQL Router 通过周期性地访问 InnoDB 集群创建的 mysql_innodb_cluster_metadata 库中的元数据获取集群成员信息，再通过 performance_schema 库获取可连接实例及其状态来实现故障迁移和读负载均衡、读写分离的路由。另一个是 MySQL Shell，它是一个新的交互式字符界面，它使用户能够通过新的管理 API，使用熟悉的 JavaScript 或 Python 语法来管理 InnoDB 集群。对 InnoDB 集群的所有管理工作都应该通过 MySQL Shell 的管理 API 进行，不建议直接对 InnoDB 集群和组复制进行操作，或者修改它们的元数据。管理 InnoDB 集群主要使用 MySQL Shell 中的两类 API，一个是 dba 类，用于准备实例和创建集群；另一个是 cluster 类，用于管理集群，

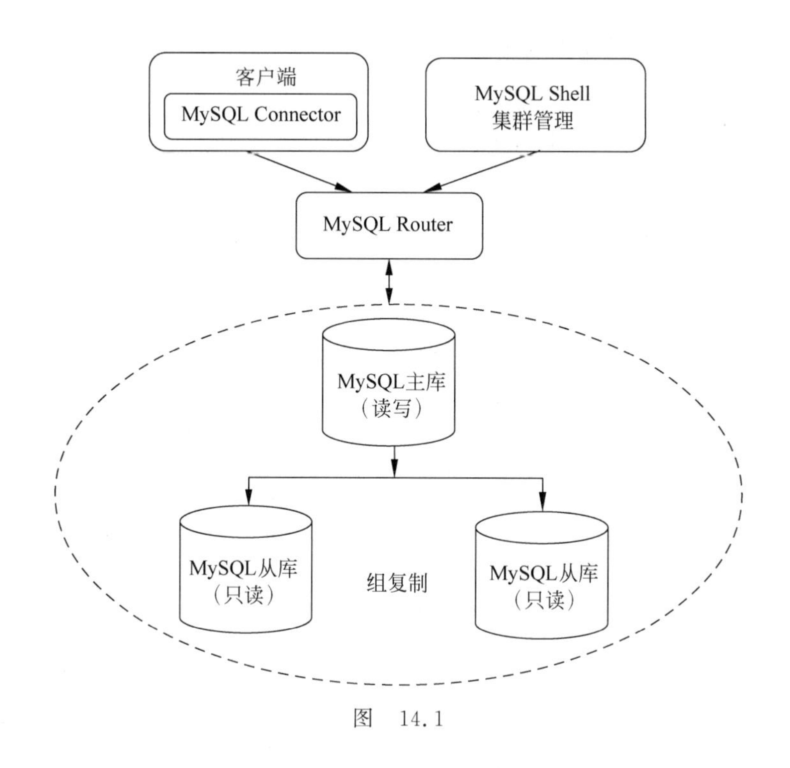

图 14.1

包括检查集群状态和解决实例故障等。

需要澄清的是，InnoDB集群和MySQL集群不是一回事，通常人们说的MySQL集群是指MySQL NDB集群，而不是InnoDB集群。这两种集群的根本区别是存储引擎的不同。另外，InnoDB集群只能进行读扩展，不能进行写扩展，因为每个更新都要写入所有的节点。而MySQL NDB集群可以通过分片的方式进行写扩展。

14.2 组复制

组复制是MySQL的一个插件，它能实现一组MySQL服务器之间的同步复制，并提供以下功能：

- 自动进行服务器故障切换。
- 当组成员离开组时，自动重新配置组。
- 解决写冲突。
- 高度可用的复制数据库。

组复制可以分为两类，一类是单主库模式，其中只有一个成员接受更新，这种模式适合于大多数应用场景；另一类是多主库模式，所有成员都接受更新。

使用MySQL Shell的AdminAPI可以方便地创建和管理组复制，它提供了两种部署组复制的方式，一种是沙箱部署，这种方式可以方便地在一台机器上部署一组实验用的集群；另一种是生产部署，这种方式可以部署和生产环境一样的集群。下面是部署沙箱集群的一个例子。

首先在同一台机器上创建三个MySQL沙箱实例，端口分别是3310、3320和3330，创建第一个沙箱实例的命令及其输出结果如下：

```
MySQL  JS > dba.deploySandboxInstance(3310)
A new MySQL sandbox instance will be created on this host in
/root/mysql-sandboxes/3310

Warning: Sandbox instances are only suitable for deploying and
running on your local machine for testing purposes and are not
accessible from external networks.

Please enter a MySQL root password for the new instance: *******

Deploying new MySQL instance...

Instance localhost: 3310 successfully deployed and started.
Use shell.connect('root@localhost: 3310') to connect to the instance.
```

创建第二个沙箱实例的命令如下：

```
MySQL  JS > dba.deploySandboxInstance(3320)
```

创建第三个沙箱实例的命令如下：

```
MySQL  JS > dba.deploySandboxInstance(3330)
```

然后连接到第一个 MySQL 实例的命令及其输出结果如下：

```
MySQL  JS > \connect root@localhost: 3310
Creating a session to 'root@localhost: 3310'
Please provide the password for 'root@localhost: 3310': *******
Save password for 'root@localhost: 3310'? [Y]es/[N]o/Ne[v]er (default No): y
Fetching schema names for autocompletion... Press ^C to stop.
Your MySQL connection id is 12
Server version: 8.0.23 MySQL Community Server - GPL
No default schema selected; type \use <schema> to set one.
```

在第一个节点上创建集群的命令及其输出结果如下：

```
MySQL  localhost: 3310 ssl  JS > var cluster = dba.createCluster('mycluster')
A new InnoDB cluster will be created on instance 'localhost: 3310'.

Validating instance configuration at localhost: 3310...
NOTE: Instance detected as a sandbox.
Please note that sandbox instances are only suitable for deploying test clusters for use within
the same host.

This instance reports its own address as 127.0.0.1: 3310

Instance configuration is suitable.
NOTE: Group Replication will communicate with other members using '127.0.0.1: 33101'. Use the
localAddress option to oveMySQL Routeride.

Creating InnoDB cluster 'mycluster' on '127.0.0.1: 3310'...

Adding Seed Instance...
```

```
Cluster successfully created. Use Cluster.addInstance() to add MySQL instances.
At least 3 instances are needed for the cluster to be able to withstand up to
one server failure.
```

再把另外两个MySQL实例加入集群中，把第二个节点加入集群的命令及其输出结果如下：

```
MySQL  localhost: 3310 ssl  JS > cluster.addInstance({user: "root", host: "localhost", port:
3320, password: "dingjia"})
...
Please select a recovery method [C]lone/[I]ncremental recovery/[A]bort (default Clone):
Validating instance configuration at localhost: 3320...
NOTE: Instance detected as a sandbox.
Please note that sandbox instances are only suitable for deploying test clusters for use within
the same host.

This instance reports its own address as 127.0.0.1: 3320

Instance configuration is suitable.
NOTE: Group Replication will communicate with other members using '127.0.0.1: 33201'. Use the
localAddress option to oveMySQL Routeride.

A new instance will be added to the InnoDB cluster. Depending on the amount of
data on the cluster this might take from a few seconds to several hours.

Adding instance to the cluster...

Monitoring recovery process of the new cluster member. Press ^C to stop monitoring and let it
continue in background.
Clone based state recovery is now in progress.

NOTE: A server restart is expected to happen as part of the clone process. If the
server does not support the RESTART command or does not come back after a
while, you may need to manually start it back.

* Waiting for clone to finish...
NOTE: 127.0.0.1: 3320 is being cloned from 127.0.0.1: 3310
** Stage DROP DATA: Completed
** Clone Transfer
    FILE COPY  ##################################
##########  100%  Compl
PAGE COPY  ####################################
########  100%  Compl
REDO COPY  ####################################
########  100%  Completed
NOTE: 127.0.0.1: 3320 is shutting down...

* Waiting for server restart... ready
* 127.0.0.1: 3320 has restarted, waiting for clone to finish...
** Stage RESTART: Completed
* Clone process has finished: 72.20 MB transfeMySQL Routered in 4 sec (18.05 MB/s)

Incremental state recovery is now in progress.
```

```
 * Waiting for distributed recovery to finish...
NOTE: '127.0.0.1: 3320' is being recovered from '127.0.0.1: 3310'
 * Distributed recovery has finished

The instance '127.0.0.1: 3320' was successfully added to the cluster.
```

把第三个节点加入集群的命令及其部分输出结果如下：

```
MySQL  localhost: 3310 ssl  JS > cluster.addInstance({user: "root", host: "localhost", port:
3330, password: "dingjia"})
...
The instance '127.0.0.1: 3330' was successfully added to the cluster.
```

这样，由 3 个 MySQL 实例组成的 InnoDB 集群就创建完成了。

14.3 MySQL Router

MySQL Router 是 MySQL 官方提供的一个轻量级中间件，它在 InnoDB 集群中位于应用和 InnoDB 组复制之间，为应用提供到 MySQL 服务器的路由，应用通过 MySQL Router 连接到 InnoDB 集群不需要指定 InnoDB 集群的细节。MySQL Router 帮助实现 InnoDB 集群的高可用、负载均衡、易扩展等特性。

14.3.1 安装

可以从 MySQL 的官方网站下载 MySQL Router，下载界面如图 14.2 所示。

图 14.2

在红帽的平台上安装 MySQL Router 的命令如下：

```
$ sudo rpm -Uvh mysql-router-community-8.0.23-1.el7.x86_64.rpm
```

14.3.2　部署 MySQL Router

启动 MySQL Router 之前，要使用 bootstrap 选项进行初始化，相应命令及其输出结果如下：

```
$ sudo mysqlrouter --bootstrap root@localhost:3310 --directory /u01/myrouter
--user=mysql
Please enter MySQL password for root:
# Bootstrapping MySQL Router instance at '/u01/myrouter'...

- Creating account(s) (only those that are needed, if any)
- Verifying account (using it to run SQL queries that would be run by Router)
- Storing account in keyring
- Adjusting permissions of generated files
- Creating configuration /u01/myrouter/mysqlrouter.conf

# MySQL Router configured for the InnoDB Cluster 'mycluster'

After this MySQL Router has been started with the generated configuration

    $ mysqlrouter -c /u01/myrouter/mysqlrouter.conf

the cluster 'mycluster' can be reached by connecting to:

## MySQL Classic protocol

- Read/Write Connections: localhost:6446
- Read/Only Connections: localhost:6447

## MySQL X protocol

- Read/Write Connections: localhost:64460
- Read/Only Connections: localhost:64470
```

MySQL Router 在初始化时会连接到 InnoDB 集群中到一个 MySQL 实例，从中获取集群的元数据并写入一些数据，这个被连接的集群成员可以是主实例，也可以是从实例，如果接入到从实例会自动切换到主实例。初始化过程中还会在指定目录下生成相关的文件，查询指定目录下的文件如下：

```
$ ls /u01/myrouter/
data   log   mysqlrouter.conf   mysqlrouter.key   run   start.sh   stop.sh
```

使用该指定目录下的启动脚本进行启动的命令及其输出结果如下：

```
$ sudo /u01/myrouter/start.sh
```

```
PID 8988 written to '/u01/myrouter/mysqlrouter.pid'
logging facility initialized, switching logging to loggers specified in configuration
```

14.3.3 使用 MySQL Router

应用连接到 MySQL Router 和连接到一般的 MySQL 实例的方法一样，其中 6646 是读写端口，6647 是只读端口。例如，连接到 MySQL Router 的读写端口检查端口和服务 id 的命令及其输出结果如下：

```
$ mysql -h 127.0.0.1 -P 6446   -e "select @@port,@@server_id"
+--------+-------------+
| @@port | @@server_id |
+--------+-------------+
|  3310  | 2078934398  |
+--------+-------------+
1 row in set (0.01 sec)
```

检查端口和服务 id，发现自动连接到了主实例，也就是可读写的实例。

连接到 MySQL Router 的只读端口检查端口和服务 id 的命令及其输出结果如下：

```
$ mysql -h 127.0.0.1 -P 6447   -e "select @@port,@@server_id"
+--------+-------------+
| @@port | @@server_id |
+--------+-------------+
|  3320  | 2000802780  |
+--------+-------------+
1 row in set (0.00 sec)
```

检查端口和服务 id，发现自动连接到了从实例，也就是只读的实例。

再次连接到 MySQL Router 的只读端口检查端口和服务 id 的命令及其输出结果如下：

```
$ mysql -h 127.0.0.1 -P 6447 -e "select @@port,@@server_id"
+--------+-------------+
| @@port | @@server_id |
+--------+-------------+
|  3330  | 3772914362  |
+--------+-------------+
1 row in set (0.00 sec)
```

发现新的连接指向了另外一个只读的实例。查看 MySQL Router 配置文件中的路由策略如下：

```
$ cat /u01/myrouter/mysqlrouter.conf
...
[routing:mycluster_rw]
bind_address = 0.0.0.0
bind_port = 6446
destinations = metadata-cache://mycluster/?role = PRIMARY
routing_strategy = first-available
```

```
protocol = classic

[routing: mycluster_ro]
bind_address = 0.0.0.0
bind_port = 6447
destinations = metadata - cache://mycluster/?role = SECONDARY
routing_strategy = round - robin - with - fallback
protocol = classic
...
```

可以看到，对于读写端口的访问策略是 first-available，对于只读端口的访问策略是 round-robin-with-fallback，从这两个访问策略的命名上就可以看出 MySQL Router 选择连接 MySQL 实例的方法。

对于因为某种原因不能让应用访问的实例，可以给实例设置隐藏的标签，相应的命令及其输出结果如下：

```
MySQL  localhost: 3310 ssl  JS >
cluster.setInstanceOption('root@localhost: 3310',"tag: _hidden",true);
```

这样设置后，MySQL Router 就不会将应用的查询引导到该实例。

14.4 管理 InnoDB 集群

14.4.1 管理沙箱实例的方法

mysqlsh 中常用的管理沙箱实例的方法如下。

- dba.deploySandboxInstance(port)：创建一些新实例。
- dba.startSandboxInstance(port)：启动一个新实例。
- dba.stopSandboxInstance(port)：有序地停止一个正在运行的实例，该方法和 dba.killSandboxInstance()不同。
- dba.killSandboxInstance(port)：立即终止一个正在运行的实例，该方法可用于模拟实例的意外终止。
- dba.deleteSandboxInstance(port)：从文件系统中删除一个实例。

14.4.2 查询状态

集群创建完成后，使用 status()方法可以查询集群的状态，相应命令及其输出结果如下：

```
MySQL  localhost: 3310 ssl  JS > cluster.status()
{
    "clusterName": "mycluster",
    "defaultReplicaSet": {
        "name": "default",
        "primary": "127.0.0.1: 3310",
```

```
            "ssl": "REQUIRED",
            "status": "OK",
            "statusText": "Cluster is ONLINE and can tolerate up to ONE failure.",
            "topology": {
                "127.0.0.1: 3310": {
                    "address": "127.0.0.1: 3310",
                    "mode": "R/W",
                    "readReplicas": {},
                    "replicationLag": null,
                    "role": "HA",
                    "status": "ONLINE",
                    "version": "8.0.23"
                },
                "127.0.0.1: 3320": {
                    "address": "127.0.0.1: 3320",
                    "mode": "R/O",
                    "readReplicas": {},
                    "replicationLag": null,
                    "role": "HA",
                    "status": "ONLINE",
                    "version": "8.0.23"
                },
                "127.0.0.1: 3330": {
                    "address": "127.0.0.1: 3330",
                    "mode": "R/O",
                    "readReplicas": {},
                    "replicationLag": null,
                    "role": "HA",
                    "status": "ONLINE",
                    "version": "8.0.23"
                }
            },
            "topologyMode": "Single - Primary"
        },
        "groupInformationSourceMember": "127.0.0.1: 3310"
    }
```

可以看到，这是一个单主模式的集群，每个成员都处于 ONLINE 的状态。单主模式是指只有一个主实例处于读写状态，其他实例处于只读状态。多主模式是指所有的实例都处于读写状态。

还可以使用 describe()方法查询集群的拓扑结构和各节点的信息，相应命令及其输出结果如下：

```
MySQL  localhost: 3310 ssl  JS > cluster.describe()
{
    "clusterName": "mycluster",
    "defaultReplicaSet": {
        "name": "default",
        "topology": [
            {
```

```
                "address": "127.0.0.1: 3310",
                "label": "127.0.0.1: 3310",
                "role": "HA"
            },
            {
                "address": "127.0.0.1: 3320",
                "label": "127.0.0.1: 3320",
                "role": "HA"
            },
            {
                "address": "127.0.0.1: 3330",
                "label": "127.0.0.1: 3330",
                "role": "HA"
            }
        ],
        "topologyMode": "Single - Primary"
    }
}
```

在连接时指定不同的端口可以连接到对应的实例，例如，连接到3320端口的命令及其输出结果如下：

```
$ mysql -P 3320 --protocol tcp -e "select @@port,@@server_id"
+---------+-------------+
| @@port  | @@server_id |
+---------+-------------+
|   3320  | 2000802780  |
+---------+-------------+
1 row in set (0.00 sec)
```

在 performance_schema.replication_group_members 中也可以查询 InnoDB 集群各个成员的状态，相应命令及其输出结果如下：

```
mysql > select member_host,member_port,member_state,member_role from performance_schema.
replication_group_members;
+-------------+-------------+--------------+-------------+
| member_host | member_port | member_state | member_role |
+-------------+-------------+--------------+-------------+
| 127.0.0.1   |        3310 | ONLINE       | PRIMARY     |
| 127.0.0.1   |        3320 | ONLINE       | SECONDARY   |
| 127.0.0.1   |        3330 | ONLINE       | SECONDARY   |
+-------------+-------------+--------------+-------------+
3 rows in set (0.00 sec)
```

14.4.3 集群中实例的异常终止

使用 dba.killSandboxInstance()可以模拟 InnoDB 集群中主实例的异常终止，相应命令及其输出结果如下：

```
MySQL  JS > dba.killSandboxInstance(3310)

Killing MySQL instance...

Instance localhost: 3310 successfully killed.
```

当端口是3310的实例终止后，连接该实例会失败，相应命令及其输出结果如下：

```
MySQL  JS > \connect root@localhost: 3310
Creating a session to 'root@localhost: 3310'
MySQL EMySQL Routeror 2003 (HY000): Can't connect to MySQL server on 'localhost' (111)
```

连接到端口是3320的实例可以成功，相应命令及其输出结果如下：

```
MySQL  JS > \connect root@localhost: 3320
Creating a session to 'root@localhost: 3320'
Please provide the password for 'root@localhost: 3320': *******
Save password for 'root@localhost: 3320'? [Y]es/[N]o/Ne[v]er (default No): y
Fetching schema names for autocompletion... Press ^C to stop.
Your MySQL connection id is 183
Server version: 8.0.23 MySQL Community Server - GPL
No default schema selected; type \use <schema> to set one.
```

使用cluster.status()可以检查实例的状态，相应命令及其输出结果如下：

```
MySQL  localhost: 3320 ssl  JS > var cluster = dba.getCluster("mycluster")
MySQL  localhost: 3320 ssl  JS > cluster.status()
{
    "clusterName": "mycluster",
    "defaultReplicaSet": {
        "name": "default",
        "primary": "127.0.0.1: 3320",
        "ssl": "REQUIRED",
        "status": "OK_NO_TOLERANCE",
        "statusText": "Cluster is NOT tolerant to any failures. 1 member is not active.",
        "topology": {
            "127.0.0.1: 3310": {
                "address": "127.0.0.1: 3310",
                "mode": "n/a",
                "readReplicas": {},
                "role": "HA",
                 "shellConnectEMySQL Routeror": "MySQL EMySQL Routeror 2003 (HY000): Can't
connect to MySQL server on '127.0.0.1' (111)",
                "status": "(MISSING)"
            },
            "127.0.0.1: 3320": {
                "address": "127.0.0.1: 3320",
                "mode": "R/W",
                "readReplicas": {},
                "replicationLag": null,
                "role": "HA",
                "status": "ONLINE",
```

```
                "version": "8.0.23"
            },
            "127.0.0.1: 3330": {
                "address": "127.0.0.1: 3330",
                "mode": "R/O",
                "readReplicas": {},
                "replicationLag": null,
                "role": "HA",
                "status": "ONLINE",
                "version": "8.0.23"
            }
        },
        "topologyMode": "Single - Primary"
    },
    "groupInformationSourceMember": "127.0.0.1: 3320"
}
```

发现第一个实例的状态是 missing，第二个实例被选举成主实例，可以接受读写。

应用端再次连接 MySQL Router 到读写端口，相应命令及其输出结果如下：

```
$ mysql  - h 127.0.0.1  - P 6446   - e "select @@port,@@server_id"
+---------+-------------+
| @@port  | @@server_id |
+---------+-------------+
|  3320   |  2000802780 |
+---------+-------------+
1 row in set (0.00 sec)
```

发现自动连接到第二个实例了，可以进行读写操作。因此应用并不需要知道 InnoDB 集群中的一个实例已经意外终止了。

使用 dba.startSandboxInstance()命令启动集群中的第一个实例，相应命令及其输出结果如下：

```
MySQL  localhost: 3320 ssl  JS > dba.startSandboxInstance(3310);

Starting MySQL instance...

Instance localhost: 3310 successfully started.
```

再次查询集群的状态，相应命令及其输出结果如下：

```
MySQL  localhost: 3320 ssl  JS > cluster.status()
{
    "clusterName": "mycluster",
    "defaultReplicaSet": {
        "name": "default",
        "primary": "127.0.0.1: 3320",
        "ssl": "REQUIRED",
        "status": "OK",
        "statusText": "Cluster is ONLINE and can tolerate up to ONE failure.",
        "topology": {
```

```
                "127.0.0.1: 3310": {
                    "address": "127.0.0.1: 3310",
                    "mode": "R/O",
                    "readReplicas": {},
                    "replicationLag": null,
                    "role": "HA",
                    "status": "ONLINE",
                    "version": "8.0.23"
                },
                "127.0.0.1: 3320": {
                    "address": "127.0.0.1: 3320",
                    "mode": "R/W",
                    "readReplicas": {},
                    "replicationLag": null,
                    "role": "HA",
                    "status": "ONLINE",
                    "version": "8.0.23"
                },
                "127.0.0.1: 3330": {
                    "address": "127.0.0.1: 3330",
                    "mode": "R/O",
                    "readReplicas": {},
                    "replicationLag": null,
                    "role": "HA",
                    "status": "ONLINE",
                    "version": "8.0.23"
                }
            },
            "topologyMode": "Single-Primary"
        },
        "groupInformationSourceMember": "127.0.0.1: 3320"
}
```

发现第一个实例已经加入了集群，但它并不是主实例，而是只读实例，主实例仍然是第二个实例。

14.4.4 移除和加入实例

将第一个实例移除的命令及其输出结果如下：

```
MySQL  localhost: 3320 ssl  JS >
cluster.removeInstance('root@localhost: 3310')
The instance will be removed from the InnoDB cluster. Depending on the instance
being the Seed or not, the Metadata session might become invalid. If so, please
start a new session to the Metadata Storage R/W instance.

Instance 'localhost: 3310' is attempting to leave the cluster...

The instance 'localhost: 3310' was successfully removed from the cluster.
```

再次检查 InnoDB 集群状态的命令及其输出结果如下：

```
MySQL  localhost: 3320 ssl   JS > cluster.status()
{
    "clusterName": "mycluster",
    "defaultReplicaSet": {
        "name": "default",
        "primary": "127.0.0.1: 3320",
        "ssl": "REQUIRED",
        "status": "OK_NO_TOLERANCE",
        "statusText": "Cluster is NOT tolerant to any failures.",
        "topology": {
            "127.0.0.1: 3320": {
                "address": "127.0.0.1: 3320",
                "mode": "R/W",
                "readReplicas": {},
                "replicationLag": null,
                "role": "HA",
                "status": "ONLINE",
                "version": "8.0.23"
            },
            "127.0.0.1: 3330": {
                "address": "127.0.0.1: 3330",
                "mode": "R/O",
                "readReplicas": {},
                "replicationLag": null,
                "role": "HA",
                "status": "ONLINE",
                "version": "8.0.23"
            }
        },
        "topologyMode": "Single - Primary"
    },
    "groupInformationSourceMember": "127.0.0.1: 3320"
}
```

可以发现，InnoDB 集群中只剩下了两个实例，从 statusText 字段可以看出这时该集群已经不能容错了。将第一个实例再次加入的命令及其输出结果如下：

```
MySQL  localhost: 3320 ssl   JS > cluster.addInstance({user: "root", host: "localhost", port: 3310, password: "dingjia"})
The safest and most convenient way to provision a new instance is through automatic clone provisioning, which will completely overwrite the state of '127.0.0.1: 3310' with a physical snapshot from an existing cluster member. To use this method by default, set the 'recoveryMethod' option to 'clone'.

The incremental state recovery may be safely used if you are sure all updates ever executed in the cluster were done with GTIDs enabled, there are no purged transactions and the new instance contains the same GTID set as the cluster or a subset of it. To use this method by default, set the 'recoveryMethod' option to 'incremental'.
```

```
Incremental state recovery was selected because it seems to be safely usable.

Validating instance configuration at localhost: 3310...
NOTE: Instance detected as a sandbox.
Please note that sandbox instances are only suitable for deploying test clusters for use within
the same host.

This instance reports its own address as 127.0.0.1: 3310

Instance configuration is suitable.
NOTE: Group Replication will communicate with other members using '127.0.0.1: 33101'. Use the
localAddress option to oveMySQL Routeride.

A new instance will be added to the InnoDB cluster. Depending on the amount of
data on the cluster this might take from a few seconds to several hours.

Adding instance to the cluster...

Monitoring recovery process of the new cluster member. Press ^C to stop monitoring and let it
continue in background.
Incremental state recovery is now in progress.

* Waiting for distributed recovery to finish...
NOTE: '127.0.0.1: 3310' is being recovered from '127.0.0.1: 3320'
* Distributed recovery has finished

The instance '127.0.0.1: 3310' was successfully added to the cluster.
```

再次检查 InnoDB 集群集群状态的命令及其输出结果如下：

```
MySQL  localhost: 3320 ssl  JS > cluster.status()
{
    "clusterName": "mycluster",
    "defaultReplicaSet": {
        "name": "default",
        "primary": "127.0.0.1: 3320",
        "ssl": "REQUIRED",
        "status": "OK",
        "statusText": "Cluster is ONLINE and can tolerate up to ONE failure.",
        "topology": {
            "127.0.0.1: 3310": {
                "address": "127.0.0.1: 3310",
                "mode": "R/O",
                "readReplicas": {},
                "replicationLag": null,
                "role": "HA",
                "status": "ONLINE",
                "version": "8.0.23"
            },
            "127.0.0.1: 3320": {
                "address": "127.0.0.1: 3320",
```

```
                "mode": "R/W",
                "readReplicas": {},
                "replicationLag": null,
                "role": "HA",
                "status": "ONLINE",
                "version": "8.0.23"
            },
            "127.0.0.1: 3330": {
                "address": "127.0.0.1: 3330",
                "mode": "R/O",
                "readReplicas": {},
                "replicationLag": null,
                "role": "HA",
                "status": "ONLINE",
                "version": "8.0.23"
            }
        },
        "topologyMode": "Single-Primary"
    },
    "groupInformationSourceMember": "127.0.0.1: 3320"
}
```

可以发现，该 InnoDB 集群又包括了 3 个实例。

14.4.5 解散 InnoDB 集群

使用 dissolve()方法可以解散 InnoDB 集群，相应的命令及其输出结果如下：

```
MySQL  JS > \connect root@localhost: 3310
Creating a session to 'root@localhost: 3310'
Fetching schema names for autocompletion... Press ^C to stop.
Your MySQL connection id is 60
Server version: 8.0.23 MySQL Community Server - GPL
No default schema selected; type \use <schema> to set one.
MySQL  localhost: 3310 ssl  JS > var cluster = dba.getCluster("mycluster")
MySQL  localhost: 3310 ssl  JS > cluster.dissolve()
The cluster still has the following registered instances:
{
    "clusterName": "mycluster",
    "defaultReplicaSet": {
        "name": "default",
        "topology": [
            {
                "address": "127.0.0.1: 3310",
                "label": "127.0.0.1: 3310",
                "role": "HA"
            },
            {
                "address": "127.0.0.1: 3320",
                "label": "127.0.0.1: 3320",
```

```
                "role": "HA"
            },
            {
                "address": "127.0.0.1: 3330",
                "label": "127.0.0.1: 3330",
                "role": "HA"
            }
        ],
        "topologyMode": "Single-Primary"
    }
}
WARNING: You are about to dissolve the whole cluster and lose the high availability features
provided by it. This operation cannot be reverted. All members will be removed from the cluster
and replication will be stopped, internal recovery user accounts and the cluster metadata will
be dropped. User data will be maintained intact in all instances.

Are you sure you want to dissolve the cluster? [y/N]: yes

Instance '127.0.0.1: 3320' is attempting to leave the cluster...
Instance '127.0.0.1: 3330' is attempting to leave the cluster...
Instance '127.0.0.1: 3310' is attempting to leave the cluster...

The cluster was successfully dissolved.
Replication was disabled but user data was left intact.
```

解散完成后，InnoDB 集群变成了 3 个独立的 MySQL 实例，数据都是完整的，相互之间的数据复制已经停止。

14.5 实验

InnoDB 集群

视频演示

(1) 创建 3 个沙箱实例。

(2) 创建一个沙箱 InnoDB 集群。

(3) 查询 InnoDB 集群的状态和拓扑结构。

第五部分

优　化

第15章

基准测试工具

基准测试工具用于检验 MySQL 在特定负载下的运行情况，本章将介绍三种常用的基准测试工具：mysqlslap、sysbench 和 TPCC_MySQL。

15.1 mysqlslap

mysqlslap 是 MySQL 官方提供的一个简单易用的压力测试工具，通过模拟多个并发客户端访问 MySQL 执行压力测试。

15.1.1 mysqlslap 运行的三个阶段

mysqlslap 的运行分为以下三个阶段。

(1) 准备阶段：创建数据库、表，可选生成数据或存储过程，该阶段只通过一个客户端连接完成。

(2) 运行阶段：运行--query 参数指定的压力测试脚本，该阶段通过多个客户端并发完成，客户端的个数由--concurrency 参数指定。

(3) 清理阶段：断开连接，删除数据，该阶段可以通过参数--no-drop 跳过，该阶段只通过一个客户端连接完成。

完成这三个阶段并不需要用户执行三条命令，而是通过执行一条 mysqlslap 命令即可完成。如下 mysqlslap 命令将创建一个 mytest 表，并向其插入记录，并发三个客户端，重复执行两次：

```
$ mysqlslap --create="create table mytest(id int not null auto_increment primary key, f1
varchar(255))" --query="insert into mytest(f1)
values(md5(rand()))" --concurrency=3 --iterations=2  --only-print
```

如上 mysqlslap 命令没有通过参数--create-schema 指定数据库，而是创建一个默认的 mysqlslap 数据库。因为指定了--only-print 参数，该命令并不会真正执行，只是打印出执行的过程如下：

```
DROP SCHEMA IF EXISTS 'mysqlslap';
CREATE SCHEMA 'mysqlslap';
use mysqlslap;
create table mytest(id int not null auto_increment primary key, f1 varchar(255));
insert into mytest(f1) values(md5(rand()));
insert into mytest(f1) values(md5(rand()));
insert into mytest(f1) values(md5(rand()));
DROP SCHEMA IF EXISTS 'mysqlslap';
CREATE SCHEMA 'mysqlslap';
use mysqlslap;
create table mytest(id int not null auto_increment primary key, f1 varchar(255));
insert into mytest(f1) values(md5(rand()));
insert into mytest(f1) values(md5(rand()));
insert into mytest(f1) values(md5(rand()));
DROP SCHEMA IF EXISTS 'mysqlslap';
```

通过分析如上的执行脚本，可以发现第一个阶段执行了 2 次运行次数由参数--iterations 指定，包括如下脚本：

```
DROP SCHEMA IF EXISTS 'mysqlslap';
CREATE SCHEMA 'mysqlslap';
use mysqlslap;
create table mytest(id int not null auto_increment primary key, f1 varchar(255));
```

第二个阶段并发运行了由参数--concurrency 指定的 3 个客户端，执行次数等于--concurrency 乘--iterations，共 6 次，只包括一条 insert 语句：

```
insert into mytest(f1) values(md5(rand()));
```

第三个阶段也执行了 2 次(运行次数由参数--iterations 指定)：

```
DROP SCHEMA IF EXISTS 'mysqlslap';
```

15.1.2 参数说明

mysqlslap 常用的参数如下。

- --concurrency：第二个阶段并发的客户端的个数。
- --iterations：重复测试的次数。
- --no-drop：不在测试结束后删除数据，但此时重复测试的次数--iterations 只能设置成 1，或者采用默认值 1。否则会出现如下错误：

```
mysqlslap: Cannot create schema mysqlslap : Can't create database 'mysqlslap'; database
```

- --create-schema：测试数据的数据库名，默认是 mysqlslap。

- --auto-generate-sql：自动生成测试表和数据，表示用 mysqlslap 工具自己生成的 SQL 脚本来测试并发压力。
- --auto-generate-sql-load-type：指定测试语句的类型。代表要测试的环境是读操作或是写操作还是两者混合的。取值包括：read，key，write，update 和 mixed(默认)。
- --auto-generate-sql-add-auto-increment：代表对生成的表自动添加 auto_increment 列。
- --number-char-cols：自动生成的测试表中包含多少个字符类型的列，默认为 1。
- --number-int-cols：自动生成的测试表中包含多少个数字类型的列，默认为 1。
- --number-of-queries：限制单个客户端执行查询的大约次数。
- --query：指定自定义脚本执行测试，可以调用自定义的一个存储过程或者 SQL 语句来执行测试。
- --commit：指定执行多少条 DML 语句后提交一次。
- --compress：如果服务器和客户端都支持压缩，则压缩信息传递。
- --concurrency：指定并发的客户端数。可指定多个值，以逗号或者--delimiter 参数指定的值作为分隔符。例如：--concurrency=100，200，500。
- --engine：指定要测试的引擎，可以有多个，用分隔符隔开。例如：-engines=myisam，innodb。
- --iterations：测试执行的迭代次数，代表要在不同并发环境下，各自运行测试多少次。
- --only-print：只打印测试语句而不实际执行。
- --detach：执行 N 条语句后断开重连。
- --debug-info：打印内存和 CPU 的相关信息。

15.1.3　使用案例

(1) 如下的 mysqlslap 命令将创建一个 mytest 数据库，创建一个 mytest 表，测试的主体是向表中插入一条记录并进行一次查询，并发 50 个客户端，重复执行 200 次：

```
$ mysqlslap  --create-schema=mytest  --delimiter=";"  --create="create table
mytest(id int not null auto_increment primary key, c1 varchar(255))"
--query="insert into mytest(c1) values(md5(rand())); select c1 from mytest;"
--concurrency=50 --iterations=200
```

(2) 如下的 mysqlslap 命令使用自动创建的 SQL 语句，创建的表中包括两个整数类型和 3 个 varchar 类型的字段：

```
$ mysqlslap --concurrency=200 --iterations=100  --number-int-cols=2
--number-char-cols=3  --auto-generate-sql
```

从性能视图中观察到如上命令自动生成如下的 SQL 语句：

```
mysql> select distinct sql_text from
performance_schema.events_statements_current where
thread_id!= ps_current_thread_id()\G
```

```
*************************** 1. row ***************************
sql_text: INSERT INTO t1 VALUES
(1233983202,789373855,'jdtPbBILRQyiu7nZN3RdI96aQOP4Y0z7dl04wbQF1OpvYdLnggGr
qMP8mjCAjB9CwQHk1jruzhhXA2BibRZPqrPSQdX9gYDTyWDr5xdxWJSgDXEozg5hNW1pzYqgx3',
'9nQEfN38KZRh5hCCL3sRkFpm4pAuL7pWvFcQXgar5203bdFwY7FHE53ImD47LtT81xaYT8Fb2Y
Wd1lQZshqyelGOzCNCxsCyimMBmWep28IvlkSTIJRw4bDvnczlLB','bvpq6cCdH6db1NCTsWgK
IdvXObOOKDRn8HdEthjAnLdochAS3qgMJclk3d0FqR2ykY5dDof119pdqz6mie816ld9NdO55R4
pr9J4pq5rrnSieQSK5SEB5rLKTlQQ4M')
1 rows in set (0.00 sec)
```

（3）如下 mysqlslap 命令将创建和执行的 SQL 语句放入文件中，文件可以包含比较复杂的 SQL 语句，这种类型的测试可以用于创建类似应用的测试，其中 create. sql 用于创建应用的测试环境，query. sql 用于并发重复执行应用的 SQL 语句：

```
$ mysqlslap --concurrency=5  --iterations=5 --query=query.sql
--create=create.sql  --delimiter=";"
```

（4）选项--no-drop 在测试完成后不删除数据，可以利用该特效快速生成一定量的数据，例如：

```
$ mysqlslap --concurrency=900 --iterations=1  --number-int-cols=2
--number-char-cols=3  --auto-generate-sql  --no-drop
```

如上命令运行完成后使用如下 mysqlshow 命令检查生成的数据如下：

```
$ mysqlshow -vv mysqlslap
Database: mysqlslap
+--------+----------+------------+
| Tables | Columns  | Total Rows |
+--------+----------+------------+
| t1     |        5 |    5499    |
+--------+----------+------------+
1 row in set.
```

采用这种方法生成的数据量并不大，如果需要生成大量的数据可以把需要测试的 SQL 语句指定为自己编写的 insert 语句，例如，如下测试向表中插入了 100 万条记录：

```
$ mysqlslap --create="create table table_a(col1 int primary key auto_increment,
col2 varchar(255))" --query="insert into table_a(col2) values(md5(rand()))"
--concurrency=10 --number-of-queries=1000000 --no-drop
```

如上命令运行完成后使用如下 mysqlshow 命令检查生成的数据如下：

```
$ mysqlshow -vv mysqlslap
Database: mysqlslap
+---------+----------+------------+
| Tables  | Columns  | Total Rows |
+---------+----------+------------+
| table_a |        2 |   1000000  |
+---------+----------+------------+
1 row in set.
```

在 mysqlslap 的运行过程中，可以使用 show processlist 监控各个连接的状态，如下是统计每类运行状态连接数的命令及其输出结果：

```
$  mysql  -e "show processlist\G"|grep State: |sort |uniq  -c|sort  -nr
    897   State: executing
      1   State: Waiting on empty queue
      1   State: Sending to client
      1   State: Opening tables
      1   State: init
      1   State: freeing items
      1   State:
```

15.2　Sysbench

15.2.1　简介

sysbench 是一个基于 LuaJIT 的脚本化多线程基准测试工具，其作者是俄罗斯人 Akopytov，它最常用于数据库基准测试，同时也可用于不涉及数据库的其他基准测试，它提供的基准测试类型如下：

- oltp 类型的数据库基准测试。
- 文件系统级基准测试。
- 简单的 cpu 基准测试。
- 内存访问基准测试。
- 基于线程的调度程序基准测试。
- POSIX 互斥锁(mutex)基准测试。

sysbench 有以下特点：

- 有关于速率和延迟的大量统计数据可用，包括延迟百分比和直方图。
- 即使有数千个并发线程，开销也很低，但它能够每秒生成和跟踪数亿个事件。
- 用户可以通过编写 lua 脚本很容易地创建新的基准测试。
- 也可使用更通用的 lua 解释器替换脚本中的/usr/bin/sysbench。

15.2.2　安装

sysbench 的安装有两种方式：一种是二进制文件安装；另一种是源码编译安装。

1. 二进制文件安装

最方便的二进制文件安装方式是使用 yum 或 apt 源安装，安装源位于：https://packagecloud.io。安装方法的说明位于：https://packagecloud.io/akopytov/sysbench/install。例如，在 RHEL/CentOS 平台上执行如下脚本即可完成安装：

```
$ curl  -s
https://packagecloud.io/install/repositories/akopytov/sysbench/script.rpm.sh | sudo bash
sudo yum  -y install sysbench
```

2. 源码编译安装

相对于二进制安装，源码安装的步骤要稍微复杂一些，但这种安装方式的优势是可以使用与 MySQL 数据库版本相对应的开发包进行编译，例如，MySQL 8.0.22 在 RHEL/CentOS 平台上的开发包是 mysql-community-devel-8.0.22-1.el7.x86_64.rpm。该包可以从 MySQL 的官方网站(https://dev.mysql.com/downloads)上下载，首先应安装该包，然后再安装相关的关联包，命令如下：

```
$ sudo yum install make automake libtool pkgconfig libaio-devel openssl-devel
```

最后安装 sysbench 软件本身，它的源码位于 GitHub 网站，可以使用如下 git clone 命令安装：

```
$ git clone https://github.com/akopytov/sysbench.git
```

也可以下载 zip 包进行安装，安装命令如下：

```
$ wget https://github.com/akopytov/sysbench/archive/master.zip
```

也可以直接从浏览器上安装，相应的下载界面如图 15.1 所示。

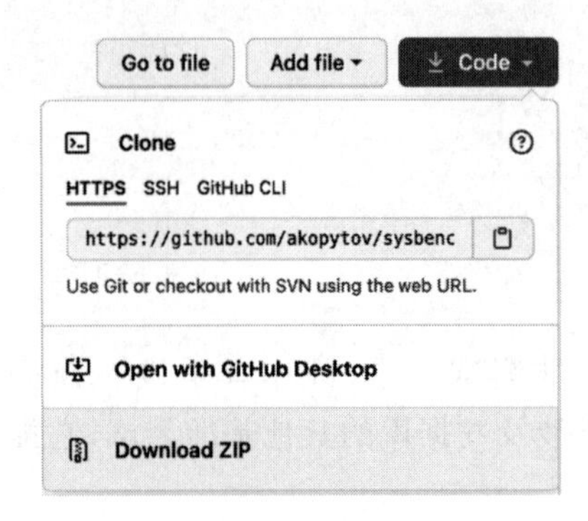

图 15.1

下载以后展开 zip 文件，展开后的文件如下：

```
$ ls
autogen.sh  config COPYING  Dockerfile  m4  missing README.md scripts src third_party
ChangeLog configure.ac debian install-sh Makefile.am mkinstalldirs rpm snap tests
```

编译安装步骤如下：

```
# ./autogen.sh
# ./configure
# make -j
# make install
```

以上步骤执行完毕，安装完成。

15.2.3 测试 MySQL

Sysbench 使用--help 选项可以查看帮助信息如下：

```
$ sysbench --help
...
Compiled-in tests:
  fileio - File I/O test
  cpu - CPU performance test
  memory - Memory functions speed test
  threads - Threads subsystem performance test
  mutex - Mutex performance test
```

```
See 'sysbench <testname> help' for a list of options for each test.
```

可以到/usr/share/sysbench 目录下查看 sysbench 自带的数据库测试脚本：

```
$ ls /usr/share/sysbench
bulk_insert.lua   oltp_delete.lua   oltp_point_select.lua   oltp_read_write.lua     oltp_
update_non_index.lua   select_random_points.lua   tests
oltp_common.lua   oltp_insert.lua   oltp_read_only.lua      oltp_update_index.lua   oltp_write
_only.lua         select_random_ranges.lua
```

以 lua 结尾的脚本可以用于对应的 MySQL 测试类型。lua 语言是一种轻量级的脚本编程语言，它的最大优势是可以嵌入其他语言中执行，这里嵌入 sysbench 中执行。对于每个 MySQL 测试类型可以用 sysbench 加测试类型名，再加 help 参数查询该测试类型的帮助信息，例如，查询 oltp_read_only 帮助信息的命令及其输出结果如下：

```
$ sysbench oltp_read_only help
sysbench 1.0.20 (using bundled LuaJIT 2.1.0-beta2)

oltp_read_only options:
  --auto_inc[=on|off]              Use AUTO_INCREMENT column as Primary Key (for MySQL), or
its alternatives in other DBMS. When disabled, use client-generated IDs [on]
  --create_secondary[=on|off]      Create a secondary index in addition to the PRIMARY KEY
[on]
  --delete_inserts=N               Number of DELETE/INSERT combinations per transaction [1]
  --distinct_ranges=N              Number of SELECT DISTINCT queries per transaction [1]
  --index_updates=N                Number of UPDATE index queries per transaction [1]
  --mysql_storage_engine=STRING Storage engine, if MySQL is used [innodb]
  --non_index_updates=N            Number of UPDATE non-index queries per transaction [1]
  --order_ranges=N                 Number of SELECT ORDER BY queries per transaction [1]
  --pgsql_variant=STRING               Use this PostgreSQL variant when running with the
PostgreSQL driver. The only currently supported variant is 'redshift'. When enabled, create_
secondary is automatically disabled, and delete_inserts is set to 0
  --point_selects=N                Number of point SELECT queries per transaction [10]
  --range_selects[=on|off]         Enable/disable all range SELECT queries [on]
  --range_size=N                   Range size for range SELECT queries [100]
  --secondary[=on|off]             Use a secondary index in place of the PRIMARY KEY [off]
  --simple_ranges=N                Number of simple range SELECT queries per transaction [1]
  --skip_trx[=on|off]              Don't start explicit transactions and execute all queries
in the AUTOCOMMIT mode [off]
  --sum_ranges=N                   Number of SELECT SUM() queries per transaction [1]
  --table_size=N                   Number of rows per table [10000]
  --tables=N                       Number of tables [1]
```

sysbench 测试分为如下三个阶段。

(1) 准备(prepare)：创建表和准备测试用的数据。

(2) 运行(run)：运行测试。

(3) 清理(cleanup)：删除测试用的表。

在运行 sysbench 之前要先创建用户和数据库并赋予权限，相应命令如下：

```
mysql> create user sbtest@'%'  identified by 'dingjia';
```

```
mysql > grant all on sbtest.* to sbtest@'%';
mysql > create database sbtest;
```

使用如下命令测试 sbtest 用户的远程连接：

```
$ mysql -usbtest -h 192.168.87.178 -pdingjia
```

使用 sysbench 的 prepare 命令创建两个表，并向每个表中插入一万条记录，相应命令及其输出结果如下：

```
$ sysbench oltp_read_only --mysql-host=192.168.87.178 \
--mysql-user=sbtest --mysql-password=dingjia \
--mysql-db=sbtest --table_size=10000 \
--tables=2 --threads=2 prepare
sysbench 1.0.20 (using bundled LuaJIT 2.1.0-beta2)

Initializing worker threads...

Creating table 'sbtest1'...
Creating table 'sbtest2'...
Inserting 10000 records into 'sbtest1'
Inserting 10000 records into 'sbtest2'
Creating a secondary index on 'sbtest1'...
Creating a secondary index on 'sbtest2'...
```

数据生成完成后使用 mysqlshow 检查 sbtest 数据库的命令及其输出结果如下：

```
$ mysqlshow -vv sbtest
Database: sbtest
+---------+----------+------------+
| Tables  | Columns  | Total Rows |
+---------+----------+------------+
| sbtest1 |        4 |      10000 |
| sbtest2 |        4 |      10000 |
+---------+----------+------------+
2 rows in set.
```

测试时间的选择可以使用两种方法：一种是使用--time 设置测试的时间；另一种使用--events 设置执行的请求次数，进行 100 秒的 oltp_read_only 类型测试的命令及其输出结果如下：

```
$ sysbench oltp_read_only --mysql-host=192.168.87.178 --mysql-user=sbtest
--mysql-password=dingjia --mysql-db=sbtest --table_size=10000 --tables=2
--threads=2 --time=100 run
sysbench 1.0.20 (using bundled LuaJIT 2.1.0-beta2)

Running the test with following options:
Number of threads: 2
Initializing random number generator from current time


Initializing worker threads...
```

```
Threads started!

SQL statistics:
    queries performed:
        read:                           638022
        write:                          0
        other:                          91146
        total:                          729168
    transactions:                       45573  (455.68 per sec.)
    queries:                            729168 (7290.90 per sec.)
    ignored errors:                     0      (0.00 per sec.)
    reconnects:                         0      (0.00 per sec.)

General statistics:
    total time:                         100.0045s
    total number of events:             45573

Latency (ms):
         min:                           4.13
         avg:                           4.39
         max:                           7.06
         95th percentile:               4.57
         sum:                           199876.24

Threads fairness:
    events (avg/stddev):           22786.5000/140.50
    execution time (avg/stddev):   99.9381/0.00
```

同样的测试换成万兆网再进行一遍：

```
$ sysbench oltp_read_only --mysql-host=10.168.85.178 --mysql-user=sbtest
--mysql-password=dingjia --mysql-db=sbtest --table_size=10000 --tables=2
--threads=2 --time=100 run
sysbench 1.0.20 (using bundled LuaJIT 2.1.0-beta2)

Running the test with following options:
Number of threads: 2
Initializing random number generator from current time


Initializing worker threads...

Threads started!

SQL statistics:
    queries performed:
        read:                           834792
        write:                          0
        other:                          119256
        total:                          954048
    transactions:                       59628  (596.22 per sec.)
    queries:                            954048 (9539.57 per sec.)
    ignored errors:                     0      (0.00 per sec.)
    reconnects:                         0      (0.00 per sec.)
```

```
General statistics:
    total time:                          100.0032s
    total number of events:              59628

Latency (ms):
         min:                            3.04
         avg:                            3.35
         max:                            9.20
         95th percentile:                3.55
         sum:                            199813.79

Threads fairness:
    events (avg/stddev):           29814.0000/288.00
    execution time (avg/stddev):   99.9069/0.00
```

表 15.1 列出了千兆网和万兆网测试结果的对比，发现仅仅将千兆网改成万兆网，性能就提高了大约 30%。

表　15.1

测试指标	千兆网	万兆网	差异/%
平均延迟	4.39ms	3.35ms	−31.0
每秒查询	7290.90 次	9539.57 次	30.8
每秒交易数	455.68 次	596.22 次	30.8

测试完成后，进行数据清理，相应命令及其输出结果如下：

```
$ sysbench oltp_read_only  --mysql-host=192.168.87.178 --mysql-user=sbtest
--mysql-password=dingjia --mysql-db=sbtest --table_size=10000 --tables=2
cleanup sysbench 1.0.20 (using bundled LuaJIT 2.1.0-beta2)

Dropping table 'sbtest1'...
Dropping table 'sbtest2'...
```

15.3　TPCC-MySQL

15.3.1　简介

TPC-C 是由 TPC(transaction processing performance council)非营利组织推出的一套联机事务处理过程(OLTP on-line transaction processing) 测试基准，它比之前的版本 TPC-A 要复杂得多，关于它的说明参见 http://www.tpc.org/tpcc/。TPCC-MySQL 是 Percona 按照 TPC-C 标准开发的，基于 MySQL 的压测工具，该工具模拟搭建一套电商的订单系统，衡量这套系统性能好坏的依据是系统的吞吐量，也就是每分钟产生的订单数(tpmc)，每产生一个订单要经过五个环节，包括下单、支付、查订单、发货、查库存等。该基准测试工具的优势是它有一套符合国际标准的业务逻辑，这样，它测试的数据可以和其他数据库进行对比，如 Oracle。

15.3.2 安装

使用如下命令下载源码安装包：

```
$ wget https://github.com/Percona-Lab/tpcc-mysql/archive/master.zip
```

编译时需要 mysql_config 工具，该工具在 MySQL 的开发包里，如果没有 MySQL 的开发包，可以到 MySQL 官方网站下载，然后使用如下命令安装：

```
$ sudo yum install mysql-community-devel-8.0.26-1.el7.x86_64.rpm
```

安装了开发包后再编译，在 src 目录里使用 make 进行编译的命令及其输出结果如下：

```
$ sudo make
cc -w -03 -g -I. 'mysql_config --include'  -c load.c
cc -w -03 -g -I. 'mysql_config --include'  -c support.c
cc load.o support.o 'mysql_config --libs_r' -lrt -o ../tpcc_load
cc -w -03 -g -I. 'mysql_config --include'  -c main.c
cc -w -03 -g -I. 'mysql_config --include'  -c spt_proc.c
cc -w -03 -g -I. 'mysql_config --include'  -c driver.c
cc -w -03 -g -I. 'mysql_config --include'  -c sequence.c
cc -w -03 -g -I. 'mysql_config --include'  -c rthist.c
cc -w -03 -g -I. 'mysql_config --include'  -c sb_percentile.c
cc -w -03 -g -I. 'mysql_config --include'  -c neword.c
cc -w -03 -g -I. 'mysql_config --include'  -c payment.c
cc -w -03 -g -I. 'mysql_config --include'  -c ordstat.c
cc -w -03 -g -I. 'mysql_config --include'  -c delivery.c
cc -w -03 -g -I. 'mysql_config --include'  -c slev.c
cc main.o spt_proc.o driver.o support.o sequence.o rthist.o sb_percentile.o neword.o payment.o
ordstat.o delivery.o slev.o 'mysql_config --libs_r' -lrt -o ../tpcc_start
```

编译后将在上一层目录中生成两个执行文件：tpcc_load 和 tpcc_start，分别用于生成数据和进行测试。

15.3.3 生成数据

在 TCPP-MySQL 安装完成并开始测试之前还要先生成用于测试的数据，步骤如下。

（1）创建数据库，命令如下：

```
$ mysqladmin create tpcc1000
```

（2）创建表，命令如下：

```
$ mysql tpcc1000 < ./create_table.sql
```

（3）加外键，命令如下：

```
$ mysql tpcc1000 < ./add_fkey_idx.sql
```

（4）生成数据。生成数据的工具 tpcc_load 的使用说明如下：

```
$ ./tpcc_load --help
```

```
*************************************
*** TPCC-mysql Data Loader        ***
*************************************
Usage: tpcc_load -h server_host -P port -d database_name -u mysql_user -p
mysql_password -w warehouses -l part -m min_wh -n max_wh
* [part]: 1=ITEMS 2=WAREHOUSE 3=CUSTOMER 4=ORDERS
```

使用如下命令生成一套包括10个仓库的电商的订单系统：

```
$ ./tpcc_load -h127.0.0.1 -d tpcc1000 -u root -p dingjia -w 10
```

生成的数据量和仓库数基本成正比，当仓库数量为10时，生成的数据量共994M；当仓库数量为100时，生成的数据量共9.1G。tpcc1000数据库中共有9个表，生成之后的EER(entity-relationship model，增强的实体关系模型)如图15.2所示。

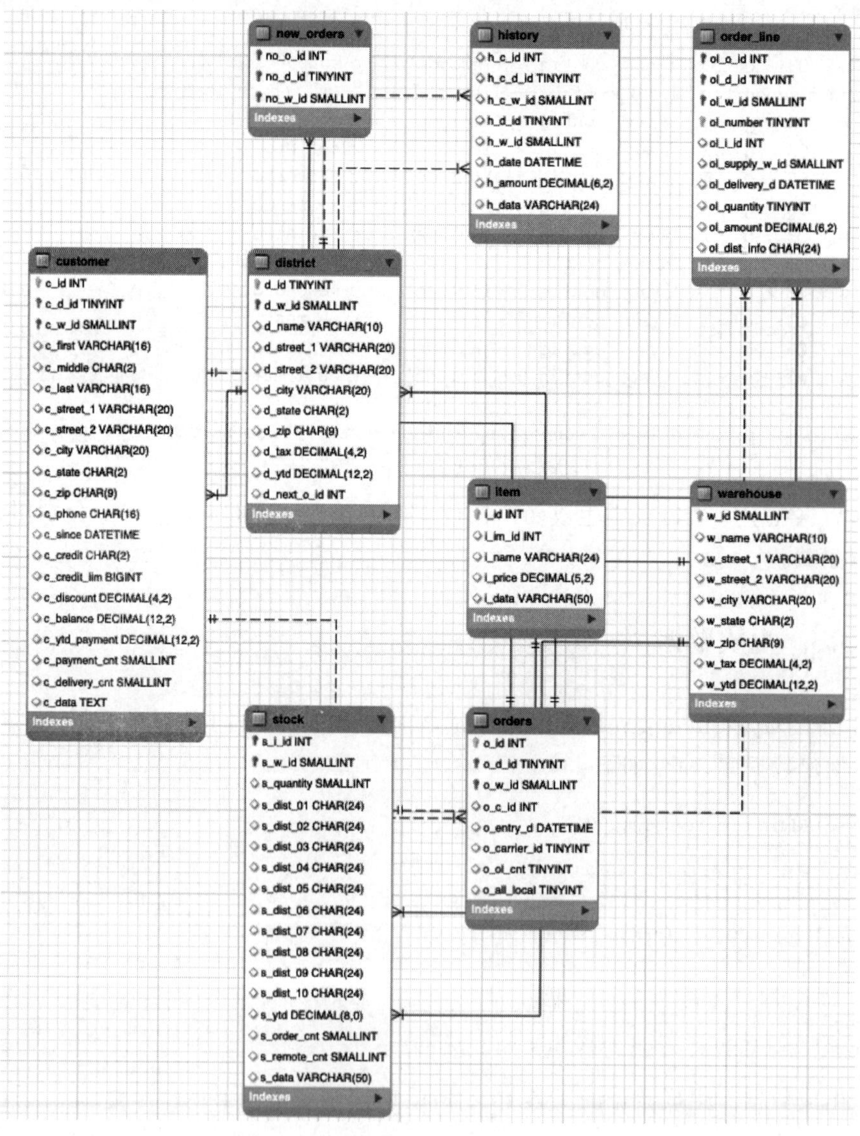

图 15.2

在 src 的上一级目录下还有两个并行生成数据的 shell 脚本，分别是 load.sh 和 load_multi_schema.sh，可以根据需要选择使用。

15.3.4 测试

测试命令 tpcc_start 的使用说明如下：

```
$ ./tpcc_start --help
***************************************
*** ###easy### TPC-C Load Generator    ***
***************************************
Usage: tpcc_start -h server_host -P port -d database_name -u mysql_user -p
mysql_password -w warehouses -c connections -r warmup_time -l running_time -i
report_interval -f report_file -t trx_file
```

使用如下命令进行 60 秒的基准测试，预热时间也是 60 秒，数据量是 10 个仓库，并发 10 个连接：

```
$ ./tpcc_start -h127.0.0.1  -d tpcc1000  -u root -p dingjia -w 10 -c 10 -r 60 -l 60
```

下面输出的是程序运行时的参数：

```
***************************************
*** ###easy### TPC-C Load Generator    ***
***************************************
option h with value '127.0.0.1'
option d with value 'tpcc1000'
option u with value 'root'
option p with value 'dingjia'
option w with value '10'
option c with value '10'
option r with value '60'
option l with value '60'
<Parameters>
      [server]: 127.0.0.1
      [port]: 3306
      [DBname]: tpcc1000
        [user]: root
        [pass]: dingjia
  [warehouse]: 10
[connection]: 10
      [rampup]: 60 (sec.)
     [measure]: 60 (sec.)
```

下面是测试进行时的输出结果：

```
RAMP-UP TIME.(60 sec.)

MEASURING START.
```

```
  10, trx: 5587, 95%: 13.919, 99%: 19.834, max_rt: 237.494, 5589|209.832, 559|11.609, 559
|241.026, 559|35.921
  20, trx: 5508, 95%: 13.647, 99%: 18.648, max_rt: 193.919, 5509|183.753, 550|1.389, 552|
214.095, 551|34.411
  30, trx: 5529, 95%: 13.965, 99%: 19.704, max_rt: 60.077, 5526|47.540, 554|1.527, 552|
96.868, 553|31.791
  40, trx: 5547, 95%: 14.024, 99%: 20.091, max_rt: 40.242, 5548|26.420, 554|1.510, 556|
56.958, 555|33.341
  50, trx: 5540, 95%: 13.803, 99%: 19.232, max_rt: 32.124, 5540|27.365, 554|1.681, 554|
64.480, 553|33.280
  60, trx: 5556, 95%: 13.762, 99%: 19.674, max_rt: 28.982, 5555|24.876, 555|1.386, 555|
60.118, 557|32.242

STOPPING THREADS.........
```

每10秒显示一次运行的结果,以第一行为例进行说明:

- 10表示第一个10秒运行的运行区间。
- "trx:5587"表示这10秒共运行了5587个订单交易。
- "95%:13.919"表示95%的交易相应时间在13.919秒之内。
- "99%:19.834"表示99%的交易相应时间在19.834秒之内。
- "max_rt:237.494"表示最慢的交易相应时间是237.494秒。
- 剩下的"5589|209.832,559|11.609,559|241.026,559|35.921"表示不同的吞吐量和在该吞吐量时最慢的相应时间。

测试完成后,下面输出的是测试结果:

```
<Raw Results>
  [0] sc: 1091 lt: 32176  rt: 0  fl: 0 avg_rt: 18.5 (5)
  [1] sc: 30918 lt: 2351  rt: 0  fl: 0 avg_rt: 6.1 (5)
  [2] sc: 3326 lt: 0  rt: 0  fl: 0 avg_rt: 1.7 (5)
  [3] sc: 3324 lt: 4  rt: 0  fl: 0 avg_rt: 63.4 (80)
  [4] sc: 110 lt: 3218  rt: 0  fl: 0 avg_rt: 48.9 (20)
in 60 sec.

<Raw Results2(sum ver.)>
  [0] sc: 1091  lt: 32177  rt: 0  fl: 0
  [1] sc: 30921  lt: 2352  rt: 0  fl: 0
  [2] sc: 3326  lt: 0  rt: 0  fl: 0
  [3] sc: 3324  lt: 4  rt: 0  fl: 0
  [4] sc: 110  lt: 3218  rt: 0  fl: 0

<Constraint Check> (all must be [OK])
[transaction percentage]
        Payment: 43.48% (>= 43.0%) [OK]
   Order-Status: 4.35% (>= 4.0%) [OK]
       Delivery: 4.35% (>= 4.0%) [OK]
    Stock-Level: 4.35% (>= 4.0%) [OK]
[response time (at least 90% passed)]
      New-Order: 3.28%  [NG] *
        Payment: 92.93%  [OK]
```

```
  Order - Status: 100.00 %   [OK]
      Delivery: 99.88 %   [OK]
   Stock - Level: 3.31 %   [NG] *

<TpmC>
                  33267.000 TpmC
```

其中,sc 表示成功；lt 表示操作延时；rt 表示重试；fl 操作失败；[0]、[1]、[2]、[3]、[4]分别代表下单、支付、查订单、发货、查库存五类操作；[OK]代表测试通过；[NG]代表 Not Good。

从上面的结果可以看出,下单的操作延时比较高。最后一行的 TpmC 值是衡量系统吞吐量的最关键指标,每分钟交易数是 33267。

15.4 实验

mysqlslap 压力测试工具

(1) 使用--only-print 选项检查 mysqlslap 的 3 个执行阶段。

(2) 检查 mysqlslap 内部运行的 SQL 和连接状态。

(3) 使用--no-drop 选项生成数据。

视频演示

第16章

实例优化

数据库优化分为 SQL 优化和实例优化。SQL 优化只对被优化的单个 SQL 有效，实例优化关注的是整个 MySQL 实例的性能，本章介绍的实例优化方法会对所有 SQL 的性能都产生影响，第 17、18 章将介绍 SQL 优化。

MySQL 实例优化的内容包括参数配置、日志文件、资源管理器的使用等。

16.1 数据库优化的重要性

数据库的优化对于应用来说是非常重要的，这是由数据库在应用中的作用和地位决定的。图 16.1 是目前行业内典型应用的架构图。

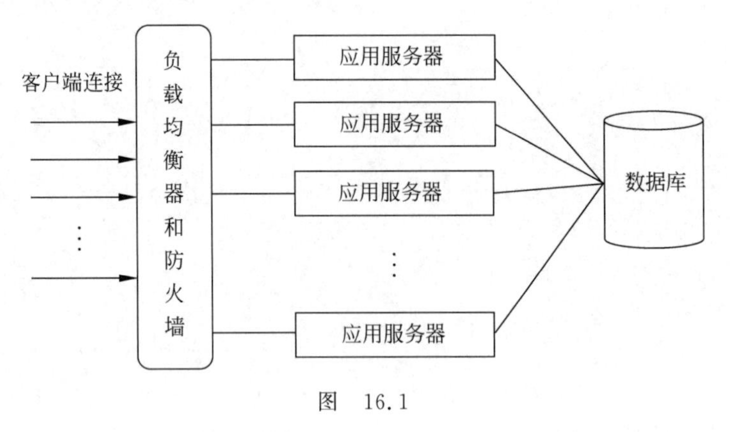

图 16.1

典型的应用数据流通常是从客户端发起连接，然后经过负载均衡器和防火墙到应用服务器(这里通常还包括 Web 服务器)，在应用服务器完成逻辑处理后，最后到数据库进行数据的存取，再原路返回。整个系统中可能出现性能瓶颈的地方是应用服务器和数据库。

应用服务器的性能特点如下：

- 应用服务器通常消耗的是 CPU 和内存，对磁盘 IO 需求很小。
- 纵向扩展（增加网络、硬盘、内存和 CPU 的处理能力）应用服务器可以线性增加性能。
- 水平扩展（增加服务器的数量）应用服务器的难度小，因为应用服务器之间通常只共享数据库。
- 水平扩展也会线性增加性能，通常服务器数量增加 N 倍，性能也会增加 N 倍。

数据库服务器的性能特点如下：

- 数据库服务器的瓶颈通常在磁盘的读写。
- 纵向扩展（增加网络、硬盘、内存和 CPU 的处理能力）数据库服务器可以线性增加性能。
- 水平扩展（增加服务器的数量）数据库服务器的难度很大，例如，将数据库服务器数量从一台增加到两台，运维的难度可能增加了 10 倍，因为涉及多台服务器之间并发时的锁和数据同步问题。
- 水平扩展可能不会带来性能的线性增加，例如，数据库服务器数量增加 1 倍，性能可能只会增加 0.5～0.8 倍。

当系统遇到性能瓶颈时，对应用服务器进行水平或纵向扩展都可以线性地增加应用层的处理能力。而数据库服务器的扩展就困难得多，纵向扩展涉及停机甚至数据迁移，而且扩展空间有限；水平扩展难度大，性能也不能线性增加。因此，当系统遇到性能瓶颈时，数据库的性能优化就显得非常重要，并且回报丰厚。目前，在项目的开发设计阶段，研发人员主要关注功能的实现，对性能缺乏足够的重视，等投产后，随着业务量和数据量的增加，性能瓶颈就暴露出来了，因此这个时候数据库优化的空间都很大，很多系统经过优化后，性能可以提高 10～100 倍。其中 SQL 语句的优化特别明显，一些需要运行数天或数小时的 SQL 语句经过优化后，可以在几分钟甚至几秒钟内完成。

16.2　系统参数的修改

MySQL 的系统参数当然是对 MySQL 性能影响最大的因素，MySQL 的系统参数保存在参数文件中，不同平台的参数文件位置不一样，Red Hat Linux 中的默认参数文件位置是/etc/my.cnf，Debian Linux 中的默认参数文件位置是/etc/mysql/my.cnf。

MySQL 的系统参数根据作用范围可以分为全局级和会话级，全局级的系统参数对实例的所有会话起作用，会话级的系统参数只对当前会话起作用，在这两个级别下修改系统参数的例子如下：

```
mysql> set session binlog_rows_query_log_events = on;
mysql> set global binlog_rows_query_log_events = on;
```

其中，session 可以省略。修改全局系统参数只对修改后连接到 MySQL 的会话生效，在修改之前已经建立的会话还保持原来的参数值，例如，如下命令为修改一个全局系统参数：

```
mysql> set global sort_buffer_size = 16 * 1024 * 1024;
```

检查该系统参数的全局级和会话级的值，相关命令及运行结果如下：

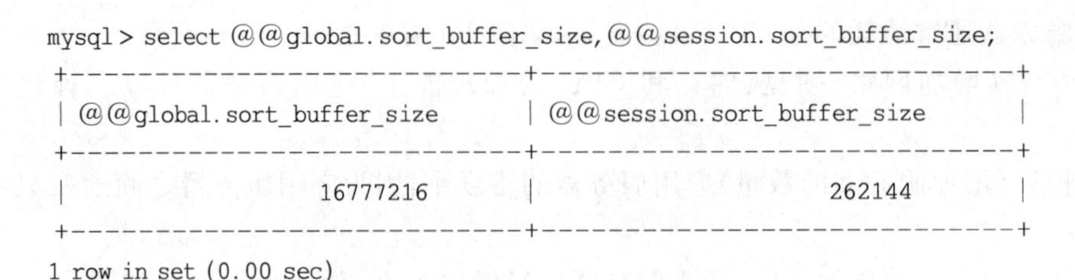

```
mysql> select @@global.sort_buffer_size,@@session.sort_buffer_size;
+---------------------------+---------------------------+
| @@global.sort_buffer_size | @@session.sort_buffer_size |
+---------------------------+---------------------------+
|          16777216         |          262144           |
+---------------------------+---------------------------+
1 row in set (0.00 sec)
```

可以发现，全局级的参数值已经改过来了，而会话级的参数值没有变。

有一些系统参数只能是全局级的，在会话级修改这类参数会出错，例如：

```
mysql> set log_error_verbosity = 2;
ERROR 1229 (HY000): Variable 'log_error_verbosity' is a GLOBAL variable and should be set with
SET GLOBAL
```

如果要查询其他会话的参数可以在 performance_schema.variables_by_thread 视图中查询，例如，查询所有会话的事务隔离级别如下：

```
mysql> select * from performance_schema.variables_by_thread where variable_name =
'transaction_isolation';
+-----------+-----------------------+-----------------+
| THREAD_ID | VARIABLE_NAME         | VARIABLE_VALUE  |
+-----------+-----------------------+-----------------+
|        60 | transaction_isolation | REPEATABLE-READ |
|        67 | transaction_isolation | SERIALIZABLE    |
+-----------+-----------------------+-----------------+
2 rows in set (0.00 sec)
```

MySQL 的系统参数还可以分为静态参数和动态参数，动态参数可以在 MySQL 运行中进行修改，静态参数在 MySQL 启动后无法修改，例如：

```
mysql> set auto_generate_certs = on;
ERROR 1238 (HY000): Variable 'auto_generate_certs' is a read only variable
```

如果把会话级系统参数设置为 DEFAULT，对应的则是全局级系统参数值。如下两个设置会话级参数的语句效果是一样的：

```
mysql> SET @@SESSION.max_join_size = DEFAULT;
mysql> SET @@SESSION.max_join_size = @@GLOBAL.max_join_size;
```

如果把全局级系统参数设置为 DEFAULT，将把系统参数恢复为 MySQL 内置的默认值，而不是像很多人认为的是参数文件里面的设置值。

在系统参数设置时，容易犯的一个错误是在 MySQL 运行时修改了参数值，但没有同时修改参数文件里面的配置，当 MySQL 重新启动后，参数文件里的旧值生效，而之前的修改丢掉了。在 MySQL 8.0 里，MySQL 推出了让参数持久化的命令，可以让在联机时修改的系统参数在重新启动后仍然生效，例如：

```
mysql> set persist max_connections = 1000;
```

或者：

```
mysql> set @@persist.max_connections = 1000;
```

如果想让系统参数在本次 MySQL 运行时不生效，只是在下次启动时生效，可以使用如下命令：

```
mysql> set persist_only back_log = 100;
```

或者：

```
mysql> set @@persist_only.back_log = 100;
```

持久化的系统参数以 JSON 格式保存在数据目录的 mysqld-auto.cnf 文件中，例如：

```
$ cat /var/lib/mysql/mysqld-auto.cnf
{ "Version" : 1 , "mysql_server" : { "max_connections" : { "Value" : "1000" , "Metadata" :
{ "Timestamp" : 1624532357790912 , "User" : "root" , "Host" : "localhost" } } , "mysql_server_
static_options" : { "back_log" : { "Value" : "100" , "Metadata" : { "Timestamp" :
1624533604896754 , "User" : "root" , "Host" : "localhost" } } } } }
```

可以通过 reset persist 命令来清除 mysqld-auto.cnf 文件中的所有配置，也可以通过 reset persist 接参数名的方式来清除某个指定的配置参数。

现在系统参数可以从多个来源进行设置，有时分不清参数值到底来自哪里，到底哪种方式的设置在起作用，这时可以查询视图 performance_schema.variables_info 找到相关信息，例如：

```
mysql> select variable_name, variable_source as source,
variable_path, set_time, set_user as user, set_host
from performance_schema.variables_info
where variable_name = 'max_connections' or variable_name = 'socket'\G
*************************** 1. row ***************************
variable_name: max_connections
       source: DYNAMIC
variable_path:
     set_time: 2021-08-06 14:30:07.128393
         user: root
     set_host: localhost
*************************** 2. row ***************************
variable_name: socket
       source: GLOBAL
variable_path:/etc/my.cnf    -- 参数文件名和路径
     set_time: NULL
         user: NULL
     set_host: NULL
2 rows in set (0.01 sec)
```

16.3 内存的分配

MySQL 运行是单进程多线程的方式，它使用的内存分为两种：一种是所有线程共用的全局共享内存；另一种是每个线程独享的内存。因此 MySQL 使用的最大内存是全局共享内存＋最大线程数×线程独享内存，计算 MySQL 在负载高峰时占用总内存的粗略方法如下：

```
mysql > select ( @@key_buffer_size + @@innodb_buffer_pool_size +
@@innodb_log_buffer_size + @@binlog_cache_size + @@max_connections *
( @@read_buffer_size + @@read_rnd_buffer_size + @@sort_buffer_size +
@@join_buffer_size + @@thread_stack + @@tmp_table_size ) ) / (1024 * 1024 * 1024) as
max_memory_gb;
```

在实际工作中，通过如上方法算出来的数值通常偏大，因为所有线程都同时用到设定内存分配最大值的情况几乎不会出现，每个线程如果只是处理简单的工作，大约只需要256KB的内存。通过查询 sys. memory_global_total 视图可以得到当前 MySQL 实例使用内存的总和。

系统参数 key_buffer_size 从字面上理解是指定索引缓存的大小，需要注意的是，它只对 MyISAM 表起作用，对 InnoDB 表无效。该参数在字面上并没有明确加上 MyISAM，这是因为它是在 MyISAM 作为 MySQL 默认存储引擎的时代产生的。由于现在通常使用的是 InnoDB 表，因此不需要调整该参数。

MySQL 读取 InnoDB 数据文件是先把数据读取到 InnoDB 缓存中，再进行后续处理，写入 InnoDB 数据文件也是把数据从 InnoDB 缓存中刷新到硬盘。因此在所有占用内存的组件中，对 MySQL 性能影响最大的是 InnoDB 缓存池，其大小由参数 innodb_buffer_pool_size 决定。MySQL 的默认配置是针对内存为 512MB 的虚拟机设计的，innodb_buffer_pool_size 默认值是 128MB，该值在生产中通常都太小，太小的缓冲池可能会导致数据页被频繁地从磁盘读取到内存，导致性能下降，但如果设置得过大，又可能会造成内存被交换到位于硬盘的内存交换分区，导致性能急剧下降，这两种情况比较起来，把 InnoDB 缓存池设置小一些对性能的负面影响相对较小。实际生产中，mysqld 进程崩溃的一个常见原因是操作系统的内存耗尽，操作系统被迫把 mysqld 进程"杀死"。当一台服务器被一个 MySQL 实例独占时，通常 innodb_buffer_pool_size 可以设置为内存的 70%左右。如果在同一台服务器上还有其他的 MySQL 或别的应用，设置 innodb_buffer_pool_size 的大小就要考虑更多因素，一个重要的因素是 InnoDB 的总数据量（包括表和索引），可以使用如下 SQL 语句查询 InnoDB 的总数据量：

```
mysql > SELECT count( * ) as TABLES, concat(round(sum(table_rows)/1000000,2),'M') num_row,
concat(round(sum(data_length)/(1024 * 1024 * 1024),2),'G') DATA,
concat(round(sum(index_length)/(1024 * 1024 * 1024),2),'G') idx,
concat(round(sum(data_length + index_length)/(1024 * 1024 * 1024),2),'G') total_size
FROM information_schema.TABLES WHERE engine = 'InnoDB';
+--------+---------+-------+-------+------------+
| TABLES | num_row | DATA  | idx   | total_size |
+--------+---------+-------+-------+------------+
|     60 | 0.90M   | 0.33G | 0.02G | 0.35G      |
+--------+---------+-------+-------+------------+
1 row in set (0.01 sec)
```

把参数 innodb_buffer_pool_size 设置成超过 InnoDB 的总数据量是没有意义的，通常设置到能容纳 InnoDB 的活跃数据量就够了，因为几乎每个数据库中都有一些很少用到的

历史数据。

一个衡量 InnoDB 缓存池效率的重要标准是它的命中率，根据如下两个 MySQL 的状态参数可以计算出 InnoDB 缓存池的命中率：

（1）Innodb_buffer_pool_read_requests：表示向 InnoDB 缓存池进行逻辑读的次数。

（2）Innodb_buffer_pool_reads：表示从物理磁盘中读取数据的次数。

所以，InnoDB 缓存池命中率的计算公式如下：

InnoDB 缓存池的命中率＝(Innodb_buffer_pool_read_requests－Innodb_buffer_pool_reads)/Innodb_buffer_pool_read_requests * 100%

通常 InnoDB 缓存池的命中率不会低于 99%，如下是一个计算 InnoDB 缓存池命中率的例子：

```
mysql> show status like  'Innodb_buffer_pool_read%s';
+----------------------------------+---------+
| Variable_name                    | Value   |
+----------------------------------+---------+
| Innodb_buffer_pool_read_requests | 1059322 |
| Innodb_buffer_pool_reads         | 6091    |
+----------------------------------+---------+
2 rows in set (0.00 sec)

mysql> select (1059322 - 6091)/1059322 * 100 'InnoDB buffer pool hit';
+------------------------+
| InnoDB buffer pool hit |
+------------------------+
|                99.4250 |
+------------------------+
1 row in set (0.01 sec)
```

另一个衡量 InnoDB 缓存池效率的指标是状态参数 Innodb_buffer_pool_reads，它代表 MySQL 不能从 InnoDB 缓存池读到需要的数据而不得不从硬盘中进行读的次数，使用如下命令可查询 MySQL 每秒从磁盘读的次数：

```
$ mysqladmin extended-status  -ri1 | grep Innodb_buffer_pool_reads
| Innodb_buffer_pool_reads                | 1476098323    |
| Innodb_buffer_pool_reads                | 734           |
| Innodb_buffer_pool_reads                | 987           |
| Innodb_buffer_pool_reads                | 595           |
| Innodb_buffer_pool_reads                | 915           |
| Innodb_buffer_pool_reads                | 1287          |
```

把查询到的值和硬盘的 I/O 能力进行对比，如果 MySQL 每秒从磁盘读的次数接近硬盘处理 I/O 的上限，那么从操作系统层查看到的 CPU 用于等待 I/O 的时间(I/O wait)会变长，这时硬盘 I/O 就成了限制系统性能的瓶颈，增大 InnoDB 缓存池可能会减少 MySQL 访问硬盘的次数，提高数据库的性能。

早期，调整 innodb_buffer_pool_size 需要重新启动 MySQL，从 MySQL 5.7 后，该参数可以动态地进行调整，例如，如下命令将该参数设置成 256MB：

```
mysql> set persist innodb_buffer_pool_size = 256 * 1024 * 1024;
```

在 MySQL 的错误日志中可以看到 MySQL 调整 InnoDB 缓存池的过程，如下所示：

```
2021-06-26T06:58:26.824890Z 6042 [Note] [MY-012398] [InnoDB] Requested to resize buffer pool. (new size: 268435456 bytes)
2021-06-26T06:58:26.866619Z 0 [Note] [MY-011880] [InnoDB] Resizing buffer pool from 134217728 to 268435456 (unit=134217728).
2021-06-26T06:58:26.892090Z 0 [Note] [MY-011880] [InnoDB] Disabling adaptive hash index.
2021-06-26T06:58:26.899180Z 0 [Note] [MY-011885] [InnoDB] disabled adaptive hash index.
2021-06-26T06:58:26.899225Z 0 [Note] [MY-011880] [InnoDB] Withdrawing blocks to be shrunken.
2021-06-26T06:58:26.899251Z 0 [Note] [MY-011880] [InnoDB] Latching whole of buffer pool.
2021-06-26T06:58:26.899294Z 0 [Note] [MY-011880] [InnoDB] buffer pool 0 : resizing with chunks 1 to 2.
2021-06-26T06:58:26.918099Z 0 [Note] [MY-011891] [InnoDB] buffer pool 0 : 1 chunks (8192 blocks) were added.
2021-06-26T06:58:26.920019Z 0 [Note] [MY-011894] [InnoDB] Completed to resize buffer pool from 134217728 to 268435456.
2021-06-26T06:58:26.920066Z 0 [Note] [MY-011895] [InnoDB] Re-enabled adaptive hash index.
2021-06-26T06:58:26.920136Z 0 [Note] [MY-011880] [InnoDB] Completed resizing buffer pool at 210626  4:28:26.
```

一个和 innodb_buffer_pool_size 相关的参数是 innodb_buffer_pool_instances，它用于设定把 InnoDB 缓存池分成几个区，当 innodb_buffer_pool_size 大于 1GB 时，这个参数才会起作用，对于大的 InnoDB 缓存池，建议把 innodb_buffer_pool_instances 设置得大一些，这样可以减少获取访问 InnoDB 缓存池时需要上锁的粒度，以提高并发度。

16.4 InnoDB 日志

InnoDB 日志保存着已经提交的数据变化，用于在崩溃恢复时把数据库的变化恢复到数据文件，除了崩溃恢复，任何情况下都不会读日志文件。向日志文件写数据的方式是顺序写，这比离散写的效率要高很多，而向数据文件写数据通常是离散写。

日志缓冲区是一个内存缓冲区，InnoDB 使用它来缓冲重做日志事件，然后再将其写入磁盘。日志缓冲区的大小由系统参数 innodb_log_buffer_size 控制，默认是 16MB，在大多数情况下是够用的。如果有大型事务或大量较小的并发事务，可以考虑增大 innodb_log_buffer_size，该参数在 MySQL 8.0 中可以动态设置。

16.4.1 日志产生量

InnoDB 的日志产生量是衡量数据库繁忙程度的重要指标，也是设置日志文件大小的依据。查询日志产生量相关信息的方法有两种。

第一种方法是查询 information_schema.innodb_metrics 或 sys.metrics 视图中的对应

计量，日志产生量相关的计量是以 log_lsn_开头的参数，这里的 lsn 代表 Log Sequence Number，直译为日志序列号，但这里的数值实际是以字节为单位的日志量大小，和序列号并没有关系。这些计量默认没有被激活，通过激活这些计量而产生的性能消耗很小，使用如下命令可以激活这些计量：

```
mysql > set global innodb_monitor_enable = 'log_lsn_%';
```

计量被激活后，一个查询结果的例子如下：

```
mysql > select name,count,status from information_schema.innodb_metrics where name like 'log_
lsn%';
+--------------------------------+-----------+---------+
| name                           | count     | STATUS  |
+--------------------------------+-----------+---------+
| log_lsn_last_flush             | 421908956 | enabled |
| log_lsn_last_checkpoint        | 404648325 | enabled |
| log_lsn_current                | 421909901 | enabled |
| log_lsn_archived               |         0 | enabled |
| log_lsn_checkpoint_age         |  17261576 | enabled |
| log_lsn_buf_dirty_pages_added  | 421909901 | enabled |
| log_lsn_buf_pool_oldest_approx | 408201960 | enabled |
| log_lsn_buf_pool_oldest_lwm    | 406104808 | enabled |
+--------------------------------+-----------+---------+
8 rows in set (0.00 sec)
```

其中，log_lsn_checkpoint_age 是当前日志产生量减去最近一次检查点的日志产生量，等于 log_lsn_current 减去 log_lsn_last_checkpoint，也就是日志文件的使用量，因为 InnoDB 引擎对日志文件的写入是循环覆盖的，而检查点之前的日志都已经写入数据文件，因此不再需要，可以被覆盖。这里看到的日志文件的使用量大约是 17MB。

第二种方法是使用 show engine innodb status 命令查询日志产生量的相关信息，这些信息在输出的 LOG 部分，该方法不需要激活 InnoDB 中的相关计量。一个查询的例子如下：

```
mysql >  show engine innodb status\G
...
---
LOG
---
Log sequence number          425640652
Log buffer assigned up to    425640652
Log buffer completed up to   425640652
Log written up to            425640652
Log flushed up to            425639974
Added dirty pages up to      425640652
Pages flushed up to          407036166
Last checkpoint at           406841423
252823 log i/o's done, 608.37 log i/o's/second
...
```

这里的 lsn 是 425640652，最近一次检查点的 lsn 是 406841423，计算出当前日志文件的使用量是这两个值之差，如下所示：

```
mysql > select round((425640652 - 406841423)/1024/1024) logsize_MB;
+------------+
| logsize_MB |
+------------+
|     18     |
+------------+
1 row in set (0.00 sec)
```

可以看出，当前日志文件的使用量大约 18MB。

16.4.2 日志文件大小

MySQL 默认在数据目录下有两个 48MB 的日志文件：ib_logfile0 和 ib_logfile1。对于繁忙的数据库来说，这样的日志文件通常太小，因为当日志文件写满时，会触发检查点，把内存中的数据写入磁盘，小的日志文件会频繁地触发检查点，增加写磁盘频率，引起系统性能的下降。大的日志文件能容纳的数据变化量大，会造成数据库在崩溃恢复时耗时较长，但新的 MySQL 版本的崩溃恢复速度已经很快，因此把日志文件设置得大一些通常不会错，甚至可以设置得和 InnoDB 缓存池一样大。另外一些备份工具要备份在备份过程中产生的重做日志，如果日志文件过小，当备份工具备份日志的速度跟不上日志产生的速度时，需要备份的日志可能已经被覆盖了，例如，XtraBackup 工具可能会遇到下面的错误：

```
xtrabackup: error: it looks like InnoDB log has wrapped around before xtrabackup could process
all records due to either log copying being too slow, or log files being too small.
```

一个合理大小的日志文件应该可以容纳数据库在高峰时 1～2 小时的数据变化。如下例子是查询一分钟产生的日志量。

设置 pager 只显示 lsn，命令如下：

```
mysql > pager grep sequence
PAGER set to 'grep sequence'
```

查询当前的 lsn，命令如下：

```
mysql > show engine innodb status \G
Log sequence number          1439955157
1 row in set (0.00 sec)
```

休眠一分钟，命令如下：

```
mysql > select sleep(60);
1 row in set (1 min 0.00 sec)
```

再次查询当前的 lsn，命令如下：

```
mysql > show engine innodb status \G
Log sequence number          1455007613
```

```
1 row in set (0.01 sec)
```

取消设置的 pager,命令如下:

```
mysql > nopager
PAGER set to stdout
```

根据一分钟的采样,可以计算出一小时产生的日志量,相关命令及其运行结果如下:

```
mysql > select round((1455007613 - 1439955157) * 60/1024/1024) "1 hour log(MB)";
```

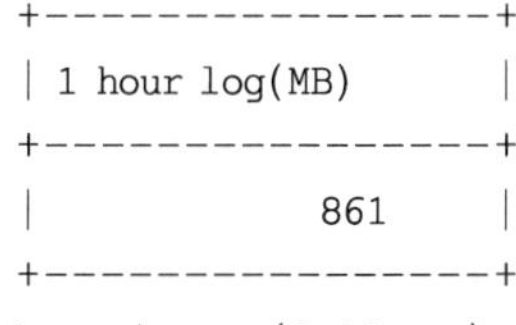

```
+-----------------+
| 1 hour log(MB)  |
+-----------------+
|             861 |
+-----------------+
1 row in set (0.00 sec)
```

可以看出,一个小时的日志量是 861MB。日志文件的大小由如下两个参数决定:

(1) innodb_log_files_in_group:表示一个组里有多少个文件,默认为 2。

(2) innodb_log_file_size:表示单个日志文件的大小,默认为 48MB。

因此,如果保持 innodb_log_files_in_group 为 2 不变,把 innodb_log_file_size 设置为 861MB,可以容纳高峰期两个小时的日志。

修改日志文件大小的方法很简单,只需要修改参数文件中 innodb_log_file_size 的设置,然后重新启动 MySQL 即可。不需要删除当前的日志文件,在启动过程中,MySQL 会发现参数值和当前日志文件的大小不一样,然后自动删除旧的日志文件,并创建新的日志文件,在 MySQL 的错误日志里会有如下记录:

```
1 [Note] [MY - 013041] [InnoDB] Resizing redo log from 2 * 50331648 to 2 * 901775360 bytes, LSN
= 1457612031
1 [Note] [MY - 013084] [InnoDB] Log background threads are being closed...
1 [Note] [MY - 012968] [InnoDB] Starting to delete and rewrite log files.
1 [Note] [MY - 013575] [InnoDB] Creating log file ./ib_logfile101
1 [Note] [MY - 013575] [InnoDB] Creating log file ./ib_logfile1
1 [Note] [MY - 012892] [InnoDB] Renaming log file ./ib_logfile101 to ./ib_logfile0
1 [Note] [MY - 012893] [InnoDB] New log files created, LSN = 1457612300
1 [Note] [MY - 013083] [InnoDB] Log background threads are being started...
```

16.4.3　innodb_dedicated_server 参数

MySQL 8.0 中新引进了一个参数 innodb_dedicated_server,该参数的默认值是 OFF,就像这个参数名所建议的一样,当 MySQL 独占当前服务器资源时,可以把该参数设置为 on,这时 MySQL 会自动探测当前服务器的内存大小并设置如下 4 个参数:

(1) innodb_buffer_pool_size。

(2) innodb_log_file_size。

(3) innodb_log_files_in_group。

(4) innodb_flush_method。

其中，前面3个参数是根据当前服务器的内存大小计算出来的，这样对运维在虚拟机或云上运行的MySQL很方便，当调整了内存的大小后，MySQL会在启动时自动调整这3个参数，省去了每次手工修改参数的工作。

innodb_buffer_pool_size根据物理内存的设置策略如表16.1所示。

表 16.1

内存大小	innodb_buffer_pool_size 的值
小于1GB	128MB
1～4GB	物理内存×0.5
大于4GB	物理内存×0.75

innodb_log_file_size 和 innodb_log_files_in_group 两个参数是根据 innodb_buffer_pool_size 计算出来的，具体策略如表16.2和表16.3所示。

表 16.2

innodb_buffer_pool_size	innodb_log_file_size
小于8GB	512MB
8～16GB	1024MB
大于16GB	2GB

表 16.3

innodb_buffer_pool_size	innodb_log_files_in_group
小于8GB	以GB为单位对 innodb_buffer_pool_size 取整
8～128GB	以GB为单位对(innodb_buffer_pool_size * 0.75)取整
大于128GB	64

innodb_flush_method 的设置策略是：如果操作系统允许设置为 O_DIRECT_NO_FSYNC，否则设置为默认值。

当参数 innodb_dedicated_server 为 ON 时，如果还显式设置了这些参数，则显式设置的这些参数会优先生效，并且在 MySQL 的错误日志中会记录如下内容：

```
0 [Warning] [MY-012360] [InnoDB]  Option innodb_dedicated_server is ignored for innodb_log_
file_size because innodb_log_file_size = 2073034752 is specified explicitly.
```

显式指定某一个值，并不会影响另外3个参数值的自动设定。

当参数 innodb_dedicated_server 为 ON 时，MySQL 每次启动时会自动探测服务器的内存并自动调整上述几个参数值。在任何时候，MySQL 都不会将自适应值保存在持久配置中，利用该参数就可以保证服务器（包括虚拟机或容器）扩展以后，MySQL 能"自动适应"，以尽量利用更多的服务器资源。

16.5 硬盘读写参数

硬盘的读写速度通常是对数据库性能影响最大的因素之一。这里介绍几个影响硬盘读

写速度的重要参数。

16.5.1 innodb_flush_log_trx_commit

innodb_flush_log_trx_commit 参数控制事务提交时写重做日志的行为方式，它有三个值，分别为 0、1 和 2。

(1) 默认值为 1，每次事务提交时都会将日志缓存中的数据写入日志文件，同时还会触发文件系统到磁盘的同步，如果发生系统崩溃，数据是零丢失，这种方式对数据是最安全的，但性能是最慢的，因为把数据从缓存同步到磁盘的成本很高。这种方式适用于对数据安全性要求高的行业，如银行业。但很多互联网的应用，对数据的安全性要求不太高，而对性能的要求很高，设置成 0 或 2 会更合适。

(2) 设置成 0 时，事务提交时不会触发写日志文件的操作，日志缓存中的数据以每秒一次的频率写入日志文件中，同时还会进行文件系统到磁盘的同步操作。

(3) 设置成 2 时，事务提交时会写日志文件，但文件系统到磁盘的同步是每秒进行一次。

0 和 2 都是每秒进行一次文件系统到磁盘的同步，因此这两种方式的性能差不多，当系统崩溃时，最多丢失 1 秒的数据。但 0 和 2 还有细微的不同，当设置成 2 时，每次事务提交都写日志文件，因此数据已经从 MySQL 的日志缓存刷新到了操作系统的文件缓存，如果只是 MySQL 崩溃，而操作系统没有崩溃，将不会丢失数据。因此 0 和 2 比较起来，通常设置为 2 比较好。

16.5.2 sync_binlog

sync_binlog 参数控制事务提交时写二进制日志的行为方式，它有三个值，分别为 0、1 和 N。

(1) 默认值为 1，每次事务提交时都会把二进制日志刷新到磁盘，这种方式对数据是最安全的，但性能是最慢的。

(2) 设置成 0 时，事务提交时不会把二进制日志刷新到磁盘，刷磁盘的动作由操作系统控制。

(3) 设置成 N(N 不等于 0 或 1)时，每进行 N 个事务提交后会进行一次把二进制日志刷新到磁盘的动作。

没有备库和使用二进制日志进行时间点恢复的需求时，可以把 sync_binlog 参数设置为 0 或 N，设置为 0 是把刷新二进制日志文件的操作交给操作系统决定，但操作系统可能会在二进制日志文件写满进行切换时才刷新磁盘文件，这样会造成数秒的延迟，在这期间事务无法提交，因此把该参数设置成 100 或 1000 之类的一个合理数值比设置成 0 好。如果使用二进制日志进行主库和备库之间的数据同步，或者使用二进制日志进行时间点恢复，并且对数据一致性要求高时，把 sync_binlog 参数设置为 1，同时要把 innodb_flush_log_trx_commit 参数也设置为 1。把这两个参数都设置成 1 对性能的负面影响很大，为了提高性能，这时使用的存储应该是带缓存的，并且设置成 Write-back，而不是 Write-through，这样数据只写入存储的缓存中即返回。但存储的缓存应该是带电池的,如果缓存不带电池,或者电池没有电,突然发生掉电时,不仅数据会丢失,而且会造成数据库损坏,无法启动,这种情况要比丢失一秒钟的数据要糟糕得多。

写二进制日志的成本比写重做日志的成本要高得多,因为重做日志的大小和文件名是固定的,重做日志循环写入日志文件。而每次写二进制日志时,文件都会进行扩展,如果写满了还要新建文件,这样每次写二进制日志不但要写数据,还要修改二进制日志文件的元数据,因此把 sync_binlog 设置成 1 比把 innodb_flush_log_trx_commit 设置成 1 对性能的负面影响还要大得多。

16.5.3 innodb_flush_method

innodb_flush_method 参数控制 MySQL 将数据刷到 InnoDB 的数据文件和日志文件的动作。在 Windows 系统上有两个选项,其一是 unbuffered,它是默认和推荐的选项;另外一个是 normal。Linux 系统上,常用的选项有以下几种:

- fsync:是默认值,使用 fsync()系统调用刷新数据文件和日志文件,数据会在操作系统的缓存中保存。
- O_DSYNC:InnoDB 使用 O_SYNC 打开和刷新日志文件,使用 fsync()刷新数据文件。
- O_DIRECT:使用 O_DIRECT 打开数据文件,使用 fsync()系统调用刷新数据文件和日志文件,数据不会在操作系统的缓存中保存。
- O_DIRECT_NO_FSYNC:使用 O_DIRECT 刷新 I/O,但写磁盘时不执行 fsync()。

通常,硬盘性能好的服务器可以设置成 O_DIRECT,这样避免在 InnoDB 缓存和操作系统缓存中存有两份数据,而且 InnoDB 缓存比操作系统缓存效率要高,因为 InnoDB 缓存是专门为 InnoDB 的数据设计的,而操作系统缓存是为通用的数据设计的。设置成 O_DIRECT_NO_FSYNC 时,因为写磁盘时不执行 fsync(),速度可能会快,但突然断电时可能会丢失数据。对于读操作大大多于写操作的应用,设置成 fsync 会比设置成 O_DIRECT 性能略好。但如何选择这些参数最终需要经过测试才能确定,测试时要注意观察状态参数 Innodb_data_fsyncs,它记录着调用 fsync()的次数。通常 fsync 和 O_DIRECT 调用 fsync()的次数差不多,O_DIRECT_NO_FSYNC 调用 fsync()的次数最少。

16.5.4 innodb_io_capacity 和 innodb_io_capacity_max

InnoDB 后台线程会进行一些 I/O 操作,例如,把缓冲池中的脏页刷新到磁盘,或将更改从更改缓冲区写入对应的二级索引。InnoDB 试图以不影响服务器正常工作的方式执行这些 I/O 操作,这需要它知道系统的 I/O 的处理能力,它根据参数 innodb_io_capacity 评估系统的 I/O 带宽。参数 innodb_io_capacity_max 值定义了系统 I/O 能力的上限,防止在 I/O 的峰值时消耗服务器的全部 I/O 带宽。

通常可以把 innodb_io_capacity 设置得低一些,但不要低到后台 I/O 滞后的程度。如果该值太高,数据将很快从缓冲池中被移除,不能充分发挥缓存的优势。但对于繁忙而且具有较高 I/O 处理能力的系统,可以设置一个较高的值来帮助服务器处理与数据快速变更相关联的后台维护工作。

innodb_io_capacity 和 innodb_io_capacity_max 两个参数的默认值如下:

```
mysql > show variables like 'innodb_io_capacity%';
+------------------------+-------+
| Variable_name          | Value |
+------------------------+-------+
| innodb_io_capacity     | 200   |
| innodb_io_capacity_max | 2000  |
+------------------------+-------+
```

如上两个参数的设定是基于系统每秒能处理的 I/O 数量(IOPS),可以把 innodb_io_capacity_max 设置成极限的 IOPS,innodb_io_capacity 设置成 innodb_io_capacity_max 的一半左右。目前业界有很多 I/O 测试软件可以测出系统的 IOPS,也可以通过硬盘配置进行估算,例如,一块 15K 转速的传统硬盘的 IOPS 的参考值大约是 200,高端固态硬盘可以达到 60 万。

状态参数 Innodb_data_fsyncs 记录着数据刷新到磁盘的次数,把 innodb_io_capacity 调大后,可以看到该状态参数也相应增加了。

16.5.5 其他参数

系统参数 max_connections 设置了允许的服务器最多连接数,防止服务器因为连接数过多而造成资源耗尽,默认是 151,该设置值在生产环境中通常偏小。该参数应当设置为经过压力测试验证后系统能承受的最多连接数。可以参考状态参数 Max_used_connections 和 Max_used_connections_time,它们记录了系统连接数曾经达到的最大值和发生时间。

系统参数 skip_name_resolve 默认为 off,这时 MySQL 每收到一个连接请求,都会进行正向和反向 DNS 解析,建议设置成 on,禁止域名解析,这样会加快客户端连接到 MySQL 服务器的速度。当 DNS 服务器运行正常时,这个优势并不明显,如果 DNS 服务器出故障,或者变慢,进行域名解析的时间可能会很长,甚至会拒绝连接。如果解析不成功,在错误日志里面会有类似下面的提示:

```
40162 [Warning] [MY-010055] [Server] IP address '192.168.87.178' could not be resolved: Name
or service not known
```

把参数 skip_name_resolve 设置成 on 也有弊端,就是只能使用 IP 进行 grant 赋权,不能使用主机名,通常主机名不会变,而 IP 改变的可能性比主机名大。因此在一个生产主机上把 skip_name_resolve 从 off 改成 on 要小心,因为原来用主机名赋予的权限不能用了。

系统参数 binlog_order_commits 默认为 on,如果把该参数设置为 off 将不能保证事务的提交顺序和写入二进制日志的顺序一致,这不会影响数据的一致性,在高并发场景下还能提升一定的吞吐量。

16.6 资源组

MySQL 8.0 中引入了资源组(resource groups)的概念,它可以设定某一类 SQL 语句所允许使用的资源(目前只包括 CPU)。在高并发的系统中,资源组可以保证关键交易的性能,例如,可以设定市场统计类的交易在白天使用较少的资源,以免影响客户的交易,在晚上可以使用较多的资源。

16.6.1 查询资源组

在 information_schema. resource_groups 视图中可以查询资源中的信息,默认有如下两个资源组:

```
mysql> select * from information_schema.resource_groups\G
*************************** 1. row ***************************
   RESOURCE_GROUP_NAME: USR_default
   RESOURCE_GROUP_TYPE: USER
RESOURCE_GROUP_ENABLED: 1
             VCPU_IDS: 0 - 3
      THREAD_PRIORITY: 0
*************************** 2. row ***************************
   RESOURCE_GROUP_NAME: SYS_default
   RESOURCE_GROUP_TYPE: SYSTEM
RESOURCE_GROUP_ENABLED: 1
             VCPU_IDS: 0 - 3
      THREAD_PRIORITY: 0
2 rows in set (0.03 sec)
```

information_schema. resource_groups 视图中的字段如下。

- RESOURCE_GROUP_NAME:资源组名称,不区分大小写,最长可达 64 个字符。
- RESOURCE_GROUP_TYPE:资源组类型,每个组都有一个类型,可以是系统(SYSTEM)或用户(USER)。系统资源组允许的优先级范围是－20～0;用户资源组允许的优先级范围是 0～19。需确保用户线程的优先级不会高于系统线程。
- RESOURCE_GROUP_ENABLED:资源组是否激活。
- VCPU_IDS:资源组可以使用的一组虚拟 CPU,虚拟 CPU 的数量由物理 CPU、核数、超线程等决定。
- THREAD_PRIORITY:分配给资源组线程的执行优先级。数值越大,优先级越低。优先级值的范围为－20(最高优先级)～19(最低优先级)。系统组和用户组的默认优先级均为 0。

information_schema. resoure_groups 视图中的每个字段对应资源组的一个属性,前面两个属性在资源组创建后不能修改,其他属性在资源组创建后可以修改。

16.6.2 管理资源组

使用 create resource group 语句可以创建资源组,创建一个 Batch 用户资源组的例子如下:

```
mysql> create resource group Batch  type = user  vcpu = 2 - 3  thread_priority = 10;
```

创建完成后可查看该batch用户资源组的信息，相关命令及运行结果如下：

```
mysql> select * from information_schema.resource_groups where resource_group_name = 'Batch'
\G
*************************** 1. row ***************************
   RESOURCE_GROUP_NAME: Batch
   RESOURCE_GROUP_TYPE: USER
RESOURCE_GROUP_ENABLED: 1
              VCPU_IDS: 2-3
       THREAD_PRIORITY: 10
```

使用alter resource group语句可以修改资源组的属性，例如，在系统高负载的时间段减少分配给资源组的CPU数量，并降低其优先级，命令如下：

```
mysql> alter resource group Batch  vcpu = 3 thread_priority = 19;
```

在系统负载较轻的情况下，增加分配给组的CPU数量，并提高其优先级，命令如下：

```
mysql> alter resource group Batch  vcpu = 0-3  thread_priority = 0;
```

注意，用户线程的优先级不能小于0，命令如下：

```
mysql> alter resource group Batch  vcpu = 3  thread_priority = -9;
ERROR 3654 (HY000): Invalid thread priority value -9 for User resource group Batch. Allowed
range is [0, 19].
```

激活Batch资源组的命令如下：

```
mysql> alter resource group Batch enable;
```

删除Batch资源组的命令如下：

```
mysql> drop resource group Batch;
```

要将线程分配给Batch资源组，执行以下命令：

```
mysql> set resource group Batch for thread_id;
```

当thread_id有多个时，中间用逗号隔开。

如果要把当前线程设定到Batch资源组中，在会话中执行以下命令：

```
mysql> set resource group batch;
```

此后，会话中的语句将使用Batch资源组的资源进行执行 。

要使用Batch组执行单个语句，需使用resource_group优化程序提示，命令如下：

```
mysql> insert /*+ resource_group(Batch) */ into t2 values(2);
```

在SQL语句里设置提示的方法可以和MySQL的中间件结合起来使用，例如，ProxySQL支持在SQL语句中增加提示。

可以在performance_schema.threads视图中的resource_group字段查询线程使用的资源组，相应的命令和输出结果如下：

```
mysql> select thread_id, resource_group from performance_schema.threads where thread_id
```

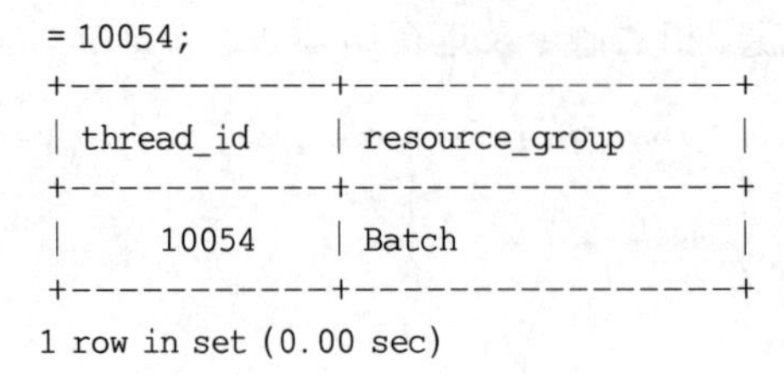

```
= 10054;
+-----------+----------------+
| thread_id | resource_group |
+-----------+----------------+
|     10054 | Batch          |
+-----------+----------------+
1 row in set (0.00 sec)
```

16.6.3 资源组的限制

目前，资源组在使用中还是有如下一些限制：

(1) 如果安装了线程池插件，则资源组不可用。

(2) 资源组在 macOS 系统中不可用，因为它不提供用于将 CPU 绑定到线程的 API。

(3) 在 FreeBSD 和 Solaris 上，忽略资源组线程优先级，尝试更改优先级会导致警告。实际上，所有线程都以优先级 0 运行。

(4) 在 Linux 系统中，需要对 mysqld 进程设置 CAP_SYS_NICE 功能，否则将忽略资源组线程优先级。CAP_SYS_NICE 可以使用 setcap 命令手动设置该功能，使用 getcap 检查功能。相应命令和输出结果如下：

```
# setcap cap_sys_nice+ep /usr/sbin/mysqld
# getcap /usr/sbin/mysqld
/usr/sbin/mysqld = cap_sys_nice+ep
```

或者使用 sudo systemctl edit mysql 在 MySQL 服务里增加如下内容：

```
[Service] AmbientCapabilities = CAP_SYS_NICE
```

然后重新启动 MySQL 服务，设置线程优先级才能生效。

(5) 在 Windows 平台上的线程优先级不用数字表示，线程有五个优先级。资源组线程优先级的范围为－20～19，其与这五个优先级的对应关系如表 16.4 所示。

表 16.4

线程优先级范围	Windows 优先级
－20～－10	THREAD_PRIORITY_HIGHEST
－9～－1	THREAD_PRIORITY_ABOVE_NORMAL
0	THREAD_PRIORITY_NORMAL
1～10	THREAD_PRIORITY_BELOW_NORMAL
10～19	THREAD_PRIORITY_LOWEST

16.7 实验

视频演示

查询日志产生量

(1) 使用 information_schema.innodb_metrics 视图找出日志产生量。

(2) 使用 show engine innodb statu 命令查询日志产生量。

第17章

SQL优化基础

本章介绍 SQL 优化的基础知识，包括 SQL 语句的执行计划、优化器、索引、表连接、统计信息、直方图和 CTE(common table express，通用表表达式)。

17.1 SQL 语句的执行计划

SQL 语句只是告诉了数据库要做什么，并没有告诉数据库如何做，查看 SQL 语句的执行计划可以使 SQL 的执行过程从“黑盒”变成“白盒”。

在 SQL 语句前面加上 EXPLAIN 即可查看 SQL 语句的执行计划，但不会实际执行这个 SQL 语句，显示 SQL 语句执行计划的格式有三种，分别是传统(TRADITIONAL)、JSON 和树形(TREE)格式。可以使用 FORMAT＝TRADITIONAL|JSON|TREE 来指定格式，默认是传统格式。

17.1.1 传统格式

传统格式提供了执行计划的概况、索引的使用等基本信息，如下是一个查询 SQL 语句执行计划的例子：

```
mysql > explain select city from city where country_id = (select country_id from country where
country = 'China')  \G
**************************** 1. row ****************************
           id: 1
  select_type: PRIMARY
        table: city
   partitions: NULL
         type: ref
possible_keys: idx_fk_country_id
```

```
          key: idx_fk_country_id
      key_len: 2
          ref: const
         rows: 53
     filtered: 100.00
        Extra: Using where
*************************** 2. row ***************************
           id: 2
  select_type: SUBQUERY
        table: country
   partitions: NULL
         type: ALL
possible_keys: NULL
          key: NULL
      key_len: NULL
          ref: NULL
         rows: 109
     filtered: 10.00
        Extra: Using where
2 rows in set, 1 warning (0.00 sec)
```

从如上执行计划里可以看到，id为2的步骤是id为1的步骤的子查询，以全表扫描的方式访问country表，找到符合country='China'条件的记录的country_id，以该country_id通过索引idx_fk_country_id访问city表。执行计划里字段的含义大部分可以从字面上推测出来，简单说明如下。

- id：一个数字标识符，显示表或子查询属于查询的哪一部分。
- select_type：每个查询块的类型。PRIMARY代表最外层的查询块；SUBQUERY代表select语句的子查询块。
- type：连接类型。ALL是全表扫描，这是成本最高的访问方式；ref是使用非唯一索引访问；const是使用主键或唯一索引访问。
- possible_keys：执行计划考虑使用的索引，如果为NULL表示不考虑使用索引。
- key：执行计划实际用到的索引，如果为NULL表示不使用索引。
- ref：索引过滤的字段，const代表常量。
- rows：估算的找到所需记录所需要读取的行数。
- filtered：返回结果的行占需要读到行(rows字段)的百分比。
- Extra：其他字段中不包含的额外说明都放在该字段。

关于执行计划的详细说明参见https://dev.mysql.com/doc/refman/8.0/en/execution-plan-information.html。

EXPLAIN语句执行完成后，会提示有一个警告信息，警告信息里包括的是优化器重写的伪SQL，该SQL语句不一定是能执行的，执行如下命令将显示前面SQL执行计划的警告信息：

```
mysql> show warnings\G
*************************** 1. row ***************************
  Level: Note
```

```
  Code: 1003
Message:/* select#1 */ select 'sakila'.'city'.'city' AS 'city' from 'sakila'.'city' where
('sakila'.'city'.'country_id' = (/* select#2 */ select 'sakila'.'country'.'country_id' from
'sakila'.'country' where ('sakila'.'country'.'country' = 'China')))
1 row in set (0.00 sec)
```

可以使用\W在MySQL客户端里打开警告信息的显示，如果使用MySQL Shell，这里的警告信息会直接显示。

17.1.2 JSON格式

JSON格式提供了以JSON格式显示的详细执行计划，该格式适合被程序调用，例如，图形工具Workbench显示的图形化的执行计划就是调用了JSON格式的接口。如下是前面SQL语句的JSON格式的执行计划：

```
mysql> explain format = json select city from city where country_id = (select country_id from
country where country = 'China') \G
*************************** 1. row ***************************
EXPLAIN: {
  "query_block": {
    "select_id": 1,
    "cost_info": {
      "query_cost": "7.55"
    },
    "table": {
      "table_name": "city",
      "access_type": "ref",
      "possible_keys": [
        "idx_fk_country_id"
      ],
      "key": "idx_fk_country_id",
      "used_key_parts": [
        "country_id"
      ],
      "key_length": "2",
      "ref": [
        "const"
      ],
      "rows_examined_per_scan": 53,
      "rows_produced_per_join": 53,
      "filtered": "100.00",
      "cost_info": {
        "read_cost": "2.25",
        "eval_cost": "5.30",
        "prefix_cost": "7.55",
        "data_read_per_join": "11K"
      },
      "used_columns": [
        "city",
```

```
          "country_id"
        ],
        "attached_condition": "('sakila'.'city'.'country_id' = (/* select#2 */ select 'sakila'.
'country'.'country_id' from 'sakila'.'country' where ('sakila'.'country'.'country' = 'China')))",
        "attached_subqueries": [
          {
            "dependent": false,
            "cacheable": true,
            "query_block": {
              "select_id": 2,
              "cost_info": {
                "query_cost": "11.15"
              },
              "table": {
                "table_name": "country",
                "access_type": "ALL",
                "rows_examined_per_scan": 109,
                "rows_produced_per_join": 10,
                "filtered": "10.00",
                "cost_info": {
                  "read_cost": "10.06",
                  "eval_cost": "1.09",
                  "prefix_cost": "11.15",
                  "data_read_per_join": "2K"
                },
                "used_columns": [
                  "country_id",
                  "country"
                ],
                "attached_condition": "('sakila'.'country'.'country' = 'China')"
              }
            }
          }
        ]
      }
    }
}
1 row in set, 1 warning (0.00 sec)
```

在 JSON 格式中 cost_info 元素提供了估算的执行成本。

17.1.3 图形方式

MySQL 的图形工具 MySQL Workbench 可以以图形方式显示 JSON 格式的执行计划。在 MySQL Workbench 里显示执行计划有如下两种方法：

(1) 在 SQL 语句没有执行时，单击 SQL 语句输入框上方的有个闪电上面带放大镜的图标就会生产该 SQL 语句的执行计划。这种方式适合于 SQL 语句执行时间长，或者 SQL 语句将修改数据的情况下使用。

(2) 执行 SQL 语句后，单击输出结果的右侧的 Execution Plan 按钮也会以图形方式显示 SQL 语句的执行计划。

例如，前面的 SQL 语句的执行计划如图 17.1 所示。

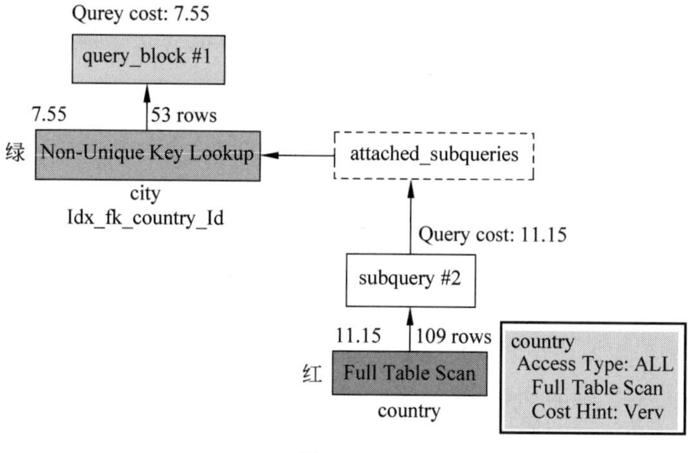

图 17.1

图 17.1 中的每个方框代表一个执行步骤，方框上方左边 Query cost 后的数字表示估算的执行成本，右边的数字代表估算的输出行数，下边是表名或索引名，其中索引名用加粗的方式显示。方框的颜色与其相应的执行成本相关，从低到高依次是蓝、绿、黄、橘、红。将光标停留在方框上还可以显示更加详细的信息，如图 17.1 中方框所示。

17.1.4 树形格式

树形格式是从 MySQL 8.0.18 开始引入的，它提供的执行计划比传统的执行计划更详细，输出格式是树形的，例如：

```
mysql> explain format = tree select city from city where country_id = (select country_id from
country where country = 'China') \G
*************************** 1. row ***************************
EXPLAIN: -> Filter: (city.country_id = (select #2))  (cost = 7.55 rows = 53)
    -> Index lookup on city using idx_fk_country_id (country_id = (select #2))  (cost = 7.55
rows = 53)
    -> Select #2 (subquery in condition; run only once)
        -> Filter: (country.country = 'China')  (cost = 11.15 rows = 11)
            -> Table scan on country  (cost = 11.15 rows = 109)

1 row in set (0.00 sec)
```

17.1.5 EXPLAIN ANALYZE

EXPLAIN ANALYZE 实际上是树形执行计划的扩展，它不但提供了执行计划，还检测并执行了 SQL 语句，提供了执行过程中的实际度量，例如：

```
mysql > EXPLAIN ANALYZE select city from city where country_id = (select country_id from country
where country = 'China') \G
**************************** 1. row ****************************
EXPLAIN: -> Filter: (city.country_id = (select #2))  (cost = 7.55 rows = 53) (actual time =
23.970..23.989 rows = 53 loops = 1)
    -> Index lookup on city using idx_fk_country_id (country_id = (select #2))  (cost = 7.55
rows = 53) (actual time = 23.967..23.981 rows = 53 loops = 1)
    -> Select #2 (subquery in condition; run only once)
        -> Filter: (country.country = 'China')  (cost = 11.15 rows = 11) (actual time =
0.450..0.515 rows = 1 loops = 1)
            -> Table scan on country  (cost = 11.15 rows = 109) (actual time = 0.405..0.467
rows = 109 loops = 1)

1 row in set (0.03 sec)
```

执行计划增加了如下 4 个实际度量：

- 获取第一行的实际时间（以毫秒为单位），是 actual time 中的第一个时间。
- 获取所有行的实际时间（以毫秒为单位），是 actual time 中的第二个时间。
- 实际读取的行数，是 actual time 后面的 rows。
- 实际循环数，是 actual time 后面的 loops。

从执行计划中可以看到，对 country 表进行过滤时，估计有 11 行，实际却只有 1 行，而其他的估计行数和实际行数是一致的。估计行数和实际行数不一致的地方往往是执行计划选择错误路径的地方，通常也是可以优化的地方。

17.1.6 EXPLAIN FOR CONNECTION

在实际工作中，如果发现一个正在执行的 SQL 语句耗时很长，这时想查询它的执行计划，通常的做法是使用 EXPLAIN 生成该 SQL 语句的执行计划，但因为统计信息等原因，生成的执行计划和正在执行的执行计划可能不完全相同，更好的做法是使用 EXPLAIN FOR CONNECTION 查询当前正在使用的执行计划。例如，如下的 SQL 语句查询出了当前的会话号：

```
mysql > select connection_id();
+-----------------+
| connection_id() |
+-----------------+
|              17 |
+-----------------+
1 row in set (0.00 sec)
```

在当前会话中执行一个慢 SQL 语句，如下所示：

```
mysql > select sleep(60), city from city where country_id = (select country_id from country
where country = 'China') \G
```

根据会话号在其他会话里查询正在执行的 SQL 语句的执行计划，相关命令及其输出结要如下：

```
mysql > EXPLAIN FOR CONNECTION  17\G
*************************** 1. row ***************************
           id: 1
  select_type: PRIMARY
        table: city
   partitions: NULL
         type: ref
possible_keys: idx_fk_country_id
          key: idx_fk_country_id
      key_len: 2
          ref: const
         rows: 53
     filtered: 100.00
        Extra: Using where
*************************** 2. row ***************************
           id: 2
  select_type: SUBQUERY
        table: country
   partitions: NULL
         type: ALL
possible_keys: NULL
          key: NULL
      key_len: NULL
          ref: NULL
         rows: 109
     filtered: 10.00
        Extra: Using where
2 rows in set (0.00 sec)
```

17.2 优化器

MySQL 的优化器负责生成 SQL 语句的执行计划。

17.2.1 优化器的开关

系统变量 optimizer_switch 可以控制优化器的行为，它的值是一组标志，每个标志有 on 或者 off 两个值，以指示相应的优化器行为是否被启用或禁用。查看当前优化器值的相关命令及其运行结果如下：

```
mysql >  select @@optimizer_switch\G
*************************** 1. row ***************************
@@optimizer_switch: index_merge = on, index_merge_union = on, index_merge_sort_union = on,
index_merge_intersection = on, engine_condition_pushdown = on, index_condition_pushdown = on,
mrr = on, mrr_cost_based = on, block_nested_loop = on, batched_key_access = off, materialization
 = on, semijoin = on, loosescan = on, firstmatch = on, duplicateweedout = on, subquery_
materialization_cost_based = on, use_index_extensions = on, condition_fanout_filter = on,
derived_merge = on, use_invisible_indexes = off, skip_scan = on, hash_join = on, subquery_to_
```

```
derived=off,prefer_ordering_index=on,hypergraph_optimizer=off,derived_condition_
pushdown=on
1 row in set (0.00 sec)
```

改变优化器的开关可以控制优化器的行为，例如，如下 SQL 语句是取两个索引的交集进行数据访问：

```
mysql> explain select * from actor where first_name='NICK' and last_name='WAHLBERG'\G
*************************** 1. row ***************************
           id: 1
  select_type: SIMPLE
        table: actor
   partitions: NULL
         type: index_merge
possible_keys: idx_actor_last_name,idx_actor_first_name
          key: idx_actor_last_name,idx_actor_first_name
      key_len: 182,182
          ref: NULL
         rows: 1
     filtered: 100.00
        Extra: Using intersect(idx_actor_last_name,idx_actor_first_name); Using where
1 row in set, 1 warning (0.01 sec)
```

修改 optimizer_switch 禁止使用 index_merge_intersection 的命令如下：

```
mysql> SET optimizer_switch='index_merge_intersection=off';
```

然后执行同样的 SQL 语句，此时它的执行计划就改变了，不再使用两个索引的交集进行数据访问，如下所示：

```
mysql> explain select * from actor where first_name='NICK' and last_name='WAHLBERG'\G
*************************** 1. row ***************************
           id: 1
  select_type: SIMPLE
        table: actor
   partitions: NULL
         type: ref
possible_keys: idx_actor_last_name,idx_actor_first_name
          key: idx_actor_last_name
      key_len: 182
          ref: const
         rows: 2
     filtered: 2.50
        Extra: Using where
1 row in set, 1 warning (0.00 sec)
```

17.2.2 提示

优化器的开关会对全局或当前会话的所有 SQL 语句起作用，而提示(hint)只用于控制单个 SQL 语句的执行计划。同时使用时，提示的优先级高于优化器开关。优化器分为两

类，一类是优化器提示，用于控制优化器的行为；另一类是索引提示，用于控制索引的使用。

1. 优化器提示

优化器提示可以在单个 SQL 语句中实现系统参数 optimizer_switch 的功能，例如，索引交集的功能可以使用如下的优化器提示实现：

```
mysql > explain  select /* + NO_INDEX_MERGE(actor) */  * from actor where first_name =
'NICK' and last_name = 'WAHLBERG'\G
*************************** 1. row ***************************
           id: 1
  select_type: SIMPLE
        table: actor
   partitions: NULL
         type: ref
possible_keys: idx_actor_last_name,idx_actor_first_name
          key: idx_actor_last_name
      key_len: 182
          ref: const
         rows: 2
     filtered: 2.50
        Extra: Using where
```

可以临时在一个 SQL 语句中设置系统变量的值，例如：

```
mysql > select /* + SET_VAR(sort_buffer_size = 16M) */ * from rental order by 1,2,3,4;
mysql > insert /* + SET_VAR(foreign_key_checks = OFF) */ into tba values('aaa');
```

如下的例子是通过优化器提示临时停止二级索引的唯一性检查，首先检查当前会话中的参数 unique_checks 的值：

```
mysql > select @@unique_checks;
+-----------------+
| @@unique_checks |
+-----------------+
|               1 |
+-----------------+
1 row in set (0.00 sec)
```

可以发现，当前会话中的参数 unique_checks 的值是 1，在 SQL 语句中使用优化器提示将 unique_checks 的值设置为 0，相关命令及其运行结果如下：

```
mysql > SELECT /* + SET_VAR(unique_checks = OFF) */ @@unique_checks;
+-----------------+
| @@unique_checks |
+-----------------+
|               0 |
+-----------------+
1 row in set (0.00 sec)
```

再次检查当前会话中的参数 unique_checks 的值，相关命令及其运行结果如下：

```
mysql > SELECT @@unique_checks;
```

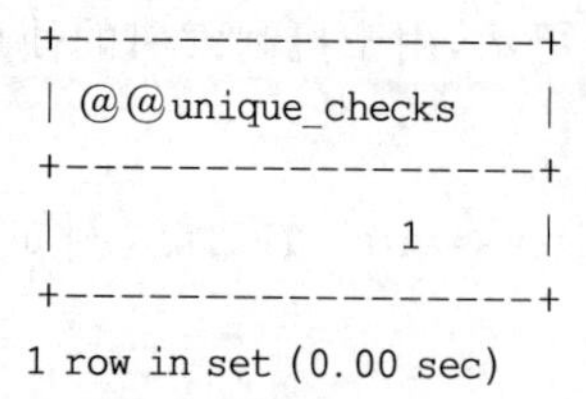

```
+-----------------+
| @@unique_checks |
+-----------------+
|               1 |
+-----------------+
1 row in set (0.00 sec)
```

可以看出，只是在优化器提示起作用的SQL语句中停止了二级索引的唯一性检查，之前和之后都没有影响。

关于优化器提示的详细信息参见 https://dev.mysql.com/doc/refman/8.0/en/optimizer-hints.html。

2. 索引提示

索引提示控制优化器对索引的使用，常用类型如下。

- USE INDEX：使用指定索引中的一个。
- USE FORCE INDEX：和USE INDEX类似，而且尽量避免全表扫描。
- IGNORE INDEX：不使用指定的索引。

索引提示的使用语法和优化器索引不同，它们直接放在指定的表名后面。

如下SQL语句是在没有使用索引提示时，优化器从两个索引中选择了rental_date：

```
mysql > explain select inventory_id from rental where rental_date between  '2005 - 05 - 27'
     AND  '2005 - 05 - 28' AND customer_id IN (433, 274, 319, 909)\G
*************************** 1. row ***************************
           id: 1
  select_type: SIMPLE
        table: rental
   partitions: NULL
         type: range
possible_keys: rental_date, idx_fk_customer_id
          key: rental_date
      key_len: 5
          ref: NULL
         rows: 166
     filtered: 0.57
        Extra: Using where; Using index
```

当可以确定使用另外一个索引效率更高时，可以使用USE INDEX指定使用另外一个索引，如下所示：

```
mysql > explain select inventory_id from rental use index(idx_fk_customer_id)
     where rental_date between  '2005 - 05 - 27' and  '2005 - 05 - 28'
     and customer_id IN (433, 274, 319, 909)\G
*************************** 1. row ***************************
           id: 1
  select_type: SIMPLE
        table: rental
   partitions: NULL
         type: range
possible_keys: idx_fk_customer_id
```

```
          key: idx_fk_customer_id
      key_len: 2
          ref: NULL
         rows: 91
     filtered: 11.11
        Extra: Using index condition; Using where
```

也可以使用 IGNORE INDEX 强制不使用 rental_date 索引，如下所示：

```
mysql > explain select inventory_id from rental ignore index(rental_date)
     where rental_date between  '2005 - 05 - 27' and  '2005 - 05 - 28'
     and customer_id IN (433, 274, 319, 909)\G
*************************** 1. row ***************************
           id: 1
  select_type: SIMPLE
        table: rental
   partitions: NULL
         type: range
possible_keys: idx_fk_customer_id
          key: idx_fk_customer_id
      key_len: 2
          ref: NULL
         rows: 91
     filtered: 11.11
        Extra: Using index condition; Using where
```

关于索引提示的详细信息参见 https://dev.mysql.com/doc/refman/8.0/en/index-hints.html。

17.2.3 成本计算

为了计算执行计划的成本，优化器使用了一个成本计算模型进行成本计算。MySQL 把各种操作的估计成本放在如下两个 MySQL 数据库的表中：

(1) engine_cost 表用于保存与特定引擎相关的操作成本，不同的引擎执行这些操作的成本不同。

(2) server_cost 表用于保存与服务相关的操作成本，不同的服务器执行这些操作的成本不同，但与引擎没有关系。

检查如上两个表的默认值如下：

```
mysql > select engine_name,cost_name,cost_value,default_value  from mysql.engine_cost;
+-------------+------------------------+------------+---------------+
| engine_name | cost_name              | cost_value | default_value |
+-------------+------------------------+------------+---------------+
| default     | io_block_read_cost     |       NULL |             1 |
| default     | memory_block_read_cost |       NULL |          0.25 |
+-------------+------------------------+------------+---------------+
2 rows in set (0.00 sec)
```

```
mysql > select cost_name,cost_value,default_value  from mysql.server_cost;
+------------------------------+------------+---------------+
| cost_name                    | cost_value | default_value |
+------------------------------+------------+---------------+
| disk_temptable_create_cost   |       NULL |            20 |
| disk_temptable_row_cost      |       NULL |           0.5 |
| key_compare_cost             |       NULL |          0.05 |
| memory_temptable_create_cost |       NULL |             1 |
| memory_temptable_row_cost    |       NULL |           0.1 |
| row_evaluate_cost            |       NULL |           0.1 |
+------------------------------+------------+---------------+
6 rows in set (0.00 sec)
```

mysql.engine_cost 表中有如下两种操作的估计成本。

- memory_block_read_cost：代表从内存中读一个块的成本，默认值是 0.25。
- io_block_read_cost：从硬盘中读一个块的成本，默认值是 1。

mysql.server_cost 表中记录了如下操作的成本。

- disk_temptable_create_cost：创建磁盘上临时表的成本，增大该值会让优化器尽量少地创建磁盘上的临时表。
- disk_temptable_row_cost：向磁盘上的临时表写入或读取一条记录的成本，增大该值会让优化器尽量少地创建磁盘上的临时表。
- key_compare_cost：进行记录比较的成本，多用在排序操作上，增大该值会提升文件排序(filesort)的成本，让优化器可能更倾向于使用索引完成排序而不是文件排序。
- memory_temptable_create_cost：创建内存中临时表的成本，增大该值会让优化器尽量少地创建内存的临时表。
- memory_temptable_row_cost：向内存中的临时表写入或读取一条记录的成本，增大该值会让优化器尽量少地创建内存的临时表。
- row_evaluate_cost：估算记录是否符合搜索条件的成本，增大该值可能让优化器更倾向于使用索引或范围访问而不是直接全表扫描。

把各种操作的估计成本放在表中的目的是方便修改，以适应计算机硬件速度的提升。初始时 engine_cost 和 server_cost 两个表中的记录 cost_value 项均为 NULL，成本值按默认值进行计算，当修改 cost_value 为非 NULL 时，成本值将按设定的值进行计算。

如下例子是向 mysql.engine_cost 插入 InnoDB 引擎的成本计量，因为该 MySQL 是安装在虚拟机上，硬盘 IO 很慢：

```
mysql > insert into mysql.engine_cost (engine_name, device_type, cost_name, cost_value,
comment) values ('InnoDB', 0, 'io_block_read_cost',2, 'Disk on virtual machine');
Query OK, 1 row affected (0.06 sec)
```

插入完成后，再次检查 engine_cost 表，相关命令及其运行结果如下：

```
mysql >  select engine_name,cost_name,cost_value,default_value  from mysql.engine_cost;
+-------------+------------------------+------------+---------------+
| engine_name | cost_name              | cost_value | default_value |
+-------------+------------------------+------------+---------------+
| default     | io_block_read_cost     |       NULL |             1 |
```

```
| InnoDB        | io_block_read_cost       |      2     |          1     |
| default       | memory_block_read_cost   |    NULL    |       0.25     |
+---------------+--------------------------+------------+----------------+
3 rows in set (0.00 sec)
```

使用如下命令把修改刷新到内存中：

```
mysql> flush optimizer_costs;
Query OK, 0 rows affected (0.0877 sec)
```

如果把 MySQL 的临时表目录 innodb_temp_tablespaces_dir 和临时文件目录 tmpdir 设置到内存中(如设备/dev/shm 或文件系统 tmpfs)，可以提高对临时表和临时文件的处理速度，例如，提高了 10 倍的处理速度，对应修改 disk_temptable_create_cost 和 disk_temptable_row_cost 成本如下：

```
mysql> update mysql.server_cost set cost_value = 2,Comment = 'Temporary tables on memory'
where cost_name = 'disk_temptable_create_cost';
Query OK, 1 row affected (0.01 sec)
Rows matched: 1  Changed: 1  Warnings: 0

mysql> update mysql.server_cost set cost_value = 0.05,Comment = 'Stored on memory disk' where
cost_name = 'disk_temptable_row_cost';
Query OK, 1 row affected (0.06 sec)
Rows matched: 1  Changed: 1  Warnings: 0
```

修改完成后，再次检查 server_cost 表，相关命令及其运行结果如下：

```
mysql> select cost_name,cost_value,default_value  from mysql.server_cost;
+------------------------------+------------+----------------+
| cost_name                    | cost_value | default_value  |
+------------------------------+------------+----------------+
| disk_temptable_create_cost   |          2 |             20 |
| disk_temptable_row_cost      |       0.05 |            0.5 |
| key_compare_cost             |       NULL |           0.05 |
| memory_temptable_create_cost |       NULL |              1 |
| memory_temptable_row_cost    |       NULL |            0.1 |
| row_evaluate_cost            |       NULL |            0.1 |
+------------------------------+------------+----------------+
6 rows in set (0.00 sec)
```

使用如下命令把修改刷新到内存中：

```
mysql> flush optimizer_costs;
Query OK, 0 rows affected (0.0877 sec)
```

变更完成后，要使用 flush optimizer_costs 命令将成本值刷新到内存中才能生效。进行这样的变更要很小心，需要反复测试，因为会影响所有 SQL 语句的执行计划。

17.2.4 优化器跟踪

优化器跟踪(optimizer trace)功能可以跟踪优化器生成执行计划的过程，准确地知道优化器选择执行路径的原因，使用优化器跟踪分为如下四步：

（1）打开优化器跟踪功能：SET optimizer_trace="enabled=on"。

（2）执行需要跟踪的 SQL 语句。

（3）查询视图 information_schema.optimizer_trace，重点关注 trace 字段中以 JSON 格式记录的优化器跟踪信息。

（4）关闭优化器跟踪功能：SET optimizer_trace="enabled=off"。

如果需要跟踪多个 SQL 语句的优化过程，可以重复第(2)步和第(3)步。优化器跟踪功能可以方便地查看优化器生成执行计划的整个过程。对于单表查询，主要关注 rows_estimation，该部分深入分析了各种执行方案的成本；对于多表连接查询，主要关注 considered_execution_plans，该部分列出了各种不同连接方式所对应的成本。

这里举一个例子，先查看如下两个 SQL 语句的执行计划：

```
mysql> explain select * from payment where customer_id<150\G
*************************** 1. row ***************************
           id: 1
  select_type: SIMPLE
        table: payment
   partitions: NULL
         type: range
possible_keys: idx_fk_customer_id
          key: idx_fk_customer_id
      key_len: 2
          ref: NULL
         rows: 4083
     filtered: 100.00
        Extra: Using index condition
1 row in set, 1 warning (0.00 sec)

mysql> explain select * from payment where customer_id<200\G
*************************** 1. row ***************************
           id: 1
  select_type: SIMPLE
        table: payment
   partitions: NULL
         type: ALL
possible_keys: idx_fk_customer_id
          key: NULL
      key_len: NULL
          ref: NULL
         rows: 16086
     filtered: 33.68
        Extra: Using where
1 row in set, 1 warning (0.00 sec)
```

从上面的执行计划里可以看到，当查询 payment 表中的客户号小于 150 的记录时使用索引，查询客户号小于 200 的记录时使用全表扫描。优化器为什么要这样选择呢？使用前面介绍的优化器跟踪功能对这两个 SQL 语句执行计划的生成过程进行跟踪，并将 trace 字段输出到文件，相应的命令如下：

```
mysql > set optimizer_trace = "enabled = on";
mysql > select * from payment where customer_id < 150\G
mysql > select trace into outfile 'payment_150'  lines terminated by '' from information_schema.
optimizer_trace;
mysql > select * from payment where customer_id < 200\G
mysql > select trace into outfile 'payment_200'  lines terminated by '' from information_schema.
optimizer_trace;
mysql > set optimizer_trace = "enabled = off";
```

跟踪这两个 SQL 语句到优化过程，分别生成了两个跟踪文件：payment_150 和 payment_200，在 payment_150 中可以看到：

```
"rows_estimation": [
              {
                "table": "'payment'",
                "range_analysis": {
                  "table_scan": {
                    "rows": 16086,
                    "cost": 1635
                  },
...
"analyzing_range_alternatives": {
                    "range_scan_alternatives": [
                      {
                        "index": "idx_fk_customer_id",
                        "ranges": [
                          "customer_id < 150"
                        ],
                        "index_dives_for_eq_ranges": true,
                        "rowid_ordered": false,
                        "using_mrr": false,
                        "index_only": false,
                        "rows": 4083,
                        "cost": 1429.3,
                        "chosen": true
                      }
                    ],
```

可以看出，全表扫描的成本是 1635，而使用 idx_fk_customer_id 索引扫描 customer_id < 150 的成本是 1429.3，索引的成本低，chosen 属性值是 true 表示优化器选择使用索引。

在 payment_200 中可以看到以下内容：

```
"rows_estimation": [
              {
                "table": "'payment'",
                "range_analysis": {
                  "table_scan": {
                    "rows": 16086,
                    "cost": 1635
```

```
                    },
...
"analyzing_range_alternatives": {
                    "range_scan_alternatives": [
                      {
                        "index": "idx_fk_customer_id",
                        "ranges": [
                          "customer_id < 200"
                        ],
                        "index_dives_for_eq_ranges": true,
                        "rowid_ordered": false,
                        "using_mrr": false,
                        "index_only": false,
                        "rows": 5417,
                        "cost": 1896.2,
                        "chosen": false,
                        "cause": "cost"
                      }
                    ],
```

可以看出，全表扫描的成本还是1635，而使用idx_fk_customer_id索引扫描customer_id＜200的成本是1896.2，索引的成本高，chosen属性值是flase表示优化器没有选择索引，而是使用全表扫描。

关于优化器跟踪的详细信息参见MySQL的内部文档https://dev.mysql.com/doc/internals/en/optimizer://tracing.html。

17.3 索引

在MySQL中，索引(index)也叫键(key)，它是对数据库表中一列或多列的值进行排序的一种结构，使用索引可快速访问数据库表中的特定信息。

17.3.1 主键和二级索引

数据库表的索引从数据存储方式上可以分为聚簇索引(clustered index)和非聚簇索引两种。

在聚簇索引中，中间节点保存指向下一层的指针，实际的数据保存在叶节点中。在InnoDB中有一种说法：一切都是索引(everything is an index in InnoDB)。InnoDB的每个表实际上是一个聚簇索引，相当于Oracle数据库中的索引组织表(index organized table)。这种把数据和索引存储在一起的方式相对于把数据和索引分开的方式节约了大量的存储空间，当按聚簇索引查找时速度也更快。InnoDB表的聚簇索引也叫主键(primary key)，用户在创建InnoDB表时应当指定一个主键，如果没有指定，MySQL会选择一个不含NULL值的唯一索引作为主键，如果这样的索引也不存在，MySQL会生成一个名为gen_clust_index的隐含聚簇索引，该索引基于InnoDB的行ID(row ID)，行ID是一个单向自增的6字节整数，可以认为它记录的是物理地址。使用如下SQL语句可以找出有该隐含聚簇索引的表：

```
mysql> select i.table_id,t.name
from information_schema.innodb_indexes i
join information_schema.innodb_tables t
on (i.table_id = t.table_id)
Where i.name = 'gen_clust_index';
+----------+------------------+
| table_id | name             |
+----------+------------------+
|     1093 | sakila/tab_a     |
|     1094 | sakila/tab_exit  |
|     1125 | tpcc1000/history |
+----------+------------------+
3 rows in set (0.01 sec)
```

当表中有一个数字类型的单列主键时，会有隐藏列_rowid 指向该主键，例如：

```
mysql> select actor_id,_rowid from sakila.actor limit 2;
+----------+--------+
| actor_id | _rowid |
+----------+--------+
|        1 |      1 |
|        2 |      2 |
+----------+--------+
2 rows in set (0.00 sec)
```

InnoDB 主键以外的索引称为二级索引(secondary indexes)。在 InnoDB 中，二级索引中不光包含二级索引的字段，还包含对应的主键字段。如果主键很长，二级索引要使用更多的空间，因此主键很短是有利的。通过二级索引访问记录，实际上进行了两次索引查找，第一次从二级索引中找到主键，第二次用该主键到聚簇索引中查找对应的行。InnoDB 的主索引和二级索引的结构如图 17.2 所示。

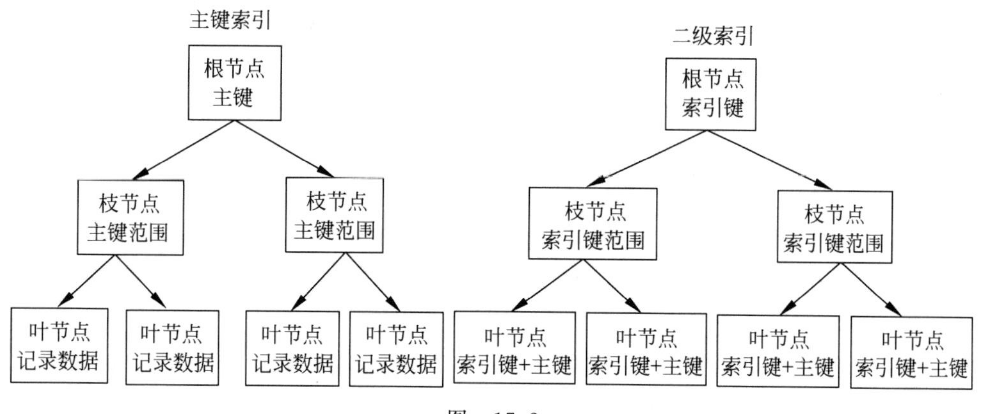

图 17.2

而非聚簇表的索引和数据是分开保存的，主键索引和二级索引的叶节点里都保存着记录的物理地址。非聚簇索引的结构如图 17.3 所示。

在大多数情况下，显式地创建一个主键通常是正确的。MySQL 8.0.13 后的版本，可以设置系统参数 sql_require_primary_key 为 on，强制创建有主键的索引，例如：

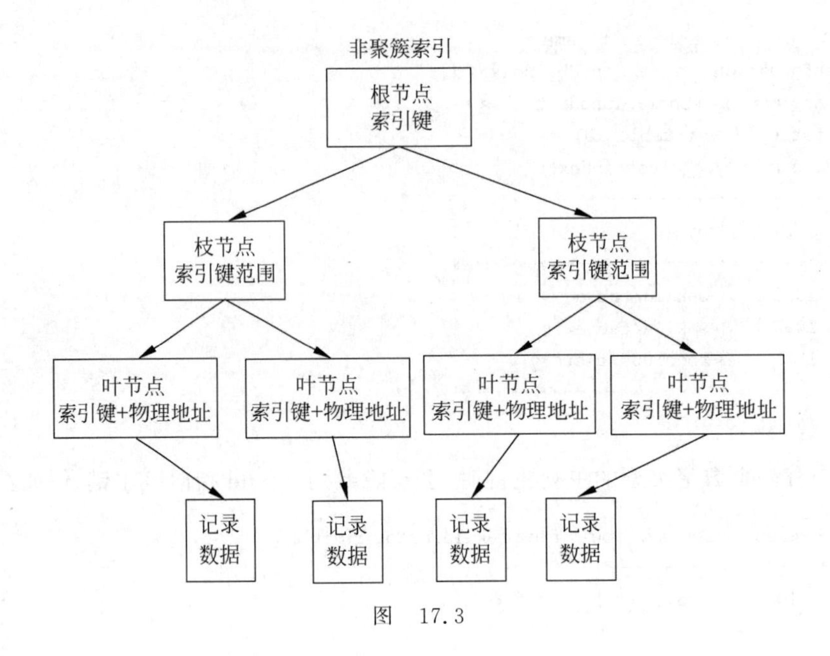

图 17.3

```
mysql > set  sql_require_primary_key = on;
Query OK, 0 rows affected (0.00 sec)

mysql > create table tab_c  (col1 char(10));
ERROR 3750 (HY000): Unable to create or change a table without a primary key, when the system
variable 'sql_require_primary_key' is set. Add a primary key to the table or unset this variable
to avoid this message. Note that tables without a primary key can cause performance problems in
row - based replication, so please consult your DBA before changing this setting.
```

使用如下 SQL 语句可以找出没有主键的表：

```
mysql > select t.table_schema as database_name,
        t.table_name
from information_schema.tables t
left join information_schema.table_constraints c
          on t.table_schema = c.table_schema
          and t.table_name = c.table_name
          and c.constraint_type = 'PRIMARY KEY'
where c.constraint_type is null
      and t.table_schema not in('mysql', 'information_schema', 'performance_schema', 'sys')
      and t.table_type = 'BASE TABLE';
```

使用 show index 命令可以查看表上的索引，使用 show extended index 命令可以看到更多索引的内部信息，如图 17.4 所示。

使用 show index 命令时可以看到，sakila.actor 表有两个索引，一个主键索引建立在字段 actor_id 上，另外一个是二级索引，建立在字段 last_name 上。使用 show extended index 命令可以看到主键实际上包括了 6 个字段，该表的 4 个字段都存放在主键上，因为主键都是聚簇索引。另外还有两个字段，一个是 DB_TRX_ID，用于存放 6 字节的事务标识；另一个是 B_ROLL_PTR，保存着指向回滚段的指针。这两个字段用于 InnoDB 的多版本控制。

```
mysql> show index from sakila.actor;
+-------+------------+--------------------+--------------+-------------+-----------+-------------+
| Table | Non_unique | Key_name           | Seq_in_index | Column_name | Collation | Cardinality |
+-------+------------+--------------------+--------------+-------------+-----------+-------------+
| actor |          0 | PRIMARY            |            1 | actor_id    | A         |         200 |
| actor |          1 | idx_actor_last_name |           1 | last_name   | A         |         121 |
+-------+------------+--------------------+--------------+-------------+-----------+-------------+
2 rows in set (0.01 sec)

mysql> show extended index from sakila.actor;
+-------+------------+--------------------+--------------+-------------+-----------+-------------+
| Table | Non_unique | Key_name           | Seq_in_index | Column_name | Collation | Cardinality |
+-------+------------+--------------------+--------------+-------------+-----------+-------------+
| actor |          0 | PRIMARY            |            1 | actor_id    | A         |         200 |
| actor |          0 | PRIMARY            |            2 | DB_TRX_ID   | A         |        NULL |
| actor |          0 | PRIMARY            |            3 | DB_ROLL_PTR | A         |        NULL |
| actor |          0 | PRIMARY            |            4 | first_name  | A         |        NULL |
| actor |          0 | PRIMARY            |            5 | last_name   | A         |        NULL |
| actor |          0 | PRIMARY            |            6 | last_update | A         |        NULL |
| actor |          1 | idx_actor_last_name |           1 | last_name   | A         |         121 |
| actor |          1 | idx_actor_last_name |           2 | actor_id    | A         |        NULL |
+-------+------------+--------------------+--------------+-------------+-----------+-------------+
8 rows in set (0.00 sec)
```

图 17.4

InnoDB 的二级索引内部也由两个字段组成，除了建立二级索引的字段，还有主键的字段，因为二级索引实际上是指向主键的索引，它的每个叶节点除了保存被索引的字段外，还保存着主键的字段，因此主键的字段越短，二级索引的存储空间越小。在使用 InnoDB 的二级索引访问数据时，实际上进行了两次索引的查找，第一次通过二级索引查找主键值，再根据该值去查找主键索引，从而找到要查找的记录。

在进行大批数据导入时，一个常用的提高效率的方法是删除二级索引，待数据导入完成后再重新创建。

17.3.2 索引合并

通常 MySQL 访问一个表时使用一个索引，当 MySQL 过滤条件中有多个字段，而且该字段并不属于同一个索引时，MySQL 可能会使用索引合并（index merge）作为访问方式。索引合并有 3 种方式，分别为交集合并（intersection merge）、联合合并（union merge）和排序联合合并（sort-union merge）。

1. 交集合并

当多个索引字段使用 and 进行连接时，可以使用交集合并，例如：

```
mysql > explain select * from rental where inventory_id = 1 and customer_id = 1 \G
*************************** 1. row ***************************
           id: 1
  select_type: SIMPLE
        table: rental
   partitions: NULL
         type: index_merge
possible_keys: idx_fk_inventory_id,idx_fk_customer_id
          key: idx_fk_inventory_id,idx_fk_customer_id
      key_len: 3,2
          ref: NULL
         rows: 1
```

```
      filtered: 100.00
         Extra: Using intersect(idx_fk_inventory_id,idx_fk_customer_id); Using where
```

可以看出，在 Extra 字段中有 Using intersect 的信息。

2. 联合合并

当多个索引字段使用 or 进行连接时，可以使用联合合并，例如：

```
mysql > explain select * from rental where inventory_id = 1 or customer_id = 1 \G
*************************** 1. row ***************************
           id: 1
  select_type: SIMPLE
        table: rental
   partitions: NULL
         type: index_merge
possible_keys: idx_fk_inventory_id,idx_fk_customer_id
          key: idx_fk_inventory_id,idx_fk_customer_id
      key_len: 3,2
          ref: NULL
         rows: 35
     filtered: 100.00
        Extra: Using union(idx_fk_inventory_id,idx_fk_customer_id); Using where
```

可以看出，在 Extra 字段中有 Using union 的信息。

3. 排序联合合并

排序联合合并与联合合并类似，只是需要先安装行 ID 进行排序后再进行联合合并，例如：

```
mysql > explain select * from rental where inventory_id < 10   or customer_id > 20000\G
*************************** 1. row ***************************
           id: 1
  select_type: SIMPLE
        table: rental
   partitions: NULL
         type: index_merge
possible_keys: idx_fk_inventory_id,idx_fk_customer_id
          key: idx_fk_inventory_id,idx_fk_customer_id
      key_len: 3,2
          ref: NULL
         rows: 26
     filtered: 100.00
        Extra: Using sort_union(idx_fk_inventory_id,idx_fk_customer_id); Using where
```

可以看出，在 Extra 字段中有 Using sort_union 的信息。

17.3.3 Multi-Range Read 优化

Multi-Range Read(MRR)优化可以减少通过二级索引进行范围扫描的磁盘随机访问，并将随机访问转换为顺序访问。这种优化方式的基本步骤如下：

(1) 按二级索引访问数据,并将数据保存到缓存中,这时缓存中的数据是按照二级索引进行排序的。

(2) 对缓存中的数据安装 RowID 进行排序,对于 InnoDB 也就是按照主键进行排序。

(3) 根据排序后的 RowID 访问数据文件。

这种访问优化方式对于性能瓶颈在传统磁盘读的 SQL 语句非常有效。缓存的大小由系统参数 read_rnd_buffer_size 决定,当读取的数据在缓存中放不下时,如上步骤要执行多次。

Multi-Range Read 优化由优化器的如下两个参数来控制:

- mrr:是否启动 Multi-Range Read 优化,默认是 on。
- mrr_cost_based:基于成本(cost based)估算决定是否启用 Multi-Range Read 优化,默认是 on。

如果将 mrr 设置为 on,mrr_cost_based 设置为 off,则总是启动 Multi-Range Read 优化,例如:

```
mysql > set @@optimizer_switch = 'mrr = on,mrr_cost_based = off';
Query OK, 0 rows affected (0.05 sec)

mysql >  explain select * from rental where rental_date between '2005 - 01 - 01' and '2005 - 01 -
31'\G
*************************** 1. row ***************************
           id: 1
  select_type: SIMPLE
        table: rental
   partitions: NULL
         type: range
possible_keys: rental_date
          key: rental_date
      key_len: 5
          ref: NULL
         rows: 1
     filtered: 100.00
        Extra: Using index condition; Using MRR
1 row in set, 1 warning (0.00 sec)
```

17.3.4　Index Condition Pushdown 优化

在没有 Index Condition Pushdown(ICP)优化的情况下,当进行索引查询时,先根据索引访问表读取记录,然后根据 where 条件进行过滤,有了 ICP 优化,访问索引时就可以根据 where 条件进行过滤,当然这需要索引中有 where 条件过滤的字段才行,该功能对其他数据库来说可能很容易实现,但对于 MySQL 数据库来说就困难一些,因为 MySQL 中有不同的存储引擎,访问索引由存储引擎负载,MySQL 服务层和存储引擎的程序代码并不是融合在一起的,它们之间通过 API 进行调用。ICP 优化方式支持 InnoDB 和 MyISAM 引擎。

当使用 ICP 优化时,执行计划的 Extra 字段中有 Using index condition 的提示,下面先创建一个索引，命令如下：

```
mysql > alter table actor add index id_first_last(first_name,last_name);
```

查询如下 SQL 语句的执行计划：

```
mysql > explain select * from actor where first_name = 'scott' and last_name like '%Yao%'\G
*************************** 1. row ***************************
           id: 1
  select_type: SIMPLE
        table: actor
   partitions: NULL
         type: ref
possible_keys: id_first_last
          key: id_first_last
      key_len: 182
          ref: const
         rows: 1
     filtered: 11.11
        Extra: Using index condition
1 row in set, 1 warning (0.00 sec)
```

可以看出，从 Extra 字段中使用了 ICP 优化。注意对索引的检索只是以第一个字段的匹配为条件，这里的第二个字段的 like 条件并不是索引的检索条件，对字符串的索引检索只能匹配最左边的字符子串。是 ICP 优化让对这两个字段的判断可以都放在索引中完成，而不用访问表。

ICP 优化的打开和关闭可以使用如下的优化器开关进行设置：

```
mysql > SET optimizer_switch = 'index_condition_pushdown = off';
mysql > SET optimizer_switch = 'index_condition_pushdown = on';
```

17.4 表连接

多表的连接在 SQL 语句里应用得也非常广泛，MySQL 里有三种表连接的方式，分别为嵌套循环、块嵌套循环和哈希连接。

17.4.1 嵌套循环

嵌套循环(nested loop)连接是最简单的一种连接方式，一直到 MySQL 5.6，它还是 MySQL 的唯一的连接方式。

嵌套循环连接中涉及两个表：一个被称为驱动表(driving table)，又叫外层表(outer table)；另一个被称为被探查表(probed table)，又叫内层表(inner table)。MySQL 首先从驱动表中提取一条记录，然后去被探查表中查找相应的匹配记录，如果有的话，就把该条记录存放到结果集中，然后再从驱动表中提取第二条记录，去被探查表中找第二条匹配的记录，依次类推，直到被探查表中的数据全部被处理完成，将结果集返回，就完成了嵌套循环连接的操作。如下是一个嵌套循环连接的例子。

检查当前线程号的命令及其输出结果如下：

```
mysql > select PS_CURRENT_THREAD_ID();
+-------------------------+
| PS_CURRENT_THREAD_ID()  |
+-------------------------+
|                     290 |
+-------------------------+
1 row in set (0.00 sec)
```

查询如下 SQL 语句的执行计划：

```
mysql > explain format = tree select count( * ) from customer inner join payment using (customer_
id) where store_id = 1\G
*************************** 1. row ***************************
EXPLAIN: -> Aggregate: count(0)
    -> Nested loop inner join  (cost = 990.87 rows = 8755)
        -> Index lookup on customer using idx_fk_store_id (store_id = 1)  (cost = 32.88 rows
= 326)
        -> Index lookup on payment using idx_fk_customer_id (customer_id = customer.customer
_id)  (cost = 0.26 rows = 27)

1 row in set (0.00 sec)
```

从如上执行计划中可以看出，执行该 SQL 语句使用了嵌套循环。

查询该嵌套循环中外表的记录数，相关命令及其运行结果如下：

```
mysql > select count( * ) from customer where store_id = 1\G
*************************** 1. row ***************************
count( * ): 326
1 row in set (0.00 sec)
```

执行这个 SQL 语句和输出结果如下：

```
mysql > select count( * ) from customer inner join payment using (customer_id) where store_id = 1
\G
*************************** 1. row ***************************
count( * ): 8748
1 row in set (0.01 sec)
```

执行该 SQL 语句会引起相关状态参数的变化，这些状态参数反映了该 SQL 语句的执行效率。查询以 Handler_read 开头的状态参数，相关命令及其运行结果如下：

```
mysql >  show status like 'Handler_read % ';
+-----------------------+-------+
| Variable_name         | Value |
+-----------------------+-------+
| Handler_read_first    | 0     |
| Handler_read_key      | 327   |
| Handler_read_last     | 0     |
| Handler_read_next     | 9074  |
```

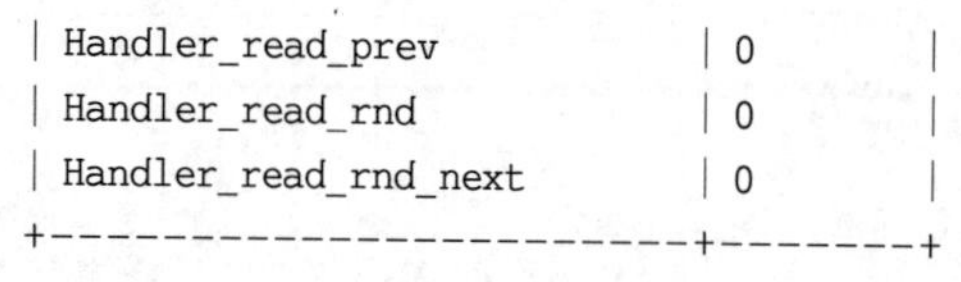

```
| Handler_read_prev          | 0      |
| Handler_read_rnd           | 0      |
| Handler_read_rnd_next      | 0      |
+----------------------------+--------+
7 rows in set (0.01 sec)
```

再按线程号在 sys.session 视图中查询该 SQL 语句执行性能的相关信息：

```
mysql > SELECT rows_examined, rows_sent, last_statement_latency AS latency FROM sys.session
WHERE thd_id = 290;
+---------------+-----------+----------+
| rows_examined | rows_sent | latency  |
+---------------+-----------+----------+
|      9074     |         1 | 6.01 ms  |
+---------------+-----------+----------+
1 row in set (0.09 sec)
```

可以看到，如上 SQL 语句检查了 9074 个记录。

把外表的记录数 326 加上输出结果的记录数 8748，相关命令及其运行结果如下：

```
mysql > select 8748 + 326;
+----------+
| 8748 + 326 |
+----------+
|   9074   |
+----------+
1 row in set (0.01 sec)
```

发现列表的记录数 326 加上输出结果的记录数 8748，正好是 9074。分析如上 SQL 语句的执行过程，发现外表是 customer，符合条件的记录有 326 个，用这 326 个记录主键访问了 payment 表中的 8748 个记录，因此 Handler_read_next 的值是 326 和 8748 的和，也就是检查的记录数。Handler_read_next 根据索引的顺序来读取下一行的值，常用于基于索引的范围扫描和 order by limit 子句中。

17.4.2 块嵌套循环

块嵌套循环(block nested loop)是嵌套循环的扩展，它从驱动表中读取数据时并不是一次读取一条记录，而是读取尽量多的记录，再将这些记录存放到连接缓存(join buffer)中，然后用这一批记录去扫描一次被探查表进行比较，这大大减少了访问被探查表的次数。

从 MySQL 8.0.20 开始，MySQL 已经不再使用块嵌套循环，由哈希连接代替。

17.4.3 哈希连接

哈希连接(hash join)是 MySQL 8.0.18 中新增的连接方式，它主要用于大数据量且连接字段没有索引时的连接，它的性能很好，有时甚至比使用索引连接的性能还好。

哈希连接主要分成两步，第一步是使用驱动表的连接字段生成哈希表，为了使哈希表能存放在内存中，通常选择小的表作为驱动表；第二步是访问被探测表的连接字段，计算哈希值并和哈希表的值进行比较，如果相符就返回结果集。

如下 SQL 语句的执行计划中就用到了哈希连接：

```
mysql > explain select count( * ) from payment, rental where payment. payment_date = rental.
return_date\G
*************************** 1. row ***************************
           id: 1
  select_type: SIMPLE
        table: rental
   partitions: NULL
         type: ALL
possible_keys: NULL
          key: NULL
      key_len: NULL
          ref: NULL
         rows: 16008
     filtered: 100.00
        Extra: NULL
*************************** 2. row ***************************
           id: 1
  select_type: SIMPLE
        table: payment
   partitions: NULL
         type: ALL
possible_keys: NULL
          key: NULL
      key_len: NULL
          ref: NULL
         rows: 16086
     filtered: 10.00
        Extra: Using where; Using join buffer (hash join)
2 rows in set, 1 warning (0.00 sec)
```

可以看到，如上 SQL 执行计划的 Extra 字段里有 hash join，说明该 SQL 执行过程中使用了哈希连接。

检查当前线程号的命令及其输出结果如下：

```
mysql >  select ps_current_thread_id();
+-------------------------+
| ps_current_thread_id()  |
+-------------------------+
|                   306   |
+-------------------------+
1 row in set (0.00 sec)
```

这里的线程号是 306，后续将使用该线程号在性能视图里查询 SQL 语句的执行性能，下面执行如下 SQL 语句：

```
mysql > select count( * )  from payment, rental where payment. payment_date = rental. return date
```

```
\G
*************************** 1. row ***************************
count( * ): 30
1 row in set (0.02 sec)
```

如下两个SQL语句分别检查了两个连接表的记录数:

```
mysql > select count( * ) from payment;
+----------+
| count( * ) |
+----------+
|    16049 |
+----------+
1 row in set (0.02 sec)

mysql > select count( * ) from rental;
+----------+
| count( * ) |
+----------+
|    16044 |
+----------+
1 row in set (0.02 sec)
```

把这两个连接表的记录数相加,相关命令及其运行结果如下:

```
mysql > select 16049 + 16044;
+-------------+
| 16049 + 16044 |
+-------------+
|       32093 |
+-------------+
1 row in set (0.00 sec)
```

再按线程号在sys.session视图中查询如下SQL语句执行性能的相关信息,相关命令及其运行结果如下:

```
mysql > select rows_examined, rows_sent, last_statement_latency as latency from sys.session
where thd_id = 306;
+---------------+-----------+----------+
| rows_examined | rows_sent | latency  |
+---------------+-----------+----------+
|         32093 |        30 | 28.90 ms |
+---------------+-----------+----------+
1 row in set (0.07 sec)
```

可以看到,如上SQL语句的检查行数正好是如上两个表的记录数之和,说明在没有索引的情况下对这两个表都只扫描了一次,效率是非常高的。

17.4.4 连接顺序的提示

与连接顺序相关的提示如下。

- JOIN_FIXED_ORDER：按 from 子句中表的排序顺序进行表连接，该提示和 SELECT STRAIGHT_JOIN 的功能一样。
- JOIN_ORDER：按提示指定的表顺序连接表。
- JOIN_PREFIX：把提示中指定的表作为连接时的第一个表，并按提示里的顺序进行连接。
- JOIN_SUFFIX：把提示中指定的表作为连接时的最后一个表，并按提示里的顺序进行连接。

17.4.5 优化器连接参数

如下两个参数影响了优化器的行为。

- optimizer_prune_level：该参数决定优化器是否会根据估计的扫描行数来决定跳过某些执行计划，默认是打开的（optimizer_prune_level=1），这时生成执行计划所花费的时间会大大缩短，但通常不会错过最优的执行计划，如果怀疑优化器错过了一个更好的执行计划，可以关闭该选项（optimizer_prune_level=0），代价是生成执行计划所花费的时间更长。
- optimizer_search_depth：该参数控制在生成执行计划时需要计算的连接表的个数，可以设置的值是 0～62，默认是 62，因为一个 SQL 语句查询块允许的连接表最多是 61，因此 62 是全排列计算。当需要连接的表比较多时，如超过 12 个表，生成执行计划所花费的时间可能是数小时，甚至数天！这时使用 show processlist 命令查询到连接的状态是 Statistics。遇到这种情况时，可以把该参数设置为 0，也就是让优化器自行决定最大搜索的深度，或者使用 JOIN_FIXED_ORDER、JOIN_ORDER、JOIN_PREFIX 和 JOIN_SUFFIX 等优化提示设定表连接的顺序。

17.5 统计信息

MySQL 统计信息是指数据库通过采样，统计出来的表、索引的相关信息，如表的记录数、聚集索引的页数、字段值的基数（cardinality）等。MySQL 在生成执行计划时，需要根据统计信息进行估算，计算出代价最低的执行计划。统计信息由存储引擎负责，MySQL 的服务器层并不保存任何统计信息，本节描述的都是 InnoDB 的统计信息。

17.5.1 自动收集统计信息

准确的表和索引的统计信息是优化器生成正确执行计划的基础，InnoDB 的统计信息分为两种：一种是持久化的统计信息，将统计信息保存到表中，在 MySQL 重启后仍然有效；另一种是临时的统计信息，统计信息保存在缓存中，MySQL 重启后丢失。第二种方式现在用得越来越少，这里不再做介绍。

收集的表的统计信息存放在 mysql 数据库的 innodb_table_stats 表中，索引的统计信息存放在 mysql 数据库的 innodb_index_stats 表中。这两个表是普通表，不是视图。这两个表可以被 update 语句修改，但尽量不要这样做，因为这样通常会造成执行计划的恶化。

InnoDB 有四个隔离级别，分别是未提交读（read uncommitted）、提交读（read committed）、可重复读（repeatable read）和序列读（serializable），默认是可重复读，但收集统计信息采用的是未提交读，也就是未提交的数据（也叫脏数据）也会被统计。这背后的逻辑是实际生产中大部分未提交的数据最终会被提交。

MySQL 有自动收集统计信息的机制，这种机制可以在全局级由系统参数控制，也可以在表级由 create table 或 alter table 选项进行控制，表 17.1 列出了控制自动收集统计信息的参数。

表 17.1

系统参数	表选项	默认值	说　明
INNODB_STATS_PERSISTENT	STATS_PERSISTENT	ON	是否把统计信息持久化
INNODB_STATS_AUTO_RECALC	STATS_AUTO_RECALC	ON	当一个表的数据变化超过 10%时是否自动收集统计信息，两次统计信息收集之间的时间间隔不能少于 10 秒
INNODB_STATS_PERSISTENT_SAMPLE_PAGES	STATS_SAMPLE_PAGES	20	统计索引时的抽样页数，该值设置得越大，收集的统计信息越准确，但收集时消耗的资源越大

应用案例：

- 当进行大批量数据导入时，可以把 INNODB_STATS_AUTO_RECALC 设置为 OFF，避免在数据导入的过程中不断地收集不准确的统计信息，在数据导入完成后再手动收集统计信息并把该参数设置为 ON。
- 对于数据量变化大的表，例如，从其他数据库导入的报告表，可以将这类表的 STATS_AUTO_RECALC 设置为 OFF，等数据量稳定后再手动进行统计信息的收集。
- 对于数据分布不规则的表，可以通过增大表的 STATS_SAMPLE_PAGES 选项，提高收集的统计信息的准确性。

下面看一个自动收集统计信息的例子。

检查当前的 innodb_stats_auto_recalc 参数，相关命令及其运行结果如下：

```
mysql> select @@innodb_stats_auto_recalc;
+----------------------------+
| @@innodb_stats_auto_recalc |
+----------------------------+
|                          1 |
+----------------------------+
1 row in set (0.01 sec)
```

参数值为 1，表示会进行自动收集统计信息。

下面检查表 tab_a 的表统计信息，相关命令及其运行结果如下：

```
mysql > select last_update, n_rows, clustered_index_size from mysql.innodb_table_stats where
table_name = 'tab_a';
```

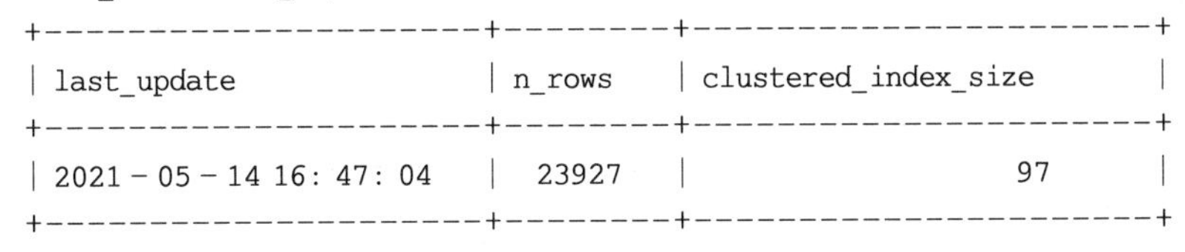

```
+----------------------+---------+-----------------------+
| last_update          | n_rows  | clustered_index_size  |
+----------------------+---------+-----------------------+
| 2021-05-14 16:47:04  |  23927  |                    97 |
+----------------------+---------+-----------------------+
1 row in set (0.00 sec)
```

下面检查表 tab_a 的索引统计信息，相关命令及其运行结果如下：

```
mysql > select last_update, stat_name, stat_value, sample_size from mysql.innodb_index_stats
where table_name = 'tab_a';
+----------------------+---------------+-------------+--------------+
| last_update          | stat_name     | stat_value  | sample_size  |
+----------------------+---------------+-------------+--------------+
| 2021-05-14 16:47:04  | n_diff_pfx01  |       23927 |           20 |
| 2021-05-14 16:47:04  | n_leaf_pages  |          59 |         NULL |
| 2021-05-14 16:47:04  | size          |          97 |         NULL |
+----------------------+---------------+-------------+--------------+
3 rows in set (0.00 sec)
```

从收集的统计信息中可以看到，表 tab_a 有 23 927 条记录(这个数字不一定准确)，占用了 97 个页，因为 MySQL 的表实际上是一个聚簇索引，因此主键和表占用的空间都是 97 个页，索引的叶节点占用了 59 页，n_diff_pfx01 是指索引中第一个字段的基数(cardinality)。

为了便于观察，把 mysql 客户端的提示符改成当前时间：

```
mysql > prompt \D >
PROMPT set to '\D > '
```

使用如下命令向表中增加超过 10%的记录：

```
Fri May 14 16:55:39 2021 > insert into tab_a select * from tab_a limit 2700;
Query OK, 2700 rows affected (0.03 sec)
Records: 2700  Duplicates: 0  Warnings: 0
```

再次检查表 tab_a 的表统计信息，相关命令及其运行结果如下：

```
Fri May 14 16:56:49 2021 > select last_update, n_rows, clustered_index_size from mysql.innodb_
table_stats where table_name = 'tab_a';
+----------------------+---------+-----------------------+
| last_update          | n_rows  | clustered_index_size  |
+----------------------+---------+-----------------------+
| 2021-05-14 16:56:49  |  26767  |                    97 |
+----------------------+---------+-----------------------+
1 row in set (0.00 sec)
```

再次检查表 tab_a 的索引统计信息，相关命令及其运行结果如下：

```
Fri May 14 16:56:53 2021 > select last_update, stat_name, stat_value, sample_size from mysql.
innodb_index_stats where table_name = 'tab_a';
+----------------------+---------------+-------------+--------------+
```

```
| last_update           | stat_name      | stat_value   | sample_size   |
+-----------------------+----------------+--------------+---------------+
| 2021-05-14 16:56:49   | n_diff_pfx01   |      26767   |          20   |
| 2021-05-14 16:56:49   | n_leaf_pages   |         66   |        NULL   |
| 2021-05-14 16:56:49   | size           |         97   |        NULL   |
+-----------------------+----------------+--------------+---------------+
3 rows in set (0.00 sec)
```

统计表 tab_a 的实际记录数，相关命令及其运行结果如下：

```
Fri May 14 16:59:41 2021 > select count(*) from tab_a;
+-----------+
| count(*)  |
+-----------+
|   28984   |
+-----------+
1 row in set (0.02 sec)
```

可以看到，当表的数据变化超过 10％时，会自动在后台进行统计信息的收集。但这里的统计信息并不特别准确，统计信息中记录有 26 767 条记录，实际上有 28 984 条记录，因为统计过程只扫描了 97 页中的 20 页，然后进行推算。

使用如下 SQL 语句把表 tab_a 的属性改成 stats_auto_recalc＝0 和 stats_sample_pages ＝ 200：

```
Fri May 14 16:57:34 2021 > alter table tab_a stats_auto_recalc = 0,
stats_sample_pages = 200;
```

再次向表 tab_a 中新增超过 10％的记录：

```
Fri May 14 17:00:59 2021 > insert into tab_a select * from tab_a limit 3000;
Query OK, 3000 rows affected (0.06 sec)
Records: 3000  Duplicates: 0  Warnings: 0
```

第 3 次检查表 tab_a 的表统计信息，相关命令及其运行结果如下：

```
Fri May 14 17:04:47 2021 > select last_update,n_rows,clustered_index_size from mysql.innodb_
table_stats where table_name = 'tab_a';
+---------------------+--------+----------------------+
| last_update         | n_rows | clustered_index_size |
+---------------------+--------+----------------------+
| 2021-05-14 16:56:49 | 26767  |                   97 |
+---------------------+--------+----------------------+
1 row in set (0.00 sec)
```

第 3 次检查表 tab_a 的索引统计信息，相关命令及其运行结果如下：

```
Fri May 14 17:04:32 2021 > select last_update,stat_name,stat_value,sample_size from mysql.
innodb_index_stats where table_name = 'tab_a';
+-----------------------+----------------+--------------+---------------+
| last_update           | stat_name      | stat_value   | sample_size   |
+-----------------------+----------------+--------------+---------------+
| 2021-05-14 16:56:49   | n_diff_pfx01   |      26767   |          20   |
```

```
| 2021-05-14 16:56:49  | n_leaf_pages  |          66 |         NULL |
| 2021-05-14 16:56:49  | size          |          97 |         NULL |
+----------------------+---------------+-------------+--------------+
3 rows in set (0.00 sec)
```

可以发现,统计信息没有发生变化,也就是不会自动进行统计信息的收集,因为表的属性 STATS_AUTO_RECALC 被设置成了 0。

使用 analyze table 命令进行统计信息的收集,相关命令及其运行结果如下:

```
Fri May 14 17:04:52 2021 > analyze table tab_a;
+------------+---------+----------+----------+
| Table      | Op      | Msg_type | Msg_text |
+------------+---------+----------+----------+
| sand.tab_a | analyze | status   | OK       |
+------------+---------+----------+----------+
1 row in set (0.01 sec)
```

第 4 次检查表 tab_a 的表统计信息,相关命令及其运行结果如下:

```
Fri May 14 17:05:57 2021 > select last_update,n_rows,clustered_index_size from mysql.innodb_
table_stats where table_name = 'tab_a';
+---------------------+--------+----------------------+
| last_update         | n_rows | clustered_index_size |
+---------------------+--------+----------------------+
| 2021-05-14 17:05:57 |  31984 |                   97 |
+---------------------+--------+----------------------+
1 row in set (0.00 sec)
```

第 4 次检查表 tab_a 的索引统计信息,相关命令及其运行结果如下:

```
Fri May 14 17:06:00 2021 > select last_update, stat_name, stat_value, sample_size from mysql.
innodb_index_stats where table_name = 'tab_a';
+----------------------+---------------+-------------+--------------+
| last_update          | stat_name     | stat_value  | sample_size  |
+----------------------+---------------+-------------+--------------+
| 2021-05-14 17:05:57  | n_diff_pfx01  |       31984 |           78 |
| 2021-05-14 17:05:57  | n_leaf_pages  |          78 |         NULL |
| 2021-05-14 17:05:57  | size          |          97 |         NULL |
+----------------------+---------------+-------------+--------------+
3 rows in set (0.00 sec)
```

再次统计表 tab_a 的实际记录数,相关命令及其统计结果如下:

```
Fri May 14 17:06:05 2021 > select count( * ) from tab_a;
+----------+
| count(*) |
+----------+
|    31984 |
+----------+
1 row in set (0.01 sec)
```

可以发现这时的统计信息很准,因为抽样页被改成了 200(STATS_SAMPLE_PAGES=

200),而表 tab_a 只有 78 页,所有页都被抽样扫描过,因此生成的统计信息就很精确。

17.5.2 手工收集统计信息

手工收集统计信息有两种方法,一种是 analyze table 命令,它可以同时收集多个表的统计信息,收集两个表统计信息的命令及其输出结果如下:

```
mysql>  analyze local table actor,rental;
+---------------+---------+----------+----------+
| Table         | Op      | Msg_type | Msg_text |
+---------------+---------+----------+----------+
| sakila.actor  | analyze | status   | OK       |
| sakila.rental | analyze | status   | OK       |
+---------------+---------+----------+----------+
2 rows in set (0.03 sec)
```

其中,local 表示该命令不会被写入二进制日志中。analyze table 命令可以方便地收集特定表的统计信息,而批量收集统计信息采用 MySQL 自带的工具 mysqlcheck 就更方便了,mysqlcheck 还可以方便地被 Linux 系统的 crontab 或 Windows 系统的任务管理器调用,收集 sakila 数据库中所有表统计信息的相关命令及其输出结果如下:

```
$ mysqlcheck --analyze sakila
sakila.actor                                    OK
sakila.actors                                   OK
sakila.address                                  OK
...
```

也可以使用--all-databases 收集所有表的统计信息。这种情况通常在进行了批量数据导入后进行。

17.5.3 查询统计信息

除了前面提到的两个保存统计信息的表,还有其他一些方式查询统计信息。

一个常用的方法是前面介绍过的 show index 命令,该命令输出的内容和视图 information_schema.statistics 差不多,例如,查询表 sakila.actor 统计信息的 SQL 语句及其输出结果如下:

```
mysql> select index_name,column_name,cardinality from information_schema.statistics where
table_name = 'actor' and table_schema = 'sakila';
+---------------------+-------------+-------------+
| INDEX_NAME          | COLUMN_NAME | CARDINALITY |
+---------------------+-------------+-------------+
| idx_actor_last_name | last_name   |         121 |
| PRIMARY             | actor_id    |         200 |
+---------------------+-------------+-------------+
2 rows in set (0.00 sec)
```

如上输出结果中的 information schema. statistics 视图里输出的最重要的信息是基数，基数是索引字段不重复值的个数。对于主键和不包含 NULL 值的唯一索引，表里的记录数就是索引的基数，因为索引中的所有值都是唯一的。从如上查询结果可以看到表 sakila. actor 两个索引的基数分别是 200 和 121，查询这两个索引字段唯一值的 SQL 语句及其输出结果如下：

```
mysql> select  count( * ) sum,count(distinct actor_id) 'primary _cardinality',count(distinct
last_name) 'last_name cardinality' from  sakila.ator;
+-----+---------------------+-----------------------+
| sum  | primary cardinality          | last_name cardinality          |
+-----+---------------------+-----------------------+
| 200  |                         200 |                           121 |
+-----+---------------------+-----------------------+
1 row in set (0.00 sec)
```

可以看到，主键的基数就是记录数，另外一个索引字段 last_name 的唯一值是 121，和该字段上的索引基数一样。

另一个和基数相关的概念是索引的选择性(selectivity)，它是指基数和表中记录总数的比值，索引的选择性越高则查询效率越好，因为高选择性的索引会在查询时过滤掉更多的行，主键和唯一索引的选择性是 1，这是最高的索引选择性。一个选择性差的索引的例子是性别，只有两个唯一值，基数是 2，因此在性别上创建索引通常不是高效的做法。计算前面两个索引选择性的 SQL 语句及其输出结果如下：

```
mysql> select count(distinct actor_id)/count( * ) 'primary selectivity',count(distinct last_
name)/count( * ) 'last_name selectivity'
from  sakila.actor;
```

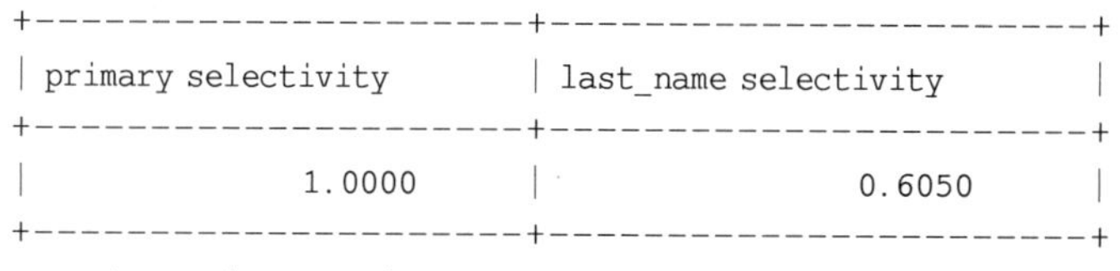

```
+---------------------+-----------------------+
| primary selectivity           | last_name selectivity          |
+---------------------+-----------------------+
|                      1.0000 |                        0.6050 |
+---------------------+-----------------------+
1 row in set (0.00 sec)
```

可以看到，主机的选择性最高是 1，二级索引的选择性要差一些。在 information_schema. statistics 视图中还可以查到索引是否可见（IS _ VISIBLE）和函数索引的表达式(EXPRESSION)。

另一个视图 information_schema. innodb_tablestats 的信息是基于 InnoDB 内部缓存中的信息，如下是一个检查表 sakila. actor 在该视图中信息的例子：

```
mysql> select * from information_schema.innodb_tablestats where name = 'sakila/actor'\G
*************************** 1. row ***************************
         TABLE_ID: 4024
             NAME: sakila/actor
STATS_INITIALIZED: Initialized
         NUM_ROWS: 200
 CLUST_INDEX_SIZE: 1
 OTHER_INDEX_SIZE: 1
 MODIFIED_COUNTER: 0
```

```
     AUTOINC: 201
   REF_COUNT: 5
1 row in set (0.01 sec)
```

其中：

- STATS_INITIALIZED 字段有两个值，分别是 Initialized 和 Uninitialized，代表该表在内存对应的架构是否被初始化。
- MODIFIED_COUNTER 字段是自上次收集统计信息后被修改的记录个数，每次收集统计信息后，该字段将会被清零。
- AUTOINC 字段是自增字段的计数，如果没有自增字段，这里会是零。
- REF_COUNT 字段是该表在缓存中的元数据被参照的次数，如果该值是零，则该表的统计信息可能被驱逐出缓存。

使用 show table status 命令可以查看相关表的统计信息，该命令的基本语法如下：

```
SHOW TABLE STATUS [{FROM | IN} db_name]  [LIKE 'pattern' | WHERE expr]
```

查询 actor 表的命令及其输出结果如下：

```
mysql> show table status like 'actor'\G
*************************** 1. row ***************************
           Name: actor
         Engine: InnoDB
        Version: 10
     Row_format: Dynamic
           Rows: 200
 Avg_row_length: 81
    Data_length: 16384
Max_data_length: 0
   Index_length: 16384
      Data_free: 0
 Auto_increment: 201
    Create_time: 2021-05-13 17:44:08
    Update_time: NULL
     Check_time: NULL
      Collation: utf8mb4_0900_ai_ci
       Checksum: NULL
 Create_options: stats_sample_pages=30 stats_auto_recalc=0 stats_persistent=1
        Comment:
1 row in set (0.00 sec)
```

show table status 的输出结果和视图 information_schema. tables 中的信息类似，在该视图里查询 actor 表的命令及其输出结果如下：

```
mysql> select * FROM information_schema.tables where table_name='actor'\G
*************************** 1. row ***************************
TABLE_CATALOG: def
TABLE_SCHEMA: sakila
TABLE_NAME: actor
TABLE_TYPE: BASE TABLE
```

```
ENGINE: InnoDB
VERSION: 10
ROW_FORMAT: Dynamic
TABLE_ROWS: 200
AVG_ROW_LENGTH: 81
DATA_LENGTH: 16384
MAX_DATA_LENGTH: 0
INDEX_LENGTH: 16384
DATA_FREE: 0
AUTO_INCREMENT: 201
CREATE_TIME: 2021-05-13 17:44:08
UPDATE_TIME: NULL
CHECK_TIME: NULL
TABLE_COLLATION: utf8mb4_0900_ai_ci
CHECKSUM: NULL
CREATE_OPTIONS: stats_sample_pages=30 stats_auto_recalc=0 stats_persistent=1
TABLE_COMMENT:
1 row in set (0.01 sec)
```

17.6　直方图

17.6.1　直方图的作用

MySQL 8.0 里引进了直方图的统计信息，用于统计字段值的分布情况。它最典型的场景是估算 where 子句中过滤谓词列的选择率，以便选择合适的执行计划。直方图的适用场景包括不是索引中第一列的列、值分布不均匀的列和在 where 子句中作为过滤条件的列。例如，在 IT 公司工作的员工，男性远多于女性，如果在性别字段上没有直方图的统计信息，优化器可能会认为男女比例相同，从而生成低效率的执行计划。再如一个作业执行完成后的状态，成功的状态通常远大于失败的状态，这些信息没有直方图统计信息的时候，MySQL 的优化器无法知道。

索引和直方图都可以用于生成正确的执行计划，但它们之间的区别却很大，例如：

- 索引可以用来减少访问所需行，直方图不能。当使用直方图进行查询时，它不会直接减少检查的行数，但可以帮助优化器选择更优化的查询计划。
- 直方图的维护成本远低于索引，对索引字段的 DML 操作要修改对应的索引，而直方图只在创建和修改时消耗资源。
- 索引需要占用大量的存储空间，直方图对存储空间的占用基本是零。
- 在判断某个范围内的行数时，索引的成本要高得多，因为索引需要当时使用索引试探(index dive)进行收集和估算，而直方图在这方面的信息是现成的。

17.6.2　桶

直方图的存放统计信息的单位是桶(bucket)，默认为 100 个，最多 1024 个。桶越多，收集统计信息的时间越长，统计信息越准确。MySQL 的直方图分为如下两类。

（1）等宽(singleton)直方图：每个桶只有一个值，保存该值和累积的频率。

（2）等高(equi-height)直方图：每个桶保存上下限、累积频率以及不同值的个数。

用户不需要指定直方图的类型，MySQL 会自动进行直方图类型的选择，当指定的桶数大于或等于桶所对应的值时，创建一个等宽直方图。否则创建一个等高直方图。

例如，在 actor 表的 first_name 字段上创建一个直方图统计信息的命令及其输出结果如下：

```
mysql > analyze table actor update histogram on first_name\G
*************************** 1. row ***************************
   Table: sakila.actor
      Op: histogram
Msg_type: status
Msg_text: Histogram statistics created for column 'first_name'.
1 row in set (0.01 sec)
```

删除如上直方图统计信息的命令及其输出结果如下：

```
mysql > analyze table actor drop histogram on first_name\G
*************************** 1. row ***************************
   Table: sakila.actor
      Op: histogram
Msg_type: status
Msg_text: Histogram statistics removed for column 'first_name'.
1 row in set (0.01 sec)
```

直方图的统计信息可以在 information_schema. column_statistics 视图中查看，其中的 histogram 字段是 JSON 格式的文档。如下为查询该视图的 SQL 语句及其输出结果：

```
mysql > select schema_name, table_name, column_name,
           histogram - >>'$."histogram - type"' as histogram_type,
           cast(histogram - >>'$."last - updated"'
               as datetime(6)) as last_updated,
           cast(histogram - >>'$."sampling - rate"'
               as decimal(4,2)) as sampling_rate,
           json_length(histogram - >'$.buckets')
               as number_of_buckets,
           cast(histogram - >'$."number - of - buckets - specified"'as unsigned)
           as number_of_buckets_specified
       from information_schema.column_statistics\G
*************************** 1. row ***************************
                SCHEMA_NAME: sakila
                 TABLE_NAME: actor
                COLUMN_NAME: first_name
             histogram_type: equi - height
               last_updated: 2021 - 08 - 13 09: 43: 33.273680
              sampling_rate: 1.00
          number_of_buckets: 100
number_of_buckets_specified: 100
1 row in set (0.00 sec)
```

可以看到这里的直方图统计信息默认使用了 100 个桶，创建了一个等高的直方图，当把桶的个数增加到 200 时创建直方图统计信息的 SQL 语句及其输出结果如下：

```
mysql > analyze table actor update histogram on first_name with 200 buckets;
*************************** 1. row ***************************
   Table: sakila.actor
      Op: histogram
Msg_type: status
Msg_text: Histogram statistics created for column 'first_name'.
1 row in set (0.01 sec)
```

再次检查生成的统计信息，相关命令及其运行结果如下：

```
mysql > select schema_name, table_name, column_name,
            histogram ->>'$."histogram-type"' as histogram_type,
            cast(histogram ->>'$."last-updated"'
                as datetime(6)) as last_updated,
            cast(histogram ->>'$."sampling-rate"'
                as decimal(4,2)) as sampling_rate,
            json_length(histogram ->'$.buckets')
                as number_of_buckets,
            cast(histogram ->'$."number-of-buckets-specified"'as unsigned)
            as number_of_buckets_specified
       from information_schema.column_statistics\G
*************************** 1. row ***************************
               SCHEMA_NAME: sakila
                TABLE_NAME: actor
               COLUMN_NAME: first_name
            histogram_type: singleton
              last_updated: 2021-08-13 09:44:36.697382
             sampling_rate: 1.00
         number_of_buckets: 130
number_of_buckets_specified: 200
1 row in set (0.00 sec)
```

可以发现，此时创建的直方图变成了等宽的，分配的 200 个桶只用了 130 个，查询该字段唯一值的个数，相关命令及其运行结果如下：

```
mysql > select count(distinct first_name) from actor;
+----------------------------+
| count(distinct first_name)   |
+----------------------------+
|                        130   |
+----------------------------+
1 row in set (0.00 sec)
```

可以发现，这时的直方图统计信息使用的桶正好也是 130 个，这说明了在等宽的直方图里一个值对应于一个桶。

17.7 CTE

MySQL 从 8.0 开始支持 CTE，CTE 是一个命名的临时结果集，它存在于单个语句的范围内，并且可以在该语句中多次引用，而且 CTE 还可以相互引用。在 MySQL 8.0 之前进行复杂查询时需要使用子查询来实现，使 SQL 语句复杂、性能低，而且可读性差。CTE 的出现简化了复杂查询语句的编写，提高了 SQL 的性能。

CTE 的基本语法如下：

```
WITH cte_name (column_list) AS (
    query
)
SELECT * FROM cte_name;
```

查询(query)中的字段数必须与 column_list 中的字段数相同。如果省略 column_list，CTE 将使用查询中的列。例如，如下 SQL 语句可查询所有的中国城市：

```
with country_cn as (
select country_id from country where country = 'China')
select city from city inner join country_cn using (country_id);
```

如上 SQL 语句比较简单，再看一个复杂点的 SQL 语句，如下是查询出租电影每个月收入的 SQL 语句及其输出结果：

```
mysql > select date_format(r.rental_date, '%y - %m - 01') as month01, sum(p.amount) as revenue
from sakila.payment p inner join sakila.rental r using (rental_id) group by month01;
+------------+----------+
| month01    | revenue  |
+------------+----------+
| 2005-05-01 |  4823.44 |
| 2005-06-01 |  9629.89 |
| 2005-07-01 | 28368.91 |
| 2005-08-01 | 24070.14 |
| 2006-02-01 |   514.18 |
+------------+----------+
5 rows in set (0.18 sec)
```

一个常见的需求是对比每个月的收入相对于上个月收入的增长情况，可以把查询每个月收入的 SQL 块写成一个 CTE，两次引用该 CTE，一次用作当前月，另一次用作上个月，对这两次引用进行左外连接，计算出相邻两个月的收入对差，相应的 SQL 语句及其输出结果如下：

```
mysql > with monthly_sales(firstofmonth, sales) as (
select date_format(r.rental_date, '%y - %m - 01') as firstofmonth, sum(p.amount) as sales from
sakila.payment p inner join sakila.rental r using (rental_id)group by firstofmonth
)
select year(cur.firstofmonth) as 'year', monthname(cur.firstofmonth) as 'month', cur.sales as
'revenue', (cur.sales - ifnull(prev.sales, 0)) as increment from monthly_sales cur left outer
join monthly_sales prev on prev.firstofmonth = (cur.firstofmonth - interval 1 month )order by
cur.firstofmonth;
+------+-----------+----------+-----------+
| year | month     | revenue  | increment |
+------+-----------+----------+-----------+
| 2005 | May       |  4823.44 |   4823.44 |
```

```
| 2005   | June       |  9629.89   |  4806.45   |
| 2005   | July       | 28368.91   | 18739.02   |
| 2005   | August     | 24070.14   | -4298.77   |
| 2006   | February   |   514.18   |   514.18   |
+------+----------+----------+-----------+
5 rows in set (0.17 sec)
```

对比一下不使用CTE时的SQL语句：

```
mysql> select year(current.month01) as 'year', monthname(current.month01) as 'month', current.
revenue, (current.revenue - ifnull(prev.revenue, 0)) as increment
from
(select date_format(r.rental_date, '%y-%m-01') as month01, sum(p.amount) as revenue from
sakila.payment p inner join sakila.rental r using (rental_id) group by month01) current
left outer join
( select date_format(r.rental_date, '%y-%m-01') as month01, sum(p.amount) as revenue from
sakila.payment p inner join sakila.rental r using (rental_id) group by month01) prev
on prev.month01 = (current.month01 - interval 1 month)
order by current.month01;
+------+----------+----------+-----------+
| year | month    | revenue  | increment |
+------+----------+----------+-----------+
| 2005 | May      |  4823.44 |  4823.44  |
| 2005 | June     |  9629.89 |  4806.45  |
| 2005 | July     | 28368.91 | 18739.02  |
| 2005 | August   | 24070.14 | -4298.77  |
| 2006 | February |   514.18 |   514.18  |
+------+----------+----------+-----------+
5 rows in set (0.31 sec)
```

可以看到，经过CTE改写后的SQL语句结构更清晰，可读性强，而且执行的速度几乎快了一倍，为什么会这样呢？对比一下这两个SQL语句的执行计划。

使用CTE改写后的SQL语句的执行计划如下：

```
mysql> explain format = tree with monthly_sales(firstofmonth, sales) as (
select date_format(r.rental_date, '%y-%m-01') as firstofmonth,sum(p.amount) as sales from
sakila.payment p inner join sakila.rental r using (rental_id)group by firstofmonth
)
select year(cur.firstofmonth) as 'year', monthname(cur.firstofmonth) as 'month', cur.sales as
'revenue', (cur.sales - ifnull(prev.sales, 0)) as increment from monthly_sales cur left outer
join monthly_sales prev on prev.firstofmonth = (cur.firstofmonth- interval 1 month )order by
cur.firstofmonth\G
*************************** 1. row ***************************
EXPLAIN: -> Nested loop left join  (cost=40140.01 rows=0)
    -> Sort: cur.firstofmonth
        -> Table scan on cur  (cost=1807.90 rows=16048)
            -> Materialize CTE monthly_sales if needed  (cost=0.00..0.00 rows=0)
                -> Table scan on <temporary>
                    -> Aggregate using temporary table
                        -> Nested loop inner join  (cost=7283.96 rows=16049)
                            -> Index scan on r using rental_date  (cost=1666.84 rows=
```

```
16008)
                                     -> Index lookup on p using fk_payment_rental (rental_id = r.rental_id)  (cost = 0.25 rows = 1)
    -> Filter: ('prev'.firstofmonth = (cur.firstofmonth - interval 1 month))  (cost = 0.25..2.50 rows = 10)
        -> Index lookup on prev using <auto_key0> (firstofmonth = (cur.firstofmonth - interval 1 month))
            -> Materialize CTE monthly_sales if needed (query plan printed elsewhere)  (cost = 0.00..0.00 rows = 0)

1 row in set (0.00 sec)
```

传统的 SQL 语句的执行计划如下：

```
mysql > explain format = tree select year(current.month01) as 'year', monthname(current.month01) as 'month', current.revenue, (current.revenue - ifnull(prev.revenue, 0)) as increment
from
(select date_format(r.rental_date, '%y-%m-01') as month01, sum(p.amount) as revenue from sakila.payment p inner join sakila.rental r using (rental_id) group by month01) current
left outer join
( select date_format(r.rental_date, '%y-%m-01') as month01, sum(p.amount) as revenue from sakila.payment p inner join sakila.rental r using (rental_id) group by month01) prev
on prev.month01 = (current.month01 - interval 1 month)
order by current.month01\G
*************************** 1. row ***************************
EXPLAIN: -> Nested loop left join  (cost = 40140.01 rows = 0)
    -> Sort: 'current'.month01
        -> Table scan on current  (cost = 1807.90 rows = 16048)
            -> Materialize  (cost = 0.00..0.00 rows = 0)
                -> Table scan on <temporary>
                    -> Aggregate using temporary table
                        -> Nested loop inner join  (cost = 7283.96 rows = 16049)
                            -> Index scan on r using rental_date  (cost = 1666.84 rows = 16008)
                            -> Index lookup on p using fk_payment_rental (rental_id = r.rental_id)  (cost = 0.25 rows = 1)
    -> Filter: ('prev'.month01 = ('current'.month01 - interval 1 month))  (cost = 0.25..2.50 rows = 10)
        -> Index lookup on prev using <auto_key0> (month01 = ('current'.month01 - interval 1 month))
            -> Materialize  (cost = 0.00..0.00 rows = 0)
                -> Table scan on <temporary>
                    -> Aggregate using temporary table
                        -> Nested loop inner join  (cost = 7283.96 rows = 16049)
                            -> Index scan on r using rental_date  (cost = 1666.84 rows = 16008)
                            -> Index lookup on p using fk_payment_rental (rental_id = r.rental_id)  (cost = 0.25 rows = 1)

1 row in set (0.00 sec)
```

可以看到，传统的SQL语句把CTE里面的SQL块两次物化（materialize）成了临时表，而使用CTE改写后的SQL语句物化CTE时使用的是Materialize CTE monthly_sales if needed，实际上只物化了一遍。

从如上例子中可以看出，复杂SQL语句使用CTE改写后，其可读性和执行效率都大大提高。

17.8　实验

统计信息收集

视频演示

（1）使用MySQL默认的自动收集统计信息的功能。

（2）关闭表的自动收集统计信息的选项。

（3）手动收集统计信息。

（参考17.5节）

第18章

SQL优化实战

第 17 章 SQL 优化基础的内容是面向优化技术进行编写。本章 SQL 优化实战的内容是面向实战进行编写，和第 17 章的内容并不重复，是 SQL 优化基础知识在实战中的扩展和应用。

要进行 SQL 优化的第一步是要先找出需要优化的 SQL，本章介绍了多种找出需要优化的 SQL 的方法，然后介绍了优化的思路和如何进行性能评估，最后分类介绍了优化的方法，包括优化索引、准确的统计信息、直方图的使用、连接优化、优化排序和表空间的碎片整理等。

18.1 找出需要优化的 SQL

当要对 MySQL 进行优化时，找到需要优化的 SQL 语句通常是第一步。目前业界有很多工具可以帮助用户快速地定位 MySQL 中消耗资源最多的 SQL 语句，如 WorkBench、MySQL Enterprise Monitor 和 Percona Monitoring & Management (PMM)等。这里介绍的是使用不需要另外安装工具的方法找出需要优化的 SQL。

18.1.1 性能视图

最需要优化的 SQL 语句并不是单次执行时间最长的 SQL 语句，而应该是总计执行时间最长的 SQL 语句，它等于执行次数乘以单次执行时间。查询 sys 和 performance_schema 数据库的性能视图可以方便地找到需要优化的 SQL 语句。例如，在视图 sys.statement_analysis 中找出总计执行时间最长的 SQL 语句：

```
mysql> select * from sys.statement_analysis limit 1\G
```

视图 sys.statement_analysis 已经按照总延迟时间降序排序，因此第一条记录就是总计用时最长的 SQL。还可以在视图 sys.statements_with_runtimes_in_95th_percentile 中查询到运行时间最长的 5%的语句。

如果要清空以前的 SQL 语句并重新进行统计，执行如下的存储过程：

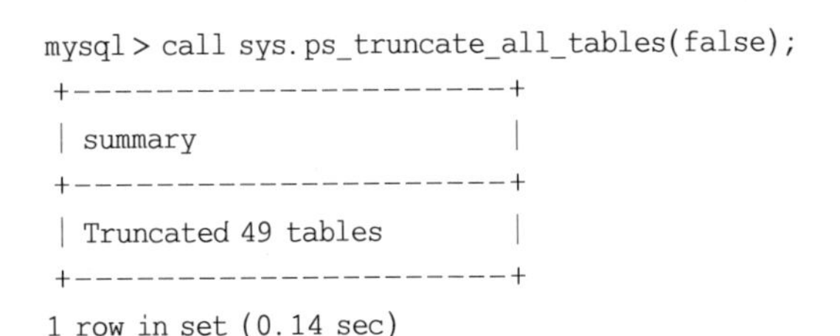

```
mysql > call sys.ps_truncate_all_tables(false);
+----------------------+
| summary              |
+----------------------+
| Truncated 49 tables  |
+----------------------+
1 row in set (0.14 sec)
```

其中，输入参数 false 表示不列出被截断的表名。实际上该存储过程截断的都是 performance_schema 数据库的表，而 sys 数据库的视图都是建立在 performance_schema 数据库上的，因此也被清空了。视图 sys.statement_analysis 的数据来自 performance_schema.events_statements_summary_by_digest，因此查询该视图可以得到更加详细的信息，在进行 SQL 性能分析时，该视图可能是最有用的视图。例如，如下 SQL 语句列出了平均执行时间最长的语句，这类 SQL 语句通常优化的空间最大：

```
mysql > select * from performance_schema.events_statements_summary_by_digest order by avg_
timer_wait desc limit 1\G
```

如下 SQL 语句列出了执行次数最多的语句，这类 SQL 语句通常对整体系统性能的影响最大：

```
mysql > select * from performance_schema.events_statements_summary_by_digest order by count_
star desc limit 1\G
```

如下 SQL 语句列出了检查行数最多的语句，这类 SQL 语句通常产生最多的硬盘 I/O：

```
mysql > select * from performance_schema.events_statements_summary_by_digest order by sum_
rows_examined desc limit 1\G
```

如下 SQL 语句列出了返回行数最多的语句，这类 SQL 语句通常占用最多网络带宽：

```
mysql > select * from performance_schema.events_statements_summary_by_digest order by sum_
rows_sent desc limit 1\G
```

18.1.2 从操作系统层监控线程

从操作系统层监控线程的运行也能找到需要优化的 SQL 语句，例如，使用 Linux 系统中的 top 加上-H 参数可以查看按繁忙程度排序的线程：

```
PID USER      PR  NI    VIRT     RES     SHR S  %CPU %MEM     TIME+  COMMAND
29345 mysql     20   0 2828020 512200  18624 R 22.9 13.2   0: 24.10 mysqld
26089 mysql     20   0 2828020 512200  18624 D  6.0 13.2   1: 57.51 mysqld
```

可以看到，最繁忙的线程号是 29345，它占用了 22.9%的 CPU 使用率。根据线程号可以从 performance_schema.threads 视图中找到正在执行的 SQL 语句，如下所示：

```
mysql> select * from performance_schema.threads where thread_os_id = 29345\G
*************************** 1. row ***************************
          THREAD_ID: 75
               NAME: thread/sql/one_connection
               TYPE: FOREGROUND
     PROCESSLIST_ID: 32
   PROCESSLIST_USER: root
   PROCESSLIST_HOST: localhost
     PROCESSLIST_DB: mysqlslap
PROCESSLIST_COMMAND: Query
   PROCESSLIST_TIME: 0
  PROCESSLIST_STATE: waiting for handler commit
   PROCESSLIST_INFO: insert into table_a(col2) values(md5(rand()))
   PARENT_THREAD_ID: NULL
               ROLE: NULL
       INSTRUMENTED: YES
            HISTORY: YES
    CONNECTION_TYPE: Socket
       THREAD_OS_ID: 29345
     RESOURCE_GROUP: USR_default
1 row in set (0.00 sec)
```

如果有必要，可以使用 PROCESSLIST_ID 字段指定的线程号“杀死”该线程或正在执行的 SQL，不能使用 THREAD_ID 字段值执行 kill 命令，不然可能会杀错了线程：

```
mysql> kill query 32;
```

或者

```
mysql> kill 32;
```

18.1.3 其他方法

1. sys 的存储过程

使用性能视图进行分析时存在的一个弊端是性能视图里保存的信息是自性能视图被重置(通常是从 MySQL 启动)以来的全量信息，当前的性能信息在这些视图里被稀释了，sys 库里有如下 4 个存储过程提供了获取增量性能信息的方法：

(1) diagnostics()存储过程会生成一个关于当前 MySQL 实例整体性能的诊断报告。

(2) statement_performance_analyzer()存储过程可以生成当前 MySQL 实例中正在运行的 SQL 语句的两个快照，并对比这两个快照，生成增量报告。

(3) ps_trace_thread()存储过程可以跟踪某个线程的执行过程，把该线程执行的所有 SQL 语句的性能信息都记录下来，并输出报告。

(4) ps_trace_statement_digest()存储过程可以根据提供的 SQL 语句摘要哈希值跟踪收集这些 SQL 语句执行过程中的性能诊断信息。

通过这 4 个存储过程也可以找到需要优化的 SQL 和它们的性能信息，这些存储过程在第 7 章已经介绍过，此处不再赘述。

2. 慢查询日志

使用慢查询日志也可以找到需要优化的SQL和它们的性能信息,该方法在第3章已经介绍过,此处不再赘述。

18.2 优化方法

实际上对SQL语句的优化是在一个项目设计阶段就要考虑的事情,但目前项目开发过程中的一个普遍现象是在设计时只考虑功能的实现,等项目投产后,随着数据量和业务量的增加,遇到了性能瓶颈,才开始考虑性能的优化。

18.2.1 优化思路

找到了最需要优化的SQL语句后,使用explain列出它们的执行计划,对执行计划进行分析,从以下几个方面找出可以优化的地方:

- SQL语句是否可以以性能更好的方式重新进行编写? 例如,子查询改成连接或者反过来、把in改成exist、使用CTE改写子查询等。
- 是否可以通过创建索引提高性能? 如果可能,执行时间缩短通常是数量级的。
- 是否正确使用了索引? 误以为使用了索引,实际上没有使用,这种情况在实际生产中并不少见。
- 是否使用了正确的数据类型? 尽量选择小的、简单的数据类型。
- 是否使用了最优的访问路径? 可以考虑索引访问、索引覆盖、合并索引等。
- 统计信息是否陈旧? 统计信息不准确会导致优化器做出错误的选择。
- 是否适合创建直方图统计信息? 直方图对生产正确的执行计划帮助很大,而且创建和维护成本也较低。
- 多表的连接是否可以优化? 例如,连接次序是否正确? 在连接的字段上是否需要创建索引? 是否可以使用哈希连接?
- 排序是否可以优化? 在排序字段上加索引、增大排序空间等。
- 优化器的成本估算和实际成本有多大差异? 优化器估算的成本只是理论成本,很多因素可能会导致估算成本和实际成本发生偏离。
- 表空间是否存在大量碎片? 经常被修改的表应定期进行整理,消除碎片。
- 应用的设置是否可能优化? 例如,应用的逻辑是否可以为了性能进行修改,数据结构呢? 这种修改可能是工作量最大的,但很多时候也是效果最好的。

在考虑这些优化方面的时候,可以不断地进行变更测试,在测试的过程中注意比较SQL语句的执行计划、执行时间、访问的行数和资源消耗的变化。在实际工作中,找到一个比MySQL自动生成的执行计划更优的执行计划的原因通常是因为我们比MySQL更了解应用的逻辑和数据特点,而不是因为我们更聪明。

18.2.2 性能评估

在SQL优化的过程中要不断地进行性能评估，最简单的评估SQL语句执行效率的方法是查看SQL语句的执行用时，mysql客户端执行每条SQL语句后都会显示这条SQL语句的执行用时，但除了这个方法之外，还有一些更精确的计量方法。

1. 性能视图

MySQL自带的performance_schema和sys数据库的很多性能视图记录了SQL语句的执行性能，例如，performance_schema.events_statements_current视图记录了当前执行SQL语句的性能，如下所示：

```
mysql > select * from performance_schema.events_statements_current where thread_id!= ps_
current_thread_id()\G
*************************** 1. row ***************************
              THREAD_ID: 261
               EVENT_ID: 85
           END_EVENT_ID: 86
             EVENT_NAME: statement/sql/select
                 SOURCE: init_net_server_extension.cc: 95
            TIMER_START: 428908093494869000
              TIMER_END: 428908101438018000
             TIMER_WAIT: 7943149000
              LOCK_TIME: 1705000000
               SQL_TEXT: select first_name, last_name from actor where actor_id in (select
actor_id from film_actor where film_id = (select film_id from film where title = 'ACADEMY
DINOSAUR'))
                 DIGEST: 8f9ed9f0cbda810c36133dc79495d9a0a8c560322e284b26146bba9bb7fa6a19
            DIGEST_TEXT: SELECT 'first_name', 'last_name' FROM 'actor' WHERE 'actor_id' IN ( SELECT
'actor_id' FROM 'film_actor' WHERE 'film_id' = ( SELECT 'film_id' FROM 'film' WHERE 'title' = ? ) )
         CURRENT_SCHEMA: sakila
            OBJECT_TYPE: NULL
          OBJECT_SCHEMA: NULL
            OBJECT_NAME: NULL
  OBJECT_INSTANCE_BEGIN: NULL
            MYSQL_ERRNO: 0
      RETURNED_SQLSTATE: NULL
           MESSAGE_TEXT: NULL
                 ERRORS: 0
               WARNINGS: 0
          ROWS_AFFECTED: 0
              ROWS_SENT: 10
          ROWS_EXAMINED: 21
CREATED_TMP_DISK_TABLES: 0
     CREATED_TMP_TABLES: 0
       SELECT_FULL_JOIN: 0
 SELECT_FULL_RANGE_JOIN: 0
           SELECT_RANGE: 0
     SELECT_RANGE_CHECK: 0
```

```
              SELECT_SCAN: 0
        SORT_MERGE_PASSES: 0
               SORT_RANGE: 0
                SORT_ROWS: 0
                SORT_SCAN: 0
            NO_INDEX_USED: 0
       NO_GOOD_INDEX_USED: 0
         NESTING_EVENT_ID: NULL
       NESTING_EVENT_TYPE: NULL
      NESTING_EVENT_LEVEL: 0
             STATEMENT_ID: 7891
1 row in set (0.01 sec)
```

performance_schema.file_summary_by_event_name 记录了按照事件分类的文件 I/O 汇总信息，如下例子反映了全表扫描的 I/O 性能。

首先重置 performance_schema.file_summary_by_event_name 视图，命令如下：

```
mysql> truncate table performance_schema.file_summary_by_event_name;
Query OK, 0 rows affected (0.01 sec)
```

然后执行一个全表扫描的语句如下所示：

```
mysql> select * from sbtest2 where c<>'a';
...
333333 rows in set (4.04 sec)
```

最后查询等待事件 wait/io/file/innodb/innodb_data_file 的性能信息，这些信息反映了全表扫描的性能，相关命令如下及其运行结果如下：

```
mysql> select event_name, count_read, avg_timer_read/1000000000.0 "Avg Read Time (ms)", sum_
number_of_bytes_read/1024/1024 "MB Read" from performance_schema.file_summary_by_event_name
where event_name = 'wait/io/file/innodb/innodb_data_file'\G
*************************** 1. row ***************************
        event_name: wait/io/file/innodb/innodb_data_file
        count_read: 4368
Avg Read Time (ms): 0.5880
           MB Read: 68.25000
1 row in set (0.00 sec)
```

2. 状态变量

MySQL 的自带了 479 个状态变量(MySQL 8.0.22)用以反映 MySQL 的运行状态，在 mysql 客户端里可以使用如下命令查询会话和全局的状态变量：

```
mysql> show session status;
mysql> show global status;
```

也可以使用 mysqladmin extended-status 查询全局的状态变量。如果要查看一段时间状态变量的变化情况，可以使用如下命令：

```
$ mysqladmin extended-status -ri60 -c3|tee my_status
```

其中，-i60 表示每 60 秒重复执行一次；-r 表示显示相邻两次查询的差值；-c3 表示重复查询 3 次；tee 命令表示把输出结果同时保存到文件。

查询 MySQL 的状态变量时，重点关注 Com_XXX 状态变量：

Com_XXX 是 SQL 语句的计数器，其中 Com 是 command 的缩写，XXX 是指 SQL 语句的类型，这些计数器包括：

- Com_begin
- Com_commit
- Com_delete
- Com_insert
- Com_select
- Com_update

查询这些 SQL 语句的计数器可以了解当前实例执行 SQL 语句的情况，可以使用如下命令查询这些计数器：

```
$ mysqladmin extended-status | grep Com_ |grep -E '
begin|commit|delete|insert|select|update'
```

Innodb_rows_XXX 是对应 SQL 语句处理 InnoDB 表行数的计数器，XXX 是指 SQL 语句的类型，这些计数器包括：

- Innodb_rows_deleted
- Innodb_rows_inserted
- Innodb_rows_read
- Innodb_rows_updated

查询这些行数的计数器可以了解当前实例处理行数的情况，可以使用如下命令查询这些计数器：

```
$ mysqladmin extended-status | grep Innodb_rows
```

查询如下两个状态变量可以知道数据库创建临时表的情况：

- Created_tmp_disk_tables
- Created_tmp_tables

如果在磁盘上创建的临时表数量较多，会遇到性能问题。有两种情况会创建位于磁盘上的临时表，一是当临时表的大小大于系统参数 tmp_table_size 或 max_heap_table_size 中的最小值时；另一种情况是表中有 blob 或 text 的字段时。第一种情况可以通过对 SQL 语句进行优化来缩小临时表的大小，第二种情况需要对表结构进行变更，设计表时应该把大字段和经常查询的小字段放在不同的表中，这和保持活跃数据独立的设计思路一致。

Handler_* 计数器统计了句柄操作，句柄 API 是 MySQL 和存储引擎之间的接口，其中 Handler_read_* 对调试 SQL 语句的性能很有用，在执行 SQL 语句之前，可以先使用如下 flush status 命令将当前状态变量重置为零：

```
mysql> flush status;
Query OK, 0 rows affected (0.01 sec)
```

然后执行一条 SQL 语句：

```
mysql > select  *  from actor where actor_id = 1;
+----------+------------+-----------+---------------------+
| actor_id | first_name | last_name | last_update         |
+----------+------------+-----------+---------------------+
|        1 | PENELOPE   | GUINESS   | 2006-02-15 04:34:33 |
+----------+------------+-----------+---------------------+
1 row in set (0.00 sec)
```

再查询状态变量 Handler_read_* 的值：

```
mysql > show session status like 'Handler_read%';
+-----------------------+-------+
| Variable_name         | Value |
+-----------------------+-------+
| Handler_read_first    | 0     |
| Handler_read_key      | 1     |
| Handler_read_last     | 0     |
| Handler_read_next     | 0     |
| Handler_read_prev     | 0     |
| Handler_read_rnd      | 0     |
| Handler_read_rnd_next | 0     |
+-----------------------+-------+
7 rows in set (0.01 sec)
```

查询最后执行的 SQL 语句的估算成本(注意：不是实际执行成本，MySQL 里没有实际执行成本)：

```
mysql > show status like 'last_query_cost';
+-----------------+----------+
| Variable_name   | Value    |
+-----------------+----------+
| Last_query_cost | 1.000000 |
+-----------------+----------+
1 row in set (0.00 sec)
```

观察状态变量值，可以看到该 SQL 语句执行了一次索引访问，估计执行成本是 1。

如果要查询其他会话的状态变量值，可以查询视图 performance_schema.status_by_thread，例如，如下 SQL 语句查询了所有会话中的状态变量 Handler_write：

```
mysql > select  *  from performance_schema.status_by_thread where variable_name = 'Handler_
write';
+-----------+---------------+----------------+
| THREAD_ID | VARIABLE_NAME | VARIABLE_VALUE |
+-----------+---------------+----------------+
|        60 | Handler_write | 94             |
|        67 | Handler_write | 477            | -- 插入记录最多的线程
|        69 | Handler_write | 101            |
+-----------+---------------+----------------+
2 rows in set (0.01 sec)
```

打开查询慢查询日志后，如果同时将系统参数 log_slow_extra 设置为 true，也会记录和慢 SQL 语句相关的状态变量。

关于具体状态变量的说明，参见 https://dev.mysql.com/doc/refman/8.0/en/server-status-variables.html。

3. explain analyze

使用 explain analyze 既可以看到估计成本，也能看到实际执行用时和访问的行数，而且每步都可以看到这些计量信息，这样对精确定位 SQL 语句执行瓶颈很有帮助。这些内容在第 17 章已经说明过，这里就不再赘述。

4. 操作系统层监控

从操作系统层也可以监控 SQL 语句的执行性能，MySQL 需要消耗操作系统的 4 种资源：CPU、磁盘、内存和网络，在 Linux 系统上可以采用监控工具包括 top、free、vmstat、iostat、mpstat、sar 和 netstat 等。

18.3 优化索引

在 MySQL 中，索引是对 SQL 语句执行性能影响最大的因素，这里介绍几种优化索引的方法。

18.3.1 正确使用索引

已经存在了索引，但因为不正确的 SQL 语句可能导致索引没有使用，例如，对于如下字符型的字段，如果使用数字类型的值进行检索时，就不会使用索引：

```
mysql > explain select * from actor where last_name = 123\G
*************************** 1. row ***************************
           id: 1
  select_type: SIMPLE
        table: actor
   partitions: NULL
         type: ALL
possible_keys: idx_actor_last_name
          key: NULL
      key_len: NULL
          ref: NULL
         rows: 200
     filtered: 10.00
        Extra: Using where
```

同样的 SQL 语句把被检索的值改用字符类型时就会使用索引，如下所示：

```
mysql > explain select * from actor where last_name = '123'\G
*************************** 1. row ***************************
           id: 1
  select_type: SIMPLE
```

```
         table: actor
    partitions: NULL
          type: ref
possible_keys: idx_actor_last_name
           key: idx_actor_last_name
       key_len: 182
           ref: const
          rows: 1
      filtered: 100.00
         Extra: NULL
```

因字段类型不同造成索引失效的情况更多是出现在表连接时，两个连接的字段从字段名上看起来相同，但实际上字段的类型却不同，从而导致无法使用索引。

不正确地使用运算符也可能造成索引失效，例如，对比如下两个 SQL 语句：

```
mysql> explain select * from actor where actor_id = 1 + 1\G
*************************** 1. row ***************************
           id: 1
  select_type: SIMPLE
        table: actor
   partitions: NULL
         type: const
possible_keys: PRIMARY
          key: PRIMARY
      key_len: 2
          ref: const
         rows: 1
     filtered: 100.00
        Extra: NULL
1 row in set, 1 warning (0.01 sec)

mysql> explain select * from actor where actor_id - 1 = 1\G
*************************** 1. row ***************************
           id: 1
  select_type: SIMPLE
        table: actor
   partitions: NULL
         type: ALL
possible_keys: NULL
          key: NULL
      key_len: NULL
          ref: NULL
         rows: 200
     filtered: 100.00
        Extra: Using where
1 row in set, 1 warning (0.00 sec)
```

字段 actor_id 上是主键索引，在第一个 SQL 语句中，对该字段的过滤为通过主键访问进行，但在第二个 SQL 语句中，由于要对字段 actor_id 进行运算后再过滤，因此只能进行全表扫描后，把所有的记录进行计算后才能进行过滤。

如下两个 SQL 语句可找出在 2005 年 6 月发生的租借电影的交易：

```
mysql > explain select * from rental where rental_date between '2005 - 06 - 01' and '2005 - 06 -
30'\G
*************************** 1. row ***************************
           id: 1
  select_type: SIMPLE
        table: rental
   partitions: NULL
         type: ALL
possible_keys: rental_date
          key: NULL
      key_len: NULL
          ref: NULL
         rows: 16008
     filtered: 14.44
        Extra: Using where
1 row in set, 1 warning (0.00 sec)

mysql > explain select * from rentalt where year(rental_date) = 2005 and month(rental_date) =
6\G
*************************** 1. row ***************************
           id: 1
  select_type: SIMPLE
        table: rental
   partitions: NULL
         type: ALL
possible_keys: NULL
          key: NULL
      key_len: NULL
          ref: NULL
         rows: 16008
     filtered: 100.00
        Extra: Using where
1 row in set, 1 warning (0.00 sec)
```

可以看到，第一种 SQL 语句可以使用 rental_date 索引，而第二种 SQL 语句不能使用索引，这也是因为运算符使用错误从而造成索引失效。

对于创建在字符串字段上的 B 树索引，需要注意，对最左边的字符子串进行匹配时可以使用索引，对中间或右边的字符子串进行匹配时则无法使用索引，对比如下两个 SQL 语句执行计划中的索引使用情况：

```
mysql > explain select * from actor where last_name like 'BALL%'\G
*************************** 1. row ***************************
           id: 1
  select_type: SIMPLE
        table: actor
   partitions: NULL
         type: range
possible_keys: idx_actor_last_name
```

```
          key: idx_actor_last_name
      key_len: 182
          ref: NULL
         rows: 1
     filtered: 100.00
        Extra: Using index condition
1 row in set, 1 warning (0.01 sec)

mysql > explain select * from actor where last_name like '%BALL%'\G
*************************** 1. row ***************************
           id: 1
  select_type: SIMPLE
        table: actor
   partitions: NULL
         type: ALL
possible_keys: NULL
          key: NULL
      key_len: NULL
          ref: NULL
         rows: 200
     filtered: 11.11
        Extra: Using where
1 row in set, 1 warning (0.00 sec)
```

18.3.2 创建索引

在一个需要被检索但又没有索引的字段上创建索引，通常是提高 SQL 语句执行效率的最简单、有效的方法。通过检查 sys 中的两个视图，可以找到需要创建的索引。

视图 schema_tables_with_full_table_scans 中包括所有没有使用高效索引的表，按扫描行数降序进行排列，例如：

```
mysql > select * from sys.schema_tables_with_full_table_scans where object_schema = 'sand';
+---------------+-------------+-------------------+-----------+
| object_schema | object_name | rows_full_scanned | latency   |
+---------------+-------------+-------------------+-----------+
| sand          | sbtest1     |           1950000 | 6.76 s    |
| sand          | tab_a       |             44084 | 257.70 ms |
+---------------+-------------+-------------------+-----------+
2 rows in set (0.01 sec)
```

可以看出，sbtest1 表被扫描了近 200 万行，用时 6 秒多。视图 statements_with_full_table_scans 中没有使用高效索引 SQL 语句，检查与 sbtest1 表相关的 SQL 语句如下：

```
mysql > select * from sys.statements_with_full_table_scans where  query like '%sbtest1%'\G
*************************** 1. row ***************************
                   query: SELECT 'pad' FROM 'sbtest1' WHERE 'c' = ?
                      db: sand
              exec_count: 1
           total_latency: 4.99 s
```

```
      no_index_used_count: 1
no_good_index_used_count: 0
        no_index_used_pct: 100
                rows_sent: 0
            rows_examined: 650000
            rows_sent_avg: 0
        rows_examined_avg: 650000
               first_seen: 2021-05-18 14:45:20.235960
                last_seen: 2021-05-18 14:45:20.235960
                   digest: 2fb9c019e982e37b73cac5bffd9137f9691159214ab8c2c99ce518eefb125bc2
...
```

可以看出，如上 SQL 语句用时 4.99 秒扫描 sbtest1 表的 65 万行，如果在字段 c 上创建索引，该 SQL 语句的执行效率将大大提高。

18.3.3 删除索引

在同一个表上创建的索引并不是越多越好，索引除了占用额外的空间外，对 DML 语句的性能也有一定的影响，因为对字段进行修改后都要修改相应字段的索引，所以不要在同一个表上创建过多的索引。冗余的索引要删除，可以从视图 sys.schema_unused_indexes 中查询没有使用的索引，这些索引可能是被删除的对象，查询 sbtest 数据库中没有使用的索引的 SQL 语句及其输出结果如下：

```
mysql> select * from sys.schema_unused_indexes where object_schema = 'sbtest';
+---------------+-------------+------------+
| object_schema | object_name | index_name |
+---------------+-------------+------------+
| sbtest        | sbtest1     | k_1        |
| sbtest        | sbtest2     | k_2        |
+---------------+-------------+------------+
2 rows in set (0.04 sec)
```

查询视图 schema_redundant_indexes 中保存着重复的索引的 SQL 语句及其输出结果如下：

```
mysql> select * from sys.schema_redundant_indexes\G
*************************** 1. row ***************************
              table_schema: sbtest
                table_name: sbtest2
     redundant_index_name: k_2
   redundant_index_columns: k
redundant_index_non_unique: 1
       dominant_index_name: id_kc
    dominant_index_columns: k,c
 dominant_index_non_unique: 1
            subpart_exists: 0
            sql_drop_index: ALTER TABLE 'sbtest'.'sbtest2' DROP INDEX 'k_2'
1 row in set (0.02 sec)
```

可以看到，id_kc 索引建立在 k 和 c 两个字段上，而 k_2 索引建立在 k 字段上，因此 k 字段上有两个索引，这里还给出了删除索引的 SQL 语句，但并不是所有的重复索引都需要删除，有些重复的索引可以提高查询的效率，还需要保留。

除了上面两个视图外，在 sys.schema_index_statistics 视图里还记录着索引被 SQL 语句使用的频率和时延，查询 actor 表索引使用情况的 SQL 语句及其输出结果如下：

```
mysql >  select * from sys.schema_index_statistics where table_name = 'actor'\G
*************************** 1. row ***************************
  table_schema: sakila
    table_name: actor
    index_name: PRIMARY
rows_selected: 34
select_latency: 1.24 ms
rows_inserted: 0
insert_latency:    0 ps
  rows_updated: 0
update_latency:    0 ps
  rows_deleted: 0
delete_latency:    0 ps
*************************** 2. row ***************************
  table_schema: sakila
    table_name: actor
    index_name: idx_actor_last_name
rows_selected: 1
select_latency: 615.35 us
rows_inserted: 0
insert_latency:    0 ps
  rows_updated: 0
update_latency:    0 ps
  rows_deleted: 0
delete_latency:    0 ps
2 rows in set (0.01 sec)
```

如上视图记录的索引使用情况也可以用来参考，和前面两个视图相结合，用来确定需要删除的索引。

18.3.4 不可见索引

在 MySQL 8.0 里，可以设置索引为不可见，这样优化器就不会使用这样的索引，带来的一个显而易见的优势是便于调试，当不能确定一个索引是否需要时，可以先把索引转换成不可见索引，并运行一段时间，从而确定该索引的确不需要，再把索引删除，如果后来发现该索引还是有用的，可以再把索引转换为可见索引，这样避免了成本高昂的创建、删除索引的动作。一个使用 actor 表 last_name 字段上的索引的 SQL 语句的执行计划如下：

```
mysql > explain select * from actor where last_name = 'BALL'\G
*************************** 1. row ***************************
           id: 1
```

```
  select_type: SIMPLE
        table: actor
   partitions: NULL
         type: ref
possible_keys: idx_actor_last_name
          key: idx_actor_last_name
      key_len: 182
          ref: const
         rows: 1
     filtered: 100.00
        Extra: NULL
1 row in set, 1 warning (0.00 sec)
```

把 last_name 字段上的索引改成不可见索引的 SQL 语句如下：

```
mysql> alter table actor alter index idx_actor_last_name invisible;
```

再次检查前面 SQL 语句的执行计划如下：

```
mysql> explain select * from actor where last_name = 'BALL'\G
*************************** 1. row ***************************
           id: 1
  select_type: SIMPLE
        table: actor
   partitions: NULL
         type: ALL
possible_keys: NULL
          key: NULL
      key_len: NULL
          ref: NULL
         rows: 200
     filtered: 0.83
        Extra: Using where
1 row in set, 1 warning (0.01 sec)
```

可以看到，在索引被改成了不可见索引后，SQL 语句将不会使用该索引。但这个索引仍然被维护着。

可以使用 show index 查询索引是否可见，也可以在视图 information_schema.statistics 中查询索引是否可见，查询某个表上的索引是否可见的 SQL 语句及其输出结果如下：

```
mysql> select index_name, is_visible from information_schema.statistics where table_schema =
'sakila' and table_name = 'actor';
+---------------------+------------+
| index_name          | is_visible |
+---------------------+------------+
| idx_actor_last_name | NO         |
| PRIMARY             | YES        |
+---------------------+------------+
2 rows in set (0.00 sec)
```

对于不可见索引可以把优化器的开关 use_invisible_indexes 设置为 on(默认是 off)，从而让优化器使用不可见索引。也可以使用提示 SET_VAR 将优化器开关 use_invisible_indexes 临时设置为 on,从而使用不可见索引,如下是使用提示让 SQL 语句使用不可见索引的执行计划:

```
mysql > explain select * /* + SET_VAR(optimizer_switch = 'use_invisible_indexes = on') */
from actor where last_name = 'BALL'\G
*************************** 1. row ***************************
           id: 1
  select_type: SIMPLE
        table: actor
   partitions: NULL
         type: ref
possible_keys: idx_actor_last_name
          key: idx_actor_last_name
      key_len: 182
          ref: const
         rows: 1
     filtered: 100.00
        Extra: NULL
1 row in set, 1 warning (0.00 sec)
```

18.3.5　在长字符串上创建索引

当在很长的字符串的字段上创建索引时,索引会变得很大而且低效,一个解决办法是使用 crc32 或 md5 函数对长字符串进行哈希计算,然后在计算的结果上创建索引。在 MySQL 5.7 以后的版本,可以创建一个自动生成的字段,例如,可以创建如下一个表:

```
mysql > create table website(
id int unsigned not null,
web varchar(100) not null,
webcrc int unsigned generated always as (crc32(web)) not null,
primary key (id)
);
```

其中,字段 webcrc 是根据字段 web 进行 crc32 哈希计算后生成的字段,如下 SQL 语句为向如上表中插入一条记录:

```
mysql > insert into website(id,web) values(1,"< https://www.scutech.com >");
```

查询如上表中的记录如下:

```
mysql > select * from website;
+----+---------------------------+------------+
| id | web                       | webcrc     |
+----+---------------------------+------------+
|  1 | < https://www.scutech.com > | 3014687870 |
+----+---------------------------+------------+
1 row in set (0.00 sec)
```

可以看到，字段 webcrc 中自动生成了 web 字段的循环冗余校验值，在该字段上创建索引，可以得到一个占用空间少，而且高效的索引。

在 MySQL 8.0.13 以后的版本中，可以直接创建函数索引，例如，创建如下一个表：

```
mysql > create table website8(
id int unsigned not null,
web varchar(100) not null,
primary key (id),
index ((crc32(web)))
);
```

在如上表中创建了一个函数索引，查询该表上索引的 SQL 语句及其输出结果如下：

```
mysql > show index from website8\G
*************************** 1. row ***************************
        Table: website8
   Non_unique: 0
     Key_name: PRIMARY
 Seq_in_index: 1
  Column_name: id
    Collation: A
  Cardinality: 0
     Sub_part: NULL
       Packed: NULL
         Null:
   Index_type: BTREE
      Comment:
Index_comment:
      Visible: YES
   Expression: NULL
*************************** 2. row ***************************
        Table: website8
   Non_unique: 1
     Key_name: functional_index
 Seq_in_index: 1
  Column_name: NULL
    Collation: A
  Cardinality: 0
     Sub_part: NULL
       Packed: NULL
         Null:
   Index_type: BTREE
      Comment:
Index_comment:
      Visible: YES
   Expression: crc32('web')
2 rows in set (0.00 sec)
```

可以看到，第一个索引是主键，第二个索引是函数索引。

解决索引字段长的另一个办法是创建前缀索引(prefix index)，前缀索引的创建语法中字段的写法是：col_name(length)，前缀索引是对字符串的前面一部分创建索引，支持的数据类型包括 char、varchar、binary 和 varbinary。创建前缀索引的关键是选择前缀的字符串的长度，长度越长，索引的选择性越高，但存储的空间也越大。

sbtest2 表中 c 字段是长度为 120 的字符串，查询在不同长度时索引选择性的 SQL 语句及其输出结果如下：

```
mysql > select
count(distinct(left(c,3)))/count( * ) sel3,
count(distinct(left(c,5)))/count( * ) sel5,
count(distinct(left(c,7)))/count( * ) sel7,
count(distinct(left(c,9)))/count( * ) sel9,
count(distinct c)/count( * ) selectivity
from sbtest1;
+--------+--------+--------+--------+-------------+
| sel3   | sel5   | sel7   | sel9   | selectivity |
+--------+--------+--------+--------+-------------+
| 0.0120 | 0.6784 | 0.9959 | 1.0000 |      1.0000 |
+--------+--------+--------+--------+-------------+
1 row in set (1.60 sec)
```

可以看到，在这个字段的前 7 位创建索引即可达到接近 1 的选择性，再增加这个索引的前缀位数，索引的选择性并不会提高，如下是创建索引的命令：

```
mysql > alter table sbtest2 add index (c(7));
```

18.3.6 使用索引改变访问路径

通过改变索引影响 SQL 语句的访问路径，往往会明显提升性能，如建立索引覆盖、多字段索引、索引合并等。但对于一些小表，很多时候全表扫描会比使用索引要快，这时就不要盲目创建索引，下面举例说明通过调整索引引起访问路径变化从而提高性能的例子。

1. 覆盖索引

覆盖索引(covering index)指一个 SQL 语句只读取索引就可以获得需要的数据，而不需要访问表。这样大大提高了 I/O 的效率，原因如下：

- 只需要访问索引，不需要访问表减少了 I/O 的次数。
- 索引通常比表小很多。
- 由于索引是按照键值顺序存储的(至少在一个页内是这样)，对于按照键值进行范围查询时使用的是顺序 I/O，相对于离散 I/O，其性能大大提高。

在使用覆盖索引时，执行计划的 Extra 字段中有 Using index 的信息，如下是一个 SQL 语句的执行计划：

```
mysql > explain select customer_id,inventory_id,rental_date from rental\G
*************************** 1. row ***************************
           id: 1
  select_type: SIMPLE
        table: rental
```

```
   partitions: NULL
         type: index
possible_keys: NULL
          key: rental_date
      key_len: 10
          ref: NULL
         rows: 16008
     filtered: 100.00
        Extra: Using index
1 row in set, 1 warning (0.01 sec)
```

如下 SQL 语句没有 where 条件,但仍然访问了 rental_date 索引,而且没有访问表,查询了 rental_date 索引的构成:

```
mysql > select column_name from information_schema.statistics where index_name = 'rental_date';
+--------------+
| column_name  |
+--------------+
| rental_date  |
| inventory_id |
| customer_id  |
+--------------+
3 rows in set (0.01 sec)
```

可以发现,该 SQL 语句要查询的 3 个字段 customer_id、inventory_id、rental_date 都包含在这个索引中了,因此只要访问该索引即可得到所有需要的数据,就没有必要再访问表了。由于二级索引实质上都包含主键,因此如果再加上主键,一样可以使用覆盖索引,如下是在输出字段中加上主键字段的 SQL 语句的执行计划:

```
mysql > explain select rental_id,customer_id,inventory_id,rental_date from rental\G
*************************** 1. row ***************************
           id: 1
  select_type: SIMPLE
        table: rental
   partitions: NULL
         type: index
possible_keys: NULL
          key: rental_date
      key_len: 10
          ref: NULL
         rows: 16008
     filtered: 100.00
        Extra: Using index
1 row in set, 1 warning (0.00 sec)
```

可以看到,一样使用了覆盖索引。但如果要访问的字段不在这个索引中,则还需要访问表,如果在上面的查询字段中再增加任意一个其他字段就不能使用覆盖索引了,例如,如下 SQL 语句将无法使用覆盖索引:

```
mysql > explain select * from  rental where rental_date = '2005 - 05 - 24 22: 53: 30' and
```

```
inventory_id = 367\G
*************************** 1. row *************************** .
           id: 1
  select_type: SIMPLE
        table: rental
   partitions: NULL
         type: ref
possible_keys: rental_date,idx_fk_inventory_id
          key: rental_date
      key_len: 8
          ref: const,const
         rows: 1
     filtered: 100.00
        Extra: NULL
1 row in set, 1 warning (0.00 sec)
```

如上 SQL 语句的执行计划的字段 Extra 里没有 Using index,因此没有使用覆盖索引。但可以对该 SQL 语句进行改写,先用一个可以使用覆盖索引的子查询查询出主键,再通过主键查找相应的记录,这种方法称为延迟关联(deferred join),改写后的 SQL 语句执行计划如下:

```
mysql > explain select * from rental where rental_id in (select rental_id from rental where
rental_date = '2005 - 05 - 24 22: 53: 30' and inventory_id = 367)\G
*************************** 1. row ***************************
           id: 1
  select_type: SIMPLE
        table: rental
   partitions: NULL
         type: ref
possible_keys: PRIMARY,rental_date,idx_fk_inventory_id
          key: rental_date
      key_len: 8
          ref: const,const
         rows: 1
     filtered: 100.00
        Extra: Using index
*************************** 2. row ***************************
           id: 1
  select_type: SIMPLE
        table: rental
   partitions: NULL
         type: eq_ref
possible_keys: PRIMARY
          key: PRIMARY
      key_len: 4
          ref: sakila.rental.rental_id
         rows: 1
     filtered: 100.00
        Extra: NULL
2 rows in set, 1 warning (0.00 sec)
```

可以看到，在第一步的查询中使用了覆盖索引。

延迟关联的方法在分页查询中对效率的提高很明显，例如，如下 SQL 语句进行排序后从 1000 行开始查询 5 行，它的执行计划如下：

```
mysql> explain analyze select * from rental order by rental_date limit 1000,5\G
*************************** 1. row ***************************
EXPLAIN: -> Limit/Offset: 5/1000 row(s)  (cost = 1625.05 rows = 5) (actual time = 17.487..
17.488 rows = 5 loops = 1)
    -> Sort: rental.rental_date, limit input to 1005 row(s) per chunk  (cost = 1625.05 rows =
16008) (actual time = 17.254..17.435 rows = 1005 loops = 1)
        -> Table scan on rental  (cost = 1625.05 rows = 16008) (actual time = 0.332..11.348
rows = 16044 loops = 1)

1 row in set (0.02 sec)
```

如上 SQL 语句要进行全表扫描和文件排序，成本很高，使用延迟关联改写后的 SQL 语句的执行计划如下：

```
mysql> explain analyze select * from rental r1 inner join (select rental_id from rental order
by rental_date limit 1000,5) r2 on r1.rental_id = r2.rental_id\G
*************************** 1. row ***************************
EXPLAIN: -> Nested loop inner join  (cost = 259.58 rows = 5) (actual time = 7.015..7.067 rows
= 5 loops = 1)
    -> Table scan on r2  (cost = 115.56 rows = 1005) (actual time = 0.002..0.003 rows = 5 loops
= 1)
        -> Materialize  (cost = 7.83 rows = 5) (actual time = 6.836..6.838 rows = 5 loops = 1)
            -> Limit/Offset: 5/1000 row(s)  (cost = 7.83 rows = 5) (actual time = 6.641..6.
644 rows = 5 loops = 1)
                -> Index scan on rental using rental_date  (cost = 7.83 rows = 1005) (actual
time = 3.762..6.568 rows = 1005 loops = 1)
    -> Single-row index lookup on r1 using PRIMARY (rental_id = r2.rental_id)  (cost = 0.25
rows = 1) (actual time = 0.044..0.044 rows = 1 loops = 5)

1 row in set (0.02 sec)
```

改写后的 SQL 充分使用了索引和主键，估算成本从 1625.05 下降到 259.58，效率大大提高。

2. 多字段索引

有时根据查询的条件，创建多字段索引会大大提高 SQL 语句的执行效率，例如如下 SQL 语句的执行计划：

```
mysql> explain select * from payment where staff_id = 1 and payment_date between '2005-06-01' and
'2005-06-20'\G
*************************** 1. row ***************************
           id: 1
  select_type: SIMPLE
        table: payment
   partitions: NULL
         type: ref
```

```
possible_keys: idx_fk_staff_id
          key: idx_fk_staff_id
      key_len: 1
          ref: const
         rows: 8057
     filtered: 11.11
        Extra: Using where
1 row in set, 1 warning (0.01 sec)
```

可以看到，在两个检索字段中，其中一个字段上有索引，可以考虑创建多字段索引。创建多字段索引时，通常将选择性高的字段放在前面，因此这里应把 payment_date 字段放在第一个，如下是创建多字段索引的 SQL 语句：

```
mysql>  alter table payment add index id_date_id(payment_date,staff_id);
```

创建索引完成后，再次查看该 SQL 语句的执行计划如下：

```
mysql> explain select * from payment where staff_id = 1 and payment_date between '2005 - 06 - 01' and
'2005 - 06 - 20'\G
*************************** 1. row ***************************
           id: 1
  select_type: SIMPLE
        table: payment
   partitions: NULL
         type: range
possible_keys: idx_fk_staff_id,id_date_id
          key: id_date_id
      key_len: 6
          ref: NULL
         rows: 1706
     filtered: 50.09
        Extra: Using index condition
1 row in set, 1 warning (0.00 sec)
```

优化器考虑了两个索引，然后选择了新建的多字段索引，从性能视图中对比一下这两个 SQL 语句的执行效率：

```
mysql> select left(sql_text,40),timer_wait,rows_examined from performance_schema.events_
statements_history
order by timer_start desc limit 3;
+------------------------------------------+--------------+---------------+
| left(sql_text,40)                        | timer_wait   | rows_examined |
+------------------------------------------+--------------+---------------+
| select * from payment where staff_id = 1 a |   4339017000 |           865 |
| alter table payment add index id_date_id | 306490524000 |             0 |
| select * from payment where staff_id = 1 a |  21490988000 |          8057 |
+------------------------------------------+--------------+---------------+
3 rows in set (0.00 sec)
```

可以看到，创建了多字段索引后，同样的 SQL 语句检查的行数减少了近 10 倍，执行速度提高到原来的 5 倍。

18.3.7 使用索引减少锁

InnoDB在访问记录时会对其进行加锁，索引可以减少InnoDB访问的记录数量，从而减少锁的数量。如果没有使用索引进行过滤，对where条件里字段的判断要到服务器层进行，InnoDB将对没有过滤的全部记录加锁。例如：

```
mysql > begin;
Query OK, 0 rows affected (0.00 sec)

mysql > select * from city where city = 'Shaoguan' for share;
+---------+----------+------------+---------------------+
| city_id | city     | country_id | last_update         |
+---------+----------+------------+---------------------+
|     469 | Shaoguan |         23 | 2006-02-15 04:45:25 |
+---------+----------+------------+---------------------+
1 row in set (0.00 sec)
```

如下SQL语句对一条记录上共享锁，然后查询锁的情况如下：

```
mysql > select index_name, lock_type, lock_mode, count(*)  from performance_schema.data_
locks group by index_name, lock_type, lock_mode;
+------------+-----------+-----------+----------+
| index_name | lock_type | lock_mode | count(*) |
+------------+-----------+-----------+----------+
| NULL       | TABLE     | IS        |        1 |
| PRIMARY    | RECORD    | S         |      602 |
+------------+-----------+-----------+----------+
2 rows in set (0.00 sec)
```

可以发现，有602个记录级别的锁，也就是对整个表的所有记录都上了锁，因为InnoDB存储引擎不能对where条件里的city='Shaoguan'进行过滤，那是服务器层的工作，它只能给所有的记录上锁，然后将所有的记录返回给服务器层。例如，如下SQL语句的执行计划：

```
mysql >  explain select * from city where city = 'Shaoguan' for share\G
*************************** 1. row ***************************
           id: 1
  select_type: SIMPLE
        table: city
   partitions: NULL
         type: ALL
possible_keys: NULL
          key: NULL
      key_len: NULL
          ref: NULL
         rows: 600
     filtered: 10.00
        Extra: Using where
1 row in set, 1 warning (0.01 sec)
```

可以看到，在Extra字段中有Using where的说明，这说明使用了where条件进行记录的过滤，而该工作是服务器层做的。在该字段上创建索引后再看该SQL语句的执行计划：

```
mysql > explain select * from city where city = 'Shaoguan' for share\G
*************************** 1. row ***************************
           id: 1
  select_type: SIMPLE
        table: city
   partitions: NULL
         type: ref
possible_keys: id_city
          key: id_city
      key_len: 202
          ref: const
         rows: 1
     filtered: 100.00
        Extra: NULL
1 row in set, 1 warning (0.00 sec)
```

可以看到，Extra字段中已经没有了Using where的说明，因为where条件里的过滤工作由InnoDB存储引擎完成了，再次查询锁的情况如下：

```
mysql > select index_name, lock_type, lock_mode, count( * )  from performance_schema.data_
locks group by index_name, lock_type, lock_mode;
+------------+-----------+---------------+----------+
| index_name | lock_type | lock_mode     | count( * ) |
+------------+-----------+---------------+----------+
| NULL       | TABLE     | IS            |        1 |
| id_city    | RECORD    | S             |        1 |
| PRIMARY    | RECORD    | S,REC_NOT_GAP |        1 |
| id_city    | RECORD    | S,GAP         |        1 |
+------------+-----------+---------------+----------+
4 rows in set (0.00 sec)
```

可以发现，只有3个记录级别的锁，大大减少了锁的竞争，因为InnoDB访问的记录减少了，锁自然也减少了。在实际工作中，如果发现大量的锁竞争，可以通过在适当字段上创建索引的方法减少锁。除了创建索引外，还有两个减少锁竞争的方法：一个是减少事务的粒度，也就是尽量避免一个事务修改大量的记录，或者持续较长的时间不提交；另一个方法是调整事务的隔离级别。

18.4 准确的统计信息

不准确的统计信息会造成优化器做出错误的选择，下面是一个模拟不准确的统计信息的例子。

首先把表改成不自动收集统计信息：

```
mysql > alter table sbtest2 stats_auto_recalc = 0;
```

然后把表中的记录增加一倍：

```
mysql > insert into sbtest2(k,c,pad) select k,c,pad from sbtest2;
```

再检查 information_schema.innodb_tablestats 视图中的 modified_counter 字段：

```
mysql > select modified_counter from information_schema.innodb_tablestats where name like
'sbtest%';
+-------------------+
| modified_counter  |
+-------------------+
|         333333    |
+-------------------+
1 row in set (0.01 sec)
```

该字段记录自上次收集统计信息后修改的行数，对比一下该表中的实际行数：

```
mysql > select count(*) from sbtest2;
+----------+
| count(*) |
+----------+
|  666666  |
+----------+
1 row in set (3.92 sec)
```

从查询中可以看到，sbtest2 表共有 666 666 条记录，自从上次收集了统计信息后，一半的记录都发生了变化，因此这个表的统计信息是不准的，再检查 mysql 数据库 innodb_table_stats 表中记录的 sbtest2 的统计信息如下：

```
mysql > select last_update,n_rows,clustered_index_size from mysql.innodb_table_stats where
table_name = 'sbtest2';
+---------------------+--------+----------------------+
| last_update         | n_rows | clustered_index_size |
+---------------------+--------+----------------------+
| 2021-05-31 20:23:23 | 328824 |                 5286 |
+---------------------+--------+----------------------+
1 row in set (0.01 sec)
```

可以看到，上次收集统计信息的时间距今已有一段时间了，记录的行数也只有大约真实行数的一半。这时可以使用 analyze table 命令手工收集统计信息如下：

```
mysql > analyze table sbtest2;
+-----------------+---------+----------+----------+
| Table           | Op      | Msg_type | Msg_text |
+-----------------+---------+----------+----------+
| sbtest.sbtest2  | analyze | status   | OK       |
+-----------------+---------+----------+----------+
1 row in set (0.10 sec)
```

再重新检查 mysql 数据库的 innodb_table_stats 表中记录的 sbtest2 的统计信息如下：

```
mysql > select last_update,n_rows,clustered_index_size from mysql.innodb_table_stats where
```

```
table_name = 'sbtest2';
+---------------------+---------+-----------------------+
| last_update         | n_rows  | clustered_index_size  |
+---------------------+---------+-----------------------+
| 2021-06-10 10:37:40 | 657576  |                  9833 |
+---------------------+---------+-----------------------+
1 row in set (0.00 sec)
```

可以发现，收集完统计信息后，统计信息中记录的行数和真实的数据已经很接近了。

收集统计信息的取样页数由系统参数 innodb_stats_persistent_sample_pages 决定，默认是 20。如果需要收集更加精确的统计信息，可以增加这个值。如果要修改单个表的收集统计信息的取样页数，可以通过 alter table table_name stats_sample_pages 来完成。

还可以通过手工修改 mysql 中两个表 innodb_table_stats 和 innodb_table_stats 的对应字段来改变统计信息，然后通过 flush table 命令把修改后的统计信息刷新到内存中，但手工修改的统计信息通常不准，最好还是使用 analyze table 命令收集统计信息。

18.5 直方图的使用

当需要过滤的字段上既没有索引也没有直方图时，优化器会根据 MySQL 代码中内置的默认规则估计过滤的比率，很大程度上是猜测，部分常用的默认规则如表 18.1 所示。

表 18.1

过滤类型	过滤比率/%
=	10
<>或≠	90
< 或 >	33.33
between	11.11
in	字段数×10 和 50 的最小值

如下是 SQL 语句的执行计划中默认过滤比率的几个例子，首先设置 pager，使执行计划只显示过滤比率：

```
mysql> pager grep filtered
PAGER set to 'grep filtered'
```

对检索字段的过滤条件是等于时，默认过滤比率为 10%：

```
mysql> explain select * from actor where first_name = 'lisa'\G
     filtered: 10.00
1 row in set, 1 warning (0.00 sec)
```

对检索字段的过滤条件是大于或小于时，默认过滤比率为三分之一：

```
mysql> explain select * from actor where first_name >'lisa'\G
     filtered: 33.33
1 row in set, 1 warning (0.00 sec)
```

对检索字段的过滤条件是不等于时，默认过滤比率为 90%：

```
mysql> explain select * from actor where first_name<>'lisa'\G
     filtered: 90.00
1 row in set, 1 warning (0.00 sec)
```

对检索字段的过滤条件是 between 时，默认过滤比率为 11.11%：

```
mysql> explain select * from actor where last_update between '2006-02-15' and
'2006-02-16'\G
     filtered: 11.11
1 row in set, 1 warning (0.00 sec)
```

下面 SQL 中，对检索字段用 in 进行过滤时，过滤比率为 20%：

```
mysql> explain select * from actor where first_name in ('lisa','THORA')\G
     filtered: 20.00
1 row in set, 1 warning (0.00 sec)
```

使用默认规则估计字段的过滤比率往往是不准确的，因此很多时候会生成错误的执行计划，这时，在字段上收集直方图统计信息可以解决这个问题。而且直方图即使不用于改变 SQL 语句的执行计划，也可以用于在执行计划中的 filt 列显示正确的过滤比例。

编写一个简单的 SQL 语句，查询在 payment 表里支付金额大于 10 元的客户号，生成该 SQL 语句的执行计划如下：

```
mysql> explain select customer_id from payment where amount>10\G
*************************** 1. row ***************************
           id: 1
  select_type: SIMPLE
        table: payment
   partitions: NULL
         type: ALL
possible_keys: NULL
          key: NULL
      key_len: NULL
          ref: NULL
         rows: 16086
     filtered: 33.33
        Extra: Using where
1 row in set, 1 warning (0.00 sec)
```

判断 amount 字段大于 10 的记录，由于该字段上没有直方图的统计信息，优化器根据代码中内置的默认值估计有三分之一的记录属于这个范围。再判断 amount 大于 100 的记录的执行计划如下：

```
mysql> explain select customer_id from payment where amount>100\G
*************************** 1. row ***************************
           id: 1
  select_type: SIMPLE
        table: payment
   partitions: NULL
         type: ALL
possible_keys: NULL
```

```
          key: NULL
      key_len: NULL
          ref: NULL
         rows: 16086
     filtered: 33.33
        Extra: Using where
1 row in set, 1 warning (0.00 sec)
```

优化器仍然估计有三分之一的记录属于这个范围，显然优化器在盲猜。为了解决这个问题，现在在 amount 字段上创建直方图的统计信息，相关命令及其输出结果如下：

```
mysql > analyze table payment update histogram on amount with 256 buckets\G
*************************** 1. row ***************************
   Table: sakila.payment
      Op: histogram
Msg_type: status
Msg_text: Histogram statistics created for column 'amount'.
1 row in set (0.31 sec)
```

再重新生成该 SQL 语句的执行计划如下：

```
mysql > explain select customer_id from payment where amount > 10\G
*************************** 1. row ***************************
           id: 1
  select_type: SIMPLE
        table: payment
   partitions: NULL
         type: ALL
possible_keys: NULL
          key: NULL
      key_len: NULL
          ref: NULL
         rows: 16086
     filtered: 0.71
        Extra: Using where
1 row in set, 1 warning (0.00 sec)
```

优化器根据直方图的统计信息估计符合这个条件的记录只占总数的 0.71%。

删除该直方图的命令如下：

```
mysql > analyze table payment drop histogram on amount\G
*************************** 1. row ***************************
   Table: sakila.payment
      Op: histogram
Msg_type: status
Msg_text: Histogram statistics removed for column 'amount'.
1 row in set (0.01 sec)
```

如下 SQL 语句为查询单次消费金额大于 10 元和在第一个店进行消费的顾客的姓名，在没有直方图时的生成的执行计划如下：

```
mysql > explain analyze select first_name, last_name from customer inner join payment using
```

```
(customer_id) where  amount > 10 and store_id = 1\G
*************************** 1. row ***************************
EXPLAIN: -> Nested loop inner join  (cost = 3100.48 rows = 2918) (actual time = 0.443..14.853
rows = 68 loops = 1)
    -> Index lookup on customer using idx_fk_store_id (store_id = 1)  (cost = 36.35 rows =
326) (actual time = 0.310..0.707 rows = 326 loops = 1)
    -> Filter: (payment.amount > 10.00)  (cost = 6.72 rows = 9) (actual time = 0.042..0.043
rows = 0 loops = 326)
        -> Index lookup on payment using idx_fk_customer_id (customer_id = customer.customer
_id)  (cost = 6.72 rows = 27) (actual time = 0.031..0.038 rows = 27 loops = 326)

1 row in set (0.01 sec)
```

可以看到，优化器先对符合在第一个店进行消费的条件进行过滤，然后再过滤消费金额大于10元的条件。在字段 amount 上有直方图统计信息之后，再次生成该 SQL 语句的执行计划如下：

```
mysql > explain analyze select first_name, last_name from customer inner join payment using
(customer_id) where  amount > 10 and store_id = 1\G
*************************** 1. row ***************************
EXPLAIN: -> Nested loop inner join  (cost = 1672.84 rows = 62) (actual time = 0.328..9.507
rows = 68 loops = 1)
    -> Filter: (payment.amount > 10.00)  (cost = 1632.85 rows = 114) (actual time = 0.224..
8.421 rows = 114 loops = 1)
        -> Table scan on payment  (cost = 1632.85 rows = 16086) (actual time = 0.191..6.482
rows = 16049 loops = 1)
    -> Filter: (customer.store_id = 1)  (cost = 0.25 rows = 1) (actual time = 0.009..0.009
rows = 1 loops = 114)
        -> Single - row index lookup on customer using PRIMARY (customer_id = payment.
customer_id)  (cost = 0.25 rows = 1) (actual time = 0.009..0.009 rows = 1 loops = 114)

1 row in set (0.02 sec)
```

可以看到，优化器将这两个过滤条件的先后次序反转过来了，因为借助直方图统计信息，优化器知道消费金额大于10元这个条件的选择性更高。从估计成本和实际执行时间都可以看出，有直方图的执行计划效率要高很多！

直方图在某些场景下可以帮助优化器生成更优的执行计划，那么在什么样的字段上考虑使用直方图呢，这里建议符合如下4个条件字段可以考虑建立直方图统计信息：

（1）值分布不均匀，优化器很难估计值的分布的字段。

（2）选择性差的字段，否则索引更适合。

（3）用于 where 子句中过滤的字段或用于连接的字段。

（4）字段值分布规律不随时间变化的字段。因为直方图统计信息不会自动收集，如果字段值分布规律发生大的变化，统计信息会失真。

实际工作中，可以使用 explain analyze 命令分析 SQL 语句的执行计划，如果估算的 rows 和实际的 rows 相差过大，可以考虑在过滤字段上创建直方图统计信息。

18.6 连接优化

连接优化是SQL优化的重要组成部分。表连接的顺序对性能的影响很大，N个表连接时不同的连接顺序的组合是N!，当N是5时，这个组合是5! = 5×4×3×2×1 = 120。不同连接顺序的性能差距可能是数量级的。

18.6.1 选择小表作为驱动表

选择小表作为驱动表可以减少访问被探测表的次数，这里的小表是指应用了查询限制条件后结果集小的表。

例如，如下SQL语句的执行计划：

```
mysql> explain format = tree select count( * ) from city inner join country using (country_id)
where city.last_update = '2006-02-15 04:45:25'\G
*************************** 1. row ***************************
EXPLAIN: -> Aggregate: count(0)
    -> Nested loop inner join  (cost = 81.75 rows = 60)
        -> Filter: (city.last_update = TIMESTAMP'2006-02-15 04:45:25')  (cost = 60.75
rows = 60)
            -> Table scan on city  (cost = 60.75 rows = 600)
        -> Single-row index lookup on country using PRIMARY (country_id = city.country_id)
  (cost = 0.25 rows = 1)

1 row in set (0.00 sec)
```

查询city表中的全部记录个数和符合过滤条件的记录个数如下：

```
mysql> select count( * ),sum(last_update = '2006-02-15 04:45:25') from city;
+----------+-------------------------------------+
| count( * ) | sum(last_update = '2006-02-15 04:45:25')  |
+----------+-------------------------------------+
|     600  |                                 600 |
+----------+-------------------------------------+
1 row in set (0.00 sec)
```

查询另一个连接表country中的记录个数如下：

```
mysql> select count( * ) from country;
+----------+
| count( * ) |
+----------+
|     109  |
+----------+
1 row in set (0.01 sec)
```

从执行计划里面可以看到，MySQL认为经过last_update字段过滤后的city表小，用它作为驱动表进行连接，但实际上所有last_update字段的值都相同，经过过滤后仍然是全集，因此 country 表比 city 表小，应该用 country 表作为驱动表，可以通过 JOIN_ORDER 提示设置表的连接顺序，修改后的 SQL 语句的执行计划如下：

```
mysql> explain format = tree select /* + JOIN_ORDER(country, city) */ count( * ) from city
inner join country using (country_id) where city.last_update = '2006 - 02 - 15 04: 45: 25'\G
*************************** 1. row ***************************
EXPLAIN: -> Aggregate: count(0)
    -> Nested loop inner join  (cost = 221.15 rows = 60)
        -> Index scan on country using PRIMARY  (cost = 11.15 rows = 109)
        -> Filter: (city.last_update = TIMESTAMP'2006 - 02 - 15 04: 45: 25')  (cost = 1.38
rows = 1)
            -> Index lookup on city using idx_fk_country_id (country_id = country.country_
id)  (cost = 1.38 rows = 6)
```

从性能视图中比较这两种连接方式的性能如下：

```
mysql> select left(sql_text, 40), rows_examined, timer_wait/1000000000 ms from performance_
schema.events_statements_history where sql_text like '%city inner join country%' and thread_
id!= ps_current_thread_id() order by timer_start desc limit 2;
+------------------------------------------+----------------+--------+
| left(sql_text, 40)                       | rows_examined  | ms     |
+------------------------------------------+----------------+--------+
| select count( * ) from city inner join cou |           1200 | 2.7704 |
| select /* + JOIN_ORDER(country, city) */ c |            709 | 2.2053 |
+------------------------------------------+----------------+--------+
2 rows in set (0.01 sec)
```

可以看到，当以 city 表为驱动表时，先访问了 city 表的 600 条记录，再用这 600 条记录的 country_id 字段查找 country 表，共访问了 600＋600＝1200 条记录；而以 country 表为驱动表时，先访问了 country 表的 109 条记录，在用这 109 条记录的 country_id 字段查找 city 表的 600 条记录，共访问了 109＋600＝709 条记录。因此小驱动表访问的记录少，执行的速度也快。

这个例子只是涉及了两个表，如果有更多的表，数据量会更大，表的连接顺序对性能的影响也更大。除了前面使用的 JOIN_ORDER 外，还有 JOIN_FIXED_ORDER、JOIN_PREFIX 和 JOIN_SUFFIX 等提示可以用于指定表的连接顺序，这些提示在指定了表的连接顺序的同时还减少了生成 SQL 语句执行计划的时间。

18.6.2 增大连接缓存

连接缓存是 MySQL 用来在连接时进行数据缓存的区域。每次连接使用一个连接缓存，因此执行一个 SQL 语句可能用到多个连接缓存，连接缓存在 SQL 语句执行之前分配，执行完成后释放。每个连接缓存的大小由系统参数 join_buffer_size 决定，对于该参数的设置，长期以来的建议是：初始值可以分配得小一些，对于需要大的连接缓存的会话和 SQL 语句可以单独进行调整，如果统一设置得很大，对于很多 SQL 语句来说是浪费的。但在 MySQL 8.0.18 以后的版本中，连接缓存的分配是根据需要进行递增分配的，join_buffer_size 只是连接缓存的上限，但外连接要分配全部的连接缓存，从 MySQL 8.0.20 以后，包括外连接对连接缓存的需求也可以进行递增的分配了，因此设置一个较大的 join_buffer_size 已经不会有什么副作用。

系统参数 join_buffer_size 默认值是 256KB，如下分别是在会话级和全局级将其设置为 1GB 的例子：

```
mysql > set join_buffer_size = 1024 * 1024 * 1024;
mysql > set global join_buffer_size = 1024 * 1024 * 1024;
```

也可以使用 set_var 提示对单个 SQL 语句调节 join_buffer_size 的大小。如下 SQL 语句为查询当前会话的系统参数 join_buffer_size：

```
mysql > select @@join_buffer_size;
+--------------------+
| @@join_buffer_size |
+--------------------+
|            262144  |
+--------------------+
1 row in set (0.00 sec)
```

当前会话的连接缓存默认是 256KB，执行一个连接 SQL 语句如下：

```
mysql > select count( * ) from sbtest1 inner join sbtest2 using (c);
+----------+
| count( * ) |
+----------+
|        0 |
+----------+
1 row in set (7.57 sec)
```

可以发现该 SQL 语句执行的用时是 7 秒多，再查询该 SQL 语句的执行计划如下：

```
mysql > explain  select count( * ) from sbtest1 inner join sbtest2 using (c)\G
*************************** 1. row ***************************
           id: 1
  select_type: SIMPLE
        table: sbtest1
   partitions: NULL
         type: ALL
possible_keys: NULL
          key: NULL
      key_len: NULL
          ref: NULL
         rows: 83333
     filtered: 100.00
        Extra: NULL
*************************** 2. row ***************************
           id: 1
  select_type: SIMPLE
        table: sbtest2
   partitions: NULL
         type: ALL
```

```
possible_keys: NULL
          key: NULL
      key_len: NULL
          ref: NULL
         rows: 657576
     filtered: 10.00
        Extra: Using where; Using join buffer (hash join)
2 rows in set, 1 warning (0.00 sec)
```

发现该 SQL 语句的执行计划使用了哈希连接，驱动表是 sbtest1，查看驱动表的状态如下：

```
mysql > show table status like 'sbtest1'\G
*************************** 1. row ***************************
           Name: sbtest1
         Engine: MyISAM
        Version: 10
     Row_format: Fixed
           Rows: 83333
 Avg_row_length: 729
    Data_length: 60749757
Max_data_length: 205195258022068223
   Index_length: 1702912
      Data_free: 0
 Auto_increment: 1000001
    Create_time: 2021 - 05 - 31 19: 10: 35
    Update_time: 2021 - 05 - 31 20: 22: 33
     Check_time: 2021 - 05 - 31 20: 22: 33
      Collation: utf8mb4_0900_ai_ci
       Checksum: NULL
 Create_options:
        Comment:
1 row in set (0.01 sec)
```

驱动表 sbtest1 的大小大约是 60MB，把根据该表的 c 字段生成的哈希表放入 256KB 的连接缓存中，显然放不下，在 SQL 语句里通过提示设置 join_buffer_size 的大小为 60MB，进行几次对比测试如下：

```
mysql > select /* +  set_var(join_buffer_size = 60M)) */ count( * ) from sbtest1 inner join
sbtest2 using (c);
+----------+
| count( * ) |
+----------+
|        0 |
+----------+
1 row in set (1.57 sec)

mysql > select count( * ) from sbtest1 inner join sbtest2 using (c);
+----------+
| count( * ) |
+----------+
```

```
|         0   |
+-----------+
1 row in set (2.21 sec)

mysql> select /*+  set_var(join_buffer_size = 60M) */ count(*) from sbtest1 inner join
sbtest2 using (c);
+-----------+
| count(*)  |
+-----------+
|         0 |
+-----------+
1 row in set (1.43 sec)

mysql> select count(*) from sbtest1 inner join sbtest2 using (c);
+-----------+
| count(*)  |
+-----------+
|         0 |
+-----------+
1 row in set (1.99 sec)
```

可以看到，增大 join_buffer_size 后，执行时间大约缩短了四分之一。

18.7 优化排序

MySQL 中对记录进行排序有两种实现方式：一种是使用索引进行排序，在执行计划的 Extra 字段中会有 Using index 的信息；另一种是不使用索引进行排序，称为文件排序(filesort)，在执行计划的 Extra 字段中会有 Using filesort 的信息。在多表连接时，如果需要保存中间排序结果然后再进行连接，Extra 字段中会有“Using temporary; Using filesort”的信息。

例如，actor 表的 last_name 字段上有索引，而 first_name 字段上没有索引，对这两个字段分别进行排序，两个执行计划如下：

```
mysql> explain select last_name from actor order by last_name\G
*************************** 1. row ***************************
           id: 1
  select_type: SIMPLE
        table: actor
   partitions: NULL
         type: index
possible_keys: NULL
          key: idx_actor_last_name
      key_len: 182
          ref: NULL
         rows: 200
     filtered: 100.00
        Extra: Using index
1 row in set, 1 warning (0.00 sec)
```

```
mysql > explain select first_name from actor order by first_name\G
*************************** 1. row ***************************
           id: 1
  select_type: SIMPLE
        table: actor
   partitions: NULL
         type: ALL
possible_keys: NULL
          key: NULL
      key_len: NULL
          ref: NULL
         rows: 200
     filtered: 100.00
        Extra: Using filesort
1 row in set, 1 warning (0.00 sec)
```

可以看到，对有索引的字段进行排序，使用了索引排序；对没有索引的字段进行排序，使用了文件排序。

18.7.1 优化文件排序

文件排序使用的内存大小由系统参数 sort_buffer_size 决定，默认是 256K，这对大数据量的排序是不够用的。在早期的版本里，sort_buffer_size 设定的是固定的排序缓存大小。从 MySQL 8.0.12 开始，sort_buffer_size 设定的是排序缓存的上限，对于排序过程中使用的内存是根据需要递增分配的，因此把它设置成一个较大的值并没有副作用。排序时需要的排序缓存可能比预计的大得多，因为 MySQL 会给每条记录都分配一个能容纳最大记录的内存，例如，varchar 类型的字符串是按照它的完整长度分配空间，对于变长的 UTF 字符集也是按最长的字节分配空间。如果排序缓存不够，MySQL 会将数据分块排序，然后进行合并，在此过程中会在磁盘上生成临时文件，因此效率会大大下降，可以从状态参数 Sort_merge_passes 中看到合并的次数，这个值最好是 0，如果过大，建议增加 sort_buffer_size 的设置。下面看一个例子。

首先重置状态变量，命令如下：

```
mysql > flush  status;
Query OK, 0 rows affected (0.01 sec)
```

执行一个排序的 SQL 语句如下：

```
mysql > select * from sbtest2 order by c;
...
666666 rows in set (8.72 sec)
```

查询与排序相关的状态变量，相关命令及其运行结果如下：

```
mysql > SHOW STATUS LIKE 'sort%';
+-------------------+--------+
| Variable_name     | Value  |
```

```
+--------------------+--------+
| Sort_merge_passes  | 193    |
| Sort_range         | 0      |
| Sort_rows          | 666666 |
| Sort_scan          | 1      |
+--------------------+--------+
4 rows in set (0.00 sec)
```

这是一个对 66 万多条记录进行的排序，因为内存里无法一次放下全部的记录，不得不进行了 193 次合并。减少查询的字段也可以减少对排序缓存的占用，例如，把检索所有的字段改成只检索 c 字段。

第二次重置状态变量，命令如下：

```
mysql > flush  status;
Query OK, 0 rows affected (0.01 sec)
```

第二次执行该排序的 SQL 语句如下：

```
mysql > select c from sbtest2 order by c;
...
666666 rows in set (7.59 sec)
```

第二次查询与排序相关的状态变量，相关命令及其运行结果如下：

```
mysql > SHOW STATUS LIKE 'sort % ';
+--------------------+--------+
| Variable_name      | Value  |
+--------------------+--------+
| Sort_merge_passes  | 165    |
| Sort_range         | 0      |
| Sort_rows          | 666666 |
| Sort_scan          | 1      |
+--------------------+--------+
4 rows in set (0.01 sec)
```

可以看到，合并的次数从 193 次减少到了 165 次，执行速度也提高了 1 秒多，增大排序缓存也可以提高执行的效率，分别在会话级别和全局级别把 sort_buffer_size 设置为 1G 的命令如下：

```
mysql > set sort_buffer_size = 1024 * 1024 * 1024;
mysql > set global sort_buffer_size = 1024 * 1024 * 1024;
```

也可以在 SQL 语句中使用提示设置 sort_buffer_size，例如：

```
mysql > select  /* +  set_var(sort_buffer_size = 1G)  */ c from sbtest2 order by c;
...
666666 rows in set (2.77 sec)
```

使用 1G 的排序缓存执行同样的 SQL 语句后的排序状态变量如下：

```
mysql > SHOW STATUS LIKE 'sort % ';
+--------------------+--------+
```

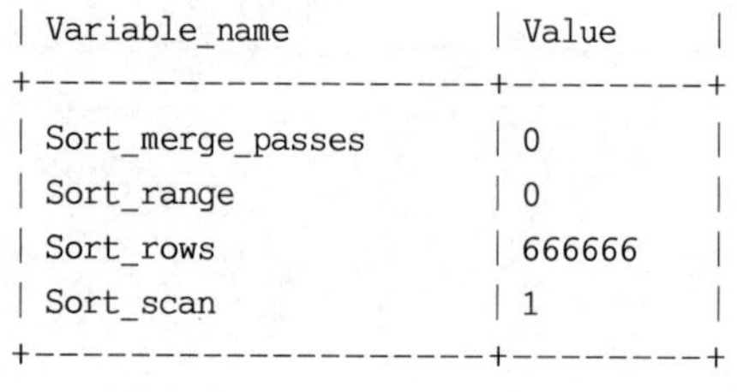

```
| Variable_name          | Value    |
+------------------------+----------+
| Sort_merge_passes      | 0        |
| Sort_range             | 0        |
| Sort_rows              | 666666   |
| Sort_scan              | 1        |
+------------------------+----------+
4 rows in set (0.01 sec)
```

可以看到，当 sort_buffer_size 设置为 1G 时，Sort_merge_passes 为 0，表示排序全部在内存中进行，原来 8 秒多的执行时间缩短到 2 秒多。下面查看性能视图，比较以上 3 个 SQL 语句的执行性能如下：

```
mysql > select left(SQL_TEXT, 40), sort_merge_passes, TIMER_WAIT/1000000000 MS from performance_schema.events_statements_history where SQL_TEXT like '%order by%' and THREAD_ID!= PS_CURRENT_THREAD_ID() order by TIMER_START desc limit 3;
+------------------------------------------+--------------------+-----------+
| left(SQL_TEXT,40)                        | sort_merge_passes  | MS        |
+------------------------------------------+--------------------+-----------+
| select /*+  set_var(sort_buffer_size = 1G |                  0 | 2768.8907 |
| select c from sbtest2 order by c         |                165 | 7590.2911 |
| select * from sbtest2 order by c         |                193 | 8716.0335 |
+------------------------------------------+--------------------+-----------+
3 rows in set (0.00 sec)
```

可以看到，增加排序缓存对性能的提升非常明显，还可以进一步考虑把系统参数 read_rnd_buffer_size 设置成一个更大的值(默认为 256K)，提升一次读取记录的数量。

注意，sort_buffer_size 和 innodb_sort_buffer_size 不同，后者是专用于指定创建 InnoDB 索引的排序缓存。

18.7.2 优化索引排序

对需要排序的字段增加索引通常可以让优化器使用索引排序，但不是绝对的，因为优化器要考虑总体的性能，例如，如下 SQL 语句的执行计划：

```
mysql > explain select * from sbtest2 order by k\G
*************************** 1. row ***************************
           id: 1
  select_type: SIMPLE
        table: sbtest2
   partitions: NULL
         type: ALL
possible_keys: NULL
          key: NULL
      key_len: NULL
          ref: NULL
         rows: 657576
     filtered: 100.00
        Extra: Using filesort
```

```
1 row in set, 1 warning (0.00 sec)
```

虽然 k 字段上有索引，优化器仍然采用了 filesort，因为该 SQL 语句查询了表里的所有字段，如果只查询 k 字段，看看它的执行计划：

```
mysql > explain select k from sbtest2 order by k\G
*************************** 1. row ***************************
           id: 1
  select_type: SIMPLE
        table: sbtest2
   partitions: NULL
         type: index
possible_keys: NULL
          key: k_2
      key_len: 4
          ref: NULL
         rows: 657576
     filtered: 100.00
        Extra: Using index
1 row in set, 1 warning (0.00 sec)
```

这时，优化器就采用了索引排序。因此在写 SQL 语句时，不要简单地使用一个星号查询所有的字段，应该把需要的字段在 select 后面逐一列出来，即使的确需要查询所有的字段，严谨的做法也是把字段逐一列出来，因为将来表结构的变化可能会增加、修改或删除字段。

18.8 表空间碎片整理

MySQL 的表在进行了多次 delete、update 和 insert 后，表空间会出现碎片。定期进行表空间整理，消除碎片可以提高访问表空间的性能。

18.8.1 检查表空间碎片

如下实验用于验证进行表空间整理后对性能的影响。

收集一个有 100 万条记录表的统计信息的命令及其输出结果如下：

```
mysql > analyze table sbtest1;
+-----------------+----------+-----------+-----------------------------+
| Table           | Op       | Msg_type  | Msg_text                    |
+-----------------+----------+-----------+-----------------------------+
| sbtest.sbtest1  | analyze  | status    | Table is already up to date |
+-----------------+----------+-----------+-----------------------------+
1 row in set (0.06 sec)
```

查询该表的状态信息如下：

```
mysql > show table status like 'sbtest1'\G
*************************** 1. row ***************************
         Name: sbtest1
```

```
         Engine: MyISAM
        Version: 10
     Row_format: Fixed
           Rows: 1000000
 Avg_row_length: 729
    Data_length: 729000000
Max_data_length: 205195258022068223
   Index_length: 20457472
      Data_free: 0
 Auto_increment: 1000001
    Create_time: 2021-05-31 18:54:22
    Update_time: 2021-05-31 18:54:43
     Check_time: 2021-05-31 18:55:05
      Collation: utf8mb4_0900_ai_ci
       Checksum: NULL
 Create_options:
        Comment:
1 row in set (0.00 sec)
```

再从操作系统层查询对应数据文件的大小如下：

```
mysql> system ls -l /var/lib/mysql/sbtest/sbtest1.*
-rw-r----- 1 mysql mysql 729000000 May 31 08:24 /var/lib/mysql/sbtest/sbtest1.MYD
-rw-r----- 1 mysql mysql  20457472 May 31 08:25 /var/lib/mysql/sbtest/sbtest1.MYI
```

命令 show table status 和从操作系统层看到的数据文件大小一致，这时的 Data_free 为零。

如下命令可删除该表三分之二的记录：

```
mysql> delete from sbtest1 where id % 3 <> 0;
Query OK, 666667 rows affected (51.72 sec)
```

重新收集该表的统计信息的命令及其输出结果如下：

```
mysql> analyze table sbtest1;
+----------------+---------+----------+----------+
| Table          | Op      | Msg_type | Msg_text |
+----------------+---------+----------+----------+
| sbtest.sbtest1 | analyze | status   | OK       |
+----------------+---------+----------+----------+
1 row in set (0.13 sec)
```

再次查看表的状态信息如下：

```
mysql> show table status like 'sbtest1'\G
*************************** 1. row ***************************
           Name: sbtest1
         Engine: MyISAM
        Version: 10
     Row_format: Fixed
           Rows: 333333
 Avg_row_length: 729
```

```
     Data_length: 729000000
 Max_data_length: 205195258022068223
    Index_length: 20457472
       Data_free: 486000243
  Auto_increment: 1000001
     Create_time: 2021-05-31 18:54:22
     Update_time: 2021-05-31 19:03:59
      Check_time: 2021-05-31 18:55:05
       Collation: utf8mb4_0900_ai_ci
        Checksum: NULL
  Create_options:
         Comment:
1 row in set (0.01 sec)
```

可以看到,这次 Data_free 不再是 0 了,计算 Data_free 和 Data_length 的比率如下:

```
mysql> select 486000243/729000000;
+---------------------+
| 486000243/729000000 |
+---------------------+
|              0.6667 |
+---------------------+
1 row in set (0.00 sec)
```

可以发现,空闲空间占用了总空间的三分之二,再次从操作系统层查询对应数据文件的大小如下:

```
mysql> system ls -l /var/lib/mysql/sbtest/sbtest1.*
-rw-r----- 1 mysql mysql 729000000 May 31 08:33 /var/lib/mysql/sbtest/sbtest1.MYD
-rw-r----- 1 mysql mysql  20457472 May 31 08:34 /var/lib/mysql/sbtest/sbtest1.MYI
```

可以发现,虽然 sbtest1 表中三分之二的记录已经被删除,但数据文件的大小还和原来一样。因为被删除的记录只是被标记成删除,它们占用的存储空间并没有被释放。

如下是进行全表扫描的 SQL 语句及其输出结果:

```
mysql> select count(*) from sbtest1 where c<>'aaa';
+----------+
| count(*) |
+----------+
|   333333 |
+----------+
1 row in set (0.82 sec)
```

可以发现,该全表扫描的 SQL 语句用时 0.82 秒,查看 sys.session 视图中的 last_statement_latency 字段也可以看到相同的用时。

18.8.2　整理表空间与性能提升

进行表空间整理的命令如下:

```
mysql> alter table sbtest1 force;
Query OK, 333333 rows affected (10.73 sec)
Records: 333333  Duplicates: 0  Warnings: 0
```

整理完成后,再次收集 sbtest1 表的统计信息的命令及其输出结果如下:

```
mysql> analyze table sbtest1;
+-----------------+---------+----------+-----------------------------+
| Table           | Op      | Msg_type | Msg_text                    |
+-----------------+---------+----------+-----------------------------+
| sbtest.sbtest1  | analyze | status   | Table is already up to date |
+-----------------+---------+----------+-----------------------------+
1 row in set (0.04 sec)
```

第 3 次查看表的状态信息如下:

```
mysql> show table status like 'sbtest1'\G
*************************** 1. row ***************************
           Name: sbtest1
         Engine: MyISAM
        Version: 10
     Row_format: Fixed
           Rows: 333333
 Avg_row_length: 729
    Data_length: 242999757
Max_data_length: 205195258022068223
   Index_length: 6820864
      Data_free: 0
 Auto_increment: 1000001
    Create_time: 2021-05-31 19:10:35
    Update_time: 2021-05-31 19:10:41
     Check_time: 2021-05-31 19:10:45
      Collation: utf8mb4_0900_ai_ci
       Checksum: NULL
 Create_options:
        Comment:
1 row in set (0.48 sec)
```

可以看到,Data_free 又变成零,而 Data_length 只剩下原来的三分之一。

第 3 次从操作系统层查询对应数据文件的大小如下:

```
mysql> system ls -l /var/lib/mysql/sbtest/sbtest1.*
-rw-r----- 1 mysql mysql 242999757 May 31 08:40 /var/lib/mysql/sbtest/sbtest1.MYD
-rw-r----- 1 mysql mysql   6820864 May 31 08:40 /var/lib/mysql/sbtest/sbtest1.MYI
```

可以发现,经过整理后,硬盘空间占用只剩下原来的三分之一,被删除的记录的硬盘空间都释放了。

再次执行全表扫描的 SQL 语句及其输出结果如下:

```
mysql> select count(*) from sbtest1 where c<>'aaa';
+----------+
| count(*) |
```

```
+----------+
|  333333  |
+----------+
1 row in set (0.29 sec)
```

可以发现，执行速度也大约提高到原来的三倍。这里使用的是 MyISAM 表进行测试，如果用 InnoDB 表，速度提高的就没有这么明显，因为 InnoDB 的数据会缓存到 InnoDB 缓存中，MyISAM 表的数据 MySQL 不进行缓存，OS 可能会缓存，因此要得到准确的测试结果，在 Linux 系统上每次测试前要使用如下命令释放系统的缓存：

```
# echo 3 > /proc/sys/vm/drop_caches
```

使用 alter table force 命令进行表空间整理和 optimize table 命令的作用一样，该命令适用于 InnoDB、MyISAM 和 ARCHIVE 三种引擎的表。但对于 InnoDB 的表，不支持 optimize table 命令，例如：

```
mysql > optimize table sbtest2\G
*************************** 1. row ***************************
   Table: sbtest.sbtest2
      Op: optimize
Msg_type: note
Msg_text: Table does not support optimize, doing recreate + analyze instead
*************************** 2. row ***************************
   Table: sbtest.sbtest2
      Op: optimize
Msg_type: status
Msg_text: OK
2 rows in set (44.63 sec)
```

可以用 alter table sbtest1 engine=innodb 代替，例如：

```
mysql > alter table sbtest2 engine = innodb;
Query OK, 0 rows affected (1 min 3.06 sec)
Records: 0  Duplicates: 0  Warnings: 0
```

18.8.3 使用 mysqlcheck 进行批量表空间优化

下面是找出表空间中可释放空间超过 10M 的最大 10 个表的 SQL 语句及其输出结果：

```
mysql > select table_name,round(data_length/1024/1024) as data_length_mb,  round(data_free/
1024/1024) as data_free_mb
from information_schema.tables  where round(data_free/1024/1024) > 10  order by data_free_
mb desc limit 10;
+------------+---------------+-------------+
| TABLE_NAME | data_length_mb | data_free_mb |
+------------+---------------+-------------+
| sbtest2    |           232 |         174 |
+------------+---------------+-------------+
1 row in set (0.02 sec)
```

可以使用 MySQL 自带的工具 mysqlcheck 的-o 选项进行表空间优化，该工具适合于在脚本中进行批量处理，可以被 Linux 系统中的 crontab 或 Windows 系统中的计划任务调用。

对单个表进行表空间优化的命令如下：

```
$ mysqlcheck -o sbtest sbtest1
```

也可以使用如下命令对某个数据库中的所有表进行表空间优化：

```
$ mysqlcheck -o sbtest
```

对整个实例中所有数据库进行表空间优化的命令如下：

```
$ mysqlcheck -o --all-databases
```

18.9 实验

视频演示

1. 覆盖索引

(1) 当索引中包括所有要查询的字段时，使用覆盖索引。

(2) 当要查询的字段不在索引中时，不能使用覆盖索引。

(3) 对不使用覆盖索引的 SQL 进行改写使其使用覆盖索引。

(4) 使用覆盖索引使 SQL 语句的性能大幅提高。

(参考 18.3.6 节)

视频演示

2. 直方图

(1) 对比字段上没有直方图统计信息和有直方图统计信息时执行计划里的过滤比率。

(2) 有直方图统计信息时生成更优的执行计划。

(参考 18.5 节)

参考文献

[1] MySQL 8.0 Reference Manual [EB/OL]. [2021-10-07]. https://dev.mysql.com/doc/refman/8.0/en/.

[2] KROGH J W. MySQL 8 Query Performance Tuning[M]. California: Apress,2020.

图书资源支持

感谢您一直以来对清华版图书的支持和爱护。为了配合本书的使用,本书提供配套的资源,有需求的读者请扫描下方的"书圈"微信公众号二维码,在图书专区下载,也可以拨打电话或发送电子邮件咨询。

如果您在使用本书的过程中遇到了什么问题,或者有相关图书出版计划,也请您发邮件告诉我们,以便我们更好地为您服务。

我们的联系方式:

地　　址:北京市海淀区双清路学研大厦A座714

邮　　编:100084

电　　话:010-83470236　010-83470237

客服邮箱:2301891038@qq.com

QQ:2301891038(请写明您的单位和姓名)

资源下载:关注公众号"书圈"下载配套资源。

资源下载、样书申请

书圈

图书案例

清华计算机学堂

观看课程直播